Introduction to Online Complexity

Introduction to Online Complexity

The New Social Physics of Extremes, Misinformation, and AI

FRANK YINGJIE HUO, PEDRO D. MANRIQUE, MINZHANG ZHENG, AND NEIL F. JOHNSON

OXFORD
UNIVERSITY PRESS

Great Clarendon Street, Oxford, OX2 6DP,
United Kingdom

Oxford University Press is a department of the University of Oxford.
It furthers the University's objective of excellence in research, scholarship,
and education by publishing worldwide. Oxford is a registered trade mark of
Oxford University Press in the UK and in certain other countries

© F. Huo, P. Manrique, M. Zhang, N. Johnson 2025

The moral rights of the authors have been asserted.

All rights reserved. No part of this publication may be reproduced, stored in a retrieval
system, transmitted, used for text and data mining, or used for training artificial intelligence,
in any form or by any means, without the prior permission in writing of Oxford University
Press, or as expressly permitted by law, by licence or under terms agreed with the appropriate
reprographics rightsorganization. Enquiries concerning reproduction outside the scope of the
above should be sent to the Rights Department, Oxford University Press, at the address above.

You must not circulate this work in any other form
and you must impose this same condition on any acquirer.

Published in the United States of America by Oxford University Press
198 Madison Avenue, New York, NY 10016, United States of America

British Library Cataloguing in Publication Data

Data available

Library of Congress Control Number: 2025938012

ISBN 9780198921011

DOI: 10.1093/oso/9780198921011.001.0001

Printed and bound by
CPI Group (UK) Ltd., Croydon, CR0 4YY

The manufacturer's authorised representative in the EU for product safety is
Oxford University Press España S.A., Parque Empresarial San Fernando de Henares,
Avenida de Castilla, 2 – 28830 Madrid (www.oup.es/enorproduct.safety@oup.com). OUP
España S.A. also acts as importer into Spain of products made by the manufacturer.)

Links to third party websites are provided by Oxford in good faith and
for information only. Oxford disclaims any responsibility for the materials
contained in any third party website referenced in this work.

To our loved ones, families, and friends

Preface

Welcome to our book! We have tried to make it a textbook that can be used by students and researchers from across disciplines, to launch into the fascinating challenge of building generative models of interacting human–technology–AI systems operating across the world's 24/7 online–offline ecosystem—including AI itself. The ultimate goal going forward is to establish some approximate 'Newton's Laws' of these future systems, including how AI itself works. These systems will undoubtedly dominate our everyday lives in the future and so it is essential to understand them and what behaviours they may exhibit—good and bad.

Future systems of humans, technology (machinery and software), and AI operating together in online–offline settings, are bound to have beneficial applications ranging from new medical procedures and cures, to reducing societal threats and enabling previously impossible space missions. Moreover, we see AI itself as a key example of a system within the scope of the theory that we present, since AI is similarly composed of a sea of evolving and interacting heterogeneous entities across scales, from tokens to Transformers. This also means that Large Language Models (LLMs) are systems to which the theory that we develop could be applied. This is an exciting project that we are currently developing and would welcome collaborators.

The central feature of heterogeneity of the interacting entities also means that this book presents new physics. The new generalizations that we derive of traditional physics, and by extension traditional chemistry and biology, could be used for a new core physics course, or a new interdisciplinary, inter-departmental, and even cross-school course. In particular, it provides an update of traditional many-body theory to the realm of non-identical particles exhibiting out-of-equilibrium behaviours. Hence the book provides an opportunity for a Many-Body Theory 2.0 course that, because of its broad applications, could attract students from other areas of Physics (e.g. Gamma Ray Burst analysis, as we mention later) and from beyond Physics Departments. For research students and researchers in any discipline, it also opens up the possibility of developing theoretical explanations of the surprises, extremes, and bursts that characterize so many real-world systems, and puts all these phenomena on a common footing. This includes biological systems, and we discuss one specific biological example later in the book in connection with control of these systems.

Our goal is to lay out the science of these future human–technology–AI systems ahead of their arrival, and hence help equip next-generation physicists and other scientists with an idea of what to expect, how such systems can be described quantitatively, and what tools could be used to design desired behaviours as well as control and mitigate undesired behaviours. Though we see the use of interacting human–technology–AI systems for good as far out-weighing their use for bad, governments and policymakers as well as the insurance industry and businesses need to assess the risk of things going wrong—and hence they need to get ahead of surprises and extreme events. This book can also serve them in that goal. For example, we include discussion of how malign activity and harms can develop and spread and how this can be mitigated. On that point of bad, nobody knows what and how future AI will be weaponized, or how a multi-dimensional online–offline war may unfold. But knowing the battlefield dynamics clarifies where and how that war will be fought. We refer to our group's laboratory for continual updates on this research: https://donlab.columbian.gwu.edu.

When reading the book, you will see that we repeatedly state the same signposts throughout the chapters. At the risk of sounding repetitive, we want to ensure that everything stays on track with a constant and clear objective and hence continuity of narrative and terminology. This book is *not* designed to provide comprehensive references or a literature review. The list of references at the end of the book includes additional sources and hence is reasonably expansive, but it is very far from exhaustive because of space restrictions. In Chapter 2, we present the robust empirical patterns in data collected so far from such human–technology–AI systems, and hence what to expect in the future. In Chapters 3–6, we present the new physics that describes this collective behaviour and the empirical patterns. It allows the interacting entities to have intrinsic heterogeneity but does not consider explicitly any internal dynamics within each entity. In Chapter 7, we add some internal dynamics for each entity in a simple way, by allowing each entity to be in a (small) number of different states to model contagion. Chapter 8 goes further by allowing each entity's internal state to exhibit adaptation and decision-making. Chapter 9 looks at how such systems can be controlled. Chapter 10 then leaves some final thoughts, focused around three exciting opportunities that we see based on current research—including the Holy Grail of understanding mechanistically how AI itself works, which we believe that this book can help answer.

We are very grateful to the prior and current PhD students and Research Scientists in our Dynamic Online Networks Laboratory at The George Washington University, who have contributed directly or indirectly to some of this work. Our website https://donlab.columbian.gwu.edu/recent-publications/ lists published versions of related ideas and results. In particular, we are very grateful to Rick Sear, Lucia Illari,

Akshay Verma, Dr Om Kant Jha, Dr Alex Dixon, and Prof. Pak Ming Hui for their important collaborations over the years, including earlier versions of some figures and some text which we have adapted from published papers, reports, and theses that we co-developed. NFJ would like to personally thank Dr Gordon Woo of Moody's for his joint work on developing a science of surprise, which greatly impacted our thinking around risk. We are also very grateful to Prof. John Horgan and Prof. Paul Gill for permission to share their mapping of fighters. Please send any thoughts or corrections etc. to neilfjohnson@me.com.

Contents

1	**New science for a new need**	**1**
1.1	Online–offline humans + machinery + software + AI = ?	1
1.2	What is 'complexity'?	5
1.3	Surprises and extremes from fusion and fission	7
2	**Empirical patterns**	**15**
2.1	Current human–machinery–software and AI examples	15
2.2	Examples of conflict mostly online	17
2.3	Examples of conflict mostly offline	22
3	**Multi-body Physics 1.0: single-species fusion and fission**	**29**
3.1	The broader scientific picture	29
3.2	What physics already says: fusion equations	30
3.3	Kernels	33
3.4	Inclusion of heterogeneity	40
3.5	Emergent phenomena from fusion	42
3.6	Gelation and shock wave	51
3.7	Generalizing to monomer addition and fission	58
	Exercises	70
4	**Multi-body Physics 2.0: multi-species fusion and fission**	**72**
4.1	Multi-dimensional fusion model	73
4.2	Scenario 1: 2-species fusion and gelation	81
4.3	Scenario 2: 3-species fusion and gelation	91
4.4	Gelation theory for general multi-species fusion systems	99
4.5	Introducing fission again: 2 species and beyond	107
4.6	Interacting shocks: a simplified approach	114
4.7	Predictions of super-shocks	119
	Exercises	121
5	**Multi-body Physics 3.0: general fusion–fission theory for time-dependent heterogeneous systems**	**123**
5.1	Motivation: later onsets and bumps grow more slowly	124
5.2	General theory of bursts	136

5.3	*D*-species fusion–fission and multi-layer random graphs	141
5.4	The x-space	148
5.5	Summary	150
	Exercises	151

6 Online wars: bad-actor AI and beyond 153
    6.1 Background   154
    6.2 What bad-actor–AI activity is likely to happen?   156
    6.3 Where will it happen?   157
    6.4 When will it happen?   159
    6.5 How can it be controlled? Battling bad-actor–AI   161
    6.6 Conclusion and final remarks   178
    Exercises   179

7 Online spreading: contagion and broadcast 180
    7.1 Classic spreading models: epidemiology   181
    7.2 Contagion: entity-to-entity   187
    7.3 Simplified approach for large-time contagion: co-existing mobility and infection dynamics   191
    7.4 Broadcast to all entities   200
    Exercises   203

8 Adding adaptation: emergence of anticrowds 204
    8.1 Extremes emerge from many-entity interactions   204
    8.2 Dynamic model of emerging extremes   208
    8.3 $N = 3$ entity example   209
    8.4 Many-entity system   212
    8.5 Rewards for winning and the 'softness' of news   224
    8.6 Impact of future social media algorithms   227
    Exercises   230

9 Controlling human–technology–AI systems 231
    9.1 Control via the generative equation   231
    9.2 Control by steering population strategies   234
    9.3 Control via link costs   251
    Exercises   262

10 Final thoughts 264

Appendix A Generative equations and their cluster dynamics 267
    A.1 One-dimensional   267

A.2	Multi-dimensional	271

Bibliography 273

Index 284

1
New science for a new need

1.1 Online–offline humans + machinery + software + AI = ?

Do you ever spend time online? What about your friends, family, and work colleagues? Of course you do—we all do. Does what you see or hear online ever influence what you think or do offline? And do your offline interactions with people and news events ever influence what you read or post online? Yes, again probably. Our offline and online worlds are now coupled, and this coupling will escalate as humans, technology (machinery, software), and new AI interact increasingly across the now-global 24/7 online–offline ecosystem.

The riots across the U.K. that followed the fatal stabbings of young girls in Southport on 29 July 2024, illustrate the complexity of this offline–online interplay—and its emergent, macroscopic consequences—which can potentially involve any of the world's 5+ billion social media users (i.e. 63 per cent of the global population as of April 2024 [1]). Humans interacting instantaneously online across both local and national scales through social media and via their smartphones, enabled the sudden emergence of offline rioting. Misinformation and disinformation about the attacker and their motives had circulated across social media communities and platforms in a largely organic way. [2] This in turn fuelled existing sentiments and resentments not only among extremists, but also members of the mainstream who feared their own kids were under threat as a result of perceived failings in immigration policies and a broader national malaise. The precise socioeconomic, political, or psychological causes will be argued over by policymakers and academics for years to come and will in any case likely vary by individual since we are all different by nature. However, the resulting impact on society is the same irrespective of the causes: days of violence and damage that further stokes already endemic online extremism and hate targeted at different races, religions, sexual orientations, political beliefs, etc. And it also further feeds into a broader yet equally dangerous breakdown of trust—and escalation of distrust—across many levels of society and across many topics, including expert advice in science (e.g. climate change) and medicine (e.g. vaccines) (Johnson et al., 2020).

[2] https://www.bbc.com/news/articles/cl4y0453nv5o

2 New science for a new need

Fig. 1.1: Current and future examples of interacting humans, technology, and AI, as envisaged by AI itself (DALL-E). These can be deemed bad or good but need to be understood in a rigorous quantitative way in order to be mitigated and controlled. Though highly schematic, the separate AI-generated panels demonstrate how real-time coupling between humans, machinery (e.g. smartphones), and software/AI (e.g. social media platforms) can have powerful offline consequences which in turn feed back to the online world. This motivates our book's goal to develop a physics that describes current and future systems that feature interacting humans, technology (machinery, software), and AI.

The U.K. is not alone in this. There are plenty of other recent examples globally of such online–offline complexity featuring offline feedback from and to the online world, including the 6 January 2021 Capitol attack in the U.S. (Verma, Sear, and Johnson, 2024). Added to this is the growing issue of the online safety of children, how online content affects mental health, online-motivated shootings or stabbings, terrorist attacks: the list goes on (Johnson *et al.*, 2019*a*). Different countries are taking different approaches to attacking such online threats through legislation, while many social media companies say they are ramping up their moderation efforts. The EU is currently leading the regulatory side through its 'Digital Services Act' and 'AI Act', with the mandate that 'Very Large Online Platforms' (e.g. Facebook) must perform risk analyses of such harms on their platform. The U.K. is adopting a similar stance with its 'Online Safety Act'. But these policies assume that large platforms hold the key, which is a massive assumption that lacks evidence. In reality, you can't win a

Fig. 1.2: This book addresses the urgent question: What is the 'physics' of large numbers of interacting heterogeneous humans, technology (machinery, software), and AI? The 'AI' entities could range from data tokens to an entire Large Language Model (LLM). During some activity (e.g. task, prompt, mission, game, economic transaction), clusters form over time (**fusion**) and may then break up (**fission**). A cluster is a subset of entities with some collective cohesion, i.e. they have become correlated, coordinated, or connected in some way. For negligible fission, a single giant cluster may form, which we call a **cohesive unit** (Chap. 2). It will appear as an out-of-the-blue **surprise** to anyone observing the system on the macroscopic scale. For non-negligible fission, clusters of all sizes may be present: if this distribution of cluster sizes is sufficiently fat-tailed (e.g. power-law) then the largest cluster can have an extreme size compared to the others. If clusters cause observable events, this means an **extreme event** will occur in a time-series of such events (e.g. conflict; Chap. 2).

battle without an accurate map of the battlefield—and our own recent work in mapping out this entire battlefield (Zheng, Sear, Illari, Restrepo, and Johnson, 2024) shows a very different picture. [3] Instead of the major platforms sitting at the centre and the smaller, less moderated, and typically more extreme platforms buzzing around the outside like a beehive, the small platforms sit at the core. This is because there are many of them and they are highly interconnected to each other and also to the larger platforms (e.g. Facebook, X, YouTube). Hence the smaller platforms act as a glue that binds together the entire online space. This is the reality of the online battlefield: harms

---

[3] see also https://donlab.columbian.gwu.edu.

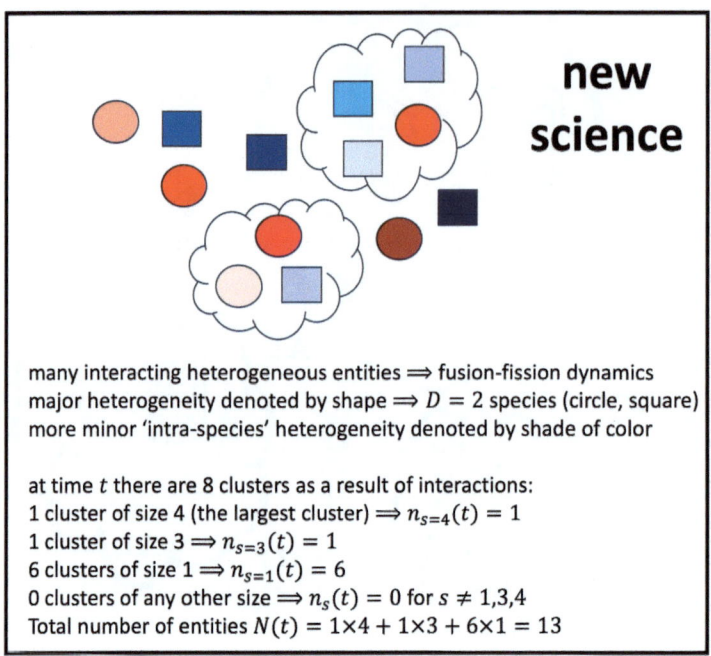

Fig. 1.3: Cartoon illustrates the cluster dynamics in Fig. 1.2 that are key to the new physics in this book. Chapters 3–6 explain the empirical patterns from existing example systems (Chap. 2) and predict that these patterns will occur in future systems.

do not lie at some supposed 'fringe'. Instead, the multiple platforms, including gaming and blockchain ones, are heavily interlinked by users creating hyperlinks between their respective in-built communities (e.g. from a Gab group to a Facebook page or YouTube channel) (Johnson et al., 2019a). These in-built communities on different platforms tend to have their own character, as do their users, and this heterogeneity needs to be accounted for. Moreover, our increased use and reliance on the online world is obviously facilitated by the wealth of different devices that we now have connected to it. Given all this, the heterogeneity of entities (Fig. 1.2) clearly needs to sit at the core of any new science aimed at describing the harms and malign activity that can develop in our current and future online–offline world.

But clearly this complex, coupled online–offline world involving interacting humans, technology (machinery, software), and now AI, could also become a massive force for good, as alluded to in Fig. 1.1's right panels. Systems containing different types of interacting humans, machinery (e.g. sensors, actuators), and software including AI, are already advancing societal needs ranging from medicine, commerce, and defence, to space exploration (Gorman et al., 2019; Huang et al., 2021). Furthermore,

Metaverse-ready social media platforms allow individuals from anywhere in the world to spontaneously aggregate into a cohesive unit (social group) (Wolfe, 2023; Palla et al., 2007) to perform activities (e.g. games, missions, economic transactions, goods exchanges) across multiple servers, assisted by physical devices (e.g. headsets or other wearables) and interactive software/AI. The interacting entities in the AI/Machine-Learning portion of this emerging ecosystem could be different categories of data tokens or topics extracted from unstructured multimodal datasets of text, images, audio, video (Sear et al., 2022a; Sear et al., 2022b)—or entire LLMs. Importantly for the modelling in this book, the communication of data between all such entities in Fig. 1.2 is now performed by extremely low latency networks which means that the interactions between these entities will be largely independent of physical separation. Hence the models and equations to be developed in this book, will not need to feature terms that describe spatial diffusion.

1.2 What is 'complexity'?

The online–offline world involving interacting humans, machinery, software, and now AI, is clearly complex. But before we proceed, we need to know what—beyond just complicated—the term 'complex' and hence complexity actually means in a scientific sense, and why it might give rise to future behaviours that are even more complicated (and hence even more interesting scientifically) than those observed currently.

To explain this, we borrow the following discussion from our earlier book *Financial Market Complexity* (Oxford University Press, 2003). When we say 'systems of interacting humans, machinery, software/AI including the example of humans online, are complex systems', we mean more than just 'complicated systems'. Making a pizza or fixing a bike puncture are both 'complicated', but neither is 'complex'. However, put these tasks together, and let the next step in one task depend on the present state of the other, and you start to incorporate at least a glimmer of complexity. Although there is no universally accepted definition of 'complexity' or 'complex system', most people would agree that any candidate complex system should have most or all of the following ingredients:

- Feedback. The nature of the feedback can change with time—for example, becoming positive one moment and negative the next—and may also change in magnitude and importance. It may operate at the macroscopic or microscopic level, or both. The presence of feedback implies that on some level, buried in the details of the dynamics, the system is 'remembering' its past and responding to it, albeit in a highly non-trivial way.
- Non-stationarity. We cannot assume that the dynamical or statistical properties observed in the system's past will remain unchanged in the system's future.

- Many interacting entities. The system contains many components or participants as in Fig. 1.2, which interact in possibly time-dependent ways. Their individual behaviours will respond to the feedback of information, which is possibly limited, from the system as a whole and/or from other entities. Since these entities may effectively be competing to win, it is unlikely that there is any such thing as a 'typical' entity.
- Adaptation. Each entity can adapt its behaviour in the hope of improving its performance. This will be particularly true for new AI entities.
- Evolution. The entire multi-body population in Fig. 1.2 evolves, driven by this ecology that interacts and adapts under the influence of feedback. The system typically remains far from equilibrium, and hence can exhibit all manner of extreme behaviours.
- Single realization. The system under study will always be a single realization, implying that standard techniques whereby averages over time are equated to averages over ensembles, may not work.
- Open system. The system is coupled to the rest of the world; hence it may be hard to distinguish between exogenous (i.e. outside) and endogenous (i.e. internal, self-generated) effects.

All these criteria are applicable to current and future online–offline scenarios of interacting humans, machinery, and software including AI as in Fig. 1.2. For example, the presence of feedback was demonstrated painfully in the global chaos caused by the 19 July 2024 technology meltdown.

This brings us to the main questions addressed in this book, of what large-scale cohesive behaviours—dangerous or desirable—could suddenly emerge from future systems of interacting humans, machinery, and software including AI; when will they emerge; and how will they evolve and be controlled? Physics is the perfect science to approach this because much of its success lies in describing real-world 'many-body' systems (i.e. systems with $N \gg 1$ interacting particles) in approximate ways which capture the key empirical patterns that emerge from the system's data. The secret sauce lies in knowing what level of approximation is just right—not too much or too little—in order to describe what would otherwise be an unsolvable set of N coupled dynamical equations. This is why we look at human–machinery–software/AI systems through this many-body physics lens—which means we look for mean-field theories that are not too strong that they end up losing the empirical patterns, and yet not too weak that the resulting equations are intractable. The significant added twist over traditional physics is that the entities in our system are **heterogeneous** and hence can have a wide range of different features (Fig. 1.2). While some differences may be small (e.g. sensors or actuators that are nominally similar but are manufactured by different companies) others can be huge (e.g. the difference between the behaviours of a human and a basic robot). Another twist—which in this case serves as a benefit mathematically—is that low-latency communications networks mean the interactions between such diverse entities will be largely independent of physical separation. Hence, as mentioned earlier, the models and equations to be developed in this book will not need to feature terms that describe spatial diffusion.

The rest of the book solves these questions by developing novel multi-dimensional generalizations of existing many-body physics—which means that this book also serves as a general textbook for complex systems in any setting. Wherever possible, we will derive analytic solutions that can reproduce—and hence offer a microscopic explanation for—empirical patterns in the data from a variety of real-world systems. We provide exercises at the end of chapters so that this can become a textbook for launching new research projects, e.g. by combining the various strands of theory that we develop and applying it to new scenarios.

1.3 Surprises and extremes from fusion and fission

There are two particular types of collective behaviour emerging from systems of humans, machinery, software/AI (Fig. 1.2), that will occupy us in this book because they both have potentially enormous real-world consequences, good or bad. These are (i) an out-of-nowhere surprise behaviour that suddenly emerges in a system that looks otherwise quiet, and (ii) in a system where things ('events') are already happening over time, an event suddenly happens with an extreme size compared to previous events, i.e. a so-called Black Swan appears. The key to both, as we explain in this book, is the interplay between the processes of **fusion and fission**. They each correspond to different regimes of behaviour of the same fusion–fission theory that we develop in Chaps 3–5.

1.3.1 Fusion

Imagine a glass of milk sitting stationary on a table. We can sit and watch this glass of milk for days and nothing happens—until it does. From out of nowhere, a milk curd suddenly appears. This is a huge cluster of molecular entities, but since we are observing the system on the macroscale of our everyday lives, we were unable to see it develop. It appeared as a surprise. If on the other hand, we had microscopic-scale vision, we would have seen this assembly of smaller clusters underway as they form and join within the liquid, i.e. a process called fusion which is a key mechanism in living systems and complex systems more generally.

8 New science for a new need

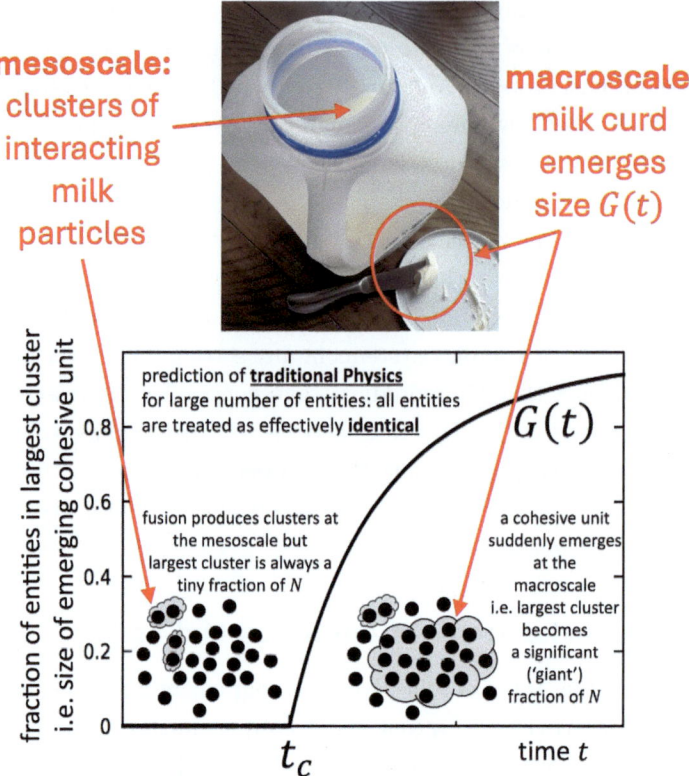

Fig. 1.4: Traditional physics view in which all the N interacting entities in a system are treated as identical. This may be reasonable for describing how a milk curd forms: but it contradicts the core feature of any real-world human–machinery–AI system which is that the entities in it have a wide range of heterogeneity, i.e. they may be *very* different from each other. In the example shown, a single giant cluster (i.e. cohesive unit) emerges, i.e. it contains a significant fraction of all the entities N. In practice whether there is a dominant cluster that truly classifies as 'giant' or simply very large, does not matter since the systems of interest involve a finite number of N entities and the sharp transition to a 'giant' cluster only strictly holds in the limit of N approaching infinity. Chapters 3–5 progressively generalize this traditional physics view to account for the heterogeneity of entities.

Figure 1.4 shows this in a cartoon way. Both the figure and our discussion massively oversimplify the chemical details involved in milk curdling, but the essential feature of fusion is correct. Since the number of microscale entities N is large, there is a sharp transition that we explain later in Chap. 3. The fundamental point is that the mesoscale clustering suddenly gives rise to a cluster that is so large, we can see it with our own eyes. The physics field defines the fraction of entities that form part of this largest cluster as $G(t)$. So the curve only rises (i.e. $G(t) > 0$ visibly) when there is a significant fraction of all N entities in the largest cluster, not a few. This means that the largest cluster, when it appears and grows, is indeed huge. As in Fig. 1.2, we call it a **cohesive unit**. But because of the different contexts in which it may be referred to in other areas of physics, chemistry, biology, and social science, we often refer to the cohesive unit in later chapters as a gel, or a giant connected component (GCC), or a shock in an inviscid fluid.

This out-of-the-blue emergence of a cohesive unit is also a surprise scientifically. Imagine taking a set of objects and adding a link between pairs of them at random. You might expect that as links get added, the largest cluster just grows steadily in size. Hence a sudden jump as seen in the milk (Fig. 1.4) would not happen. But this argument underestimates the fact that as the fusion process proceeds in time by assembling smaller-scale clusters, at some stage it will join together clusters that are already sizeable. This means that a very large cluster will suddenly appear in an abrupt, nonlinear way; as though its growth was suddenly switched on, as observed visually in Fig. 1.4. The NetLogo simulation of this in Fig. 1.5 is freely available for anyone to explore. [4]

Such a cohesive unit may be desirable. For example, we might welcome the sudden, spontaneous emergence of cooperation and coordination in some future human–AI–robot surgical team or space mission. But it may be undesirable and even highly dangerous, particularly if such systems belong to adversaries, e.g. emergence of AI-assisted online extremism (Zheng et al., 2024; Manrique et al., 2023; Johnson et al., 2016). A damaging cohesive unit may also arise without intentional malice, e.g. the global chaos caused by the 19 July 2024 technology meltdown.

On a technical point, our use of the term 'cohesion' here is consistent with the huge body of existing physics on polymers, networks, and also social systems (Jusup et al., 2022; Artime et al., 2024; Ji et al., 2023; Palla et al., 2007; Stockmayer, 1943; Stockmayer, 1944; Flory, 1953; Ziff et al., 1982; Hendriks et al., 1983; Wattis, 2006; Drake, 1972; Lushnikov, 2006; Nelson et al., 2020; Krapivsky et al., 2010; Barabási, 2016; Newman, 2018; Soulier and Halpin-Healy, 2003; Schweitzer and Andres, 2022; Niu et al., 2017; Dodds et al., 2011; Fagan et al., 2024; Battiston et al., 2021), i.e. sudden, spontaneous emergence and growth of some macroscopic unit (cohesive unit) containing a non-negligible fraction of the entire multi-species population. Importantly, cohesion does not mean the objects in question are necessarily spatially near each other. Instead, low-latency communications networks mean the interactions between such diverse entities will be largely independent of physical separation. In more traditional physics settings, this cohesive unit represents the well-known gel, or giant connected

---

[4]see http://www.netlogoweb.org.

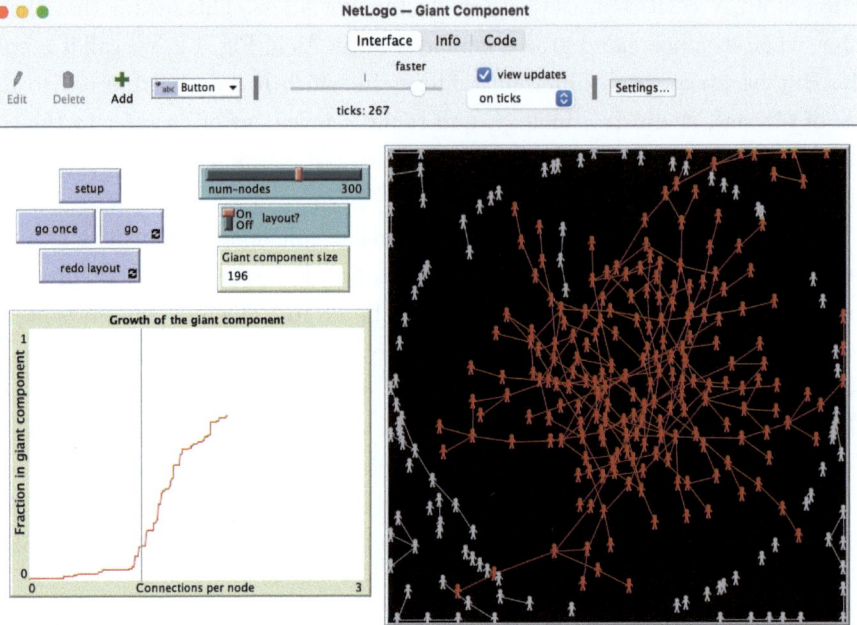

Fig. 1.5: Free NetLogo software allows readers to explore in real time the appearance and growth of the cohesive unit (i.e. giant connected component, GCC, in network language, or a gel in chemical language) which was shown in Fig. 1.4 for the simple case of identical entities. For the mesoscopic (i.e. cluster) theory that we develop in Chapter 3 onwards, it does not matter precisely what links are where—only that all the entities in the cluster are indirectly connected, i.e. it is a connected cluster. Hence we show clusters as 'clouds' in Figs 1.2–1.4 and we do not need to discuss the microscopics of what links are where.

component (GCC), or it could even be a cartoon of the correlations in some dynamically emerging, highly entangled quantum state (Liu *et al.*, 2022). We mention this again briefly in Chap. 10.

Figure 1.6 (left panel) provides a glimpse of the successively richer $G(t)$ shapes that we will derive in Chaps 3–5. Figure 1.6 right panel shows the derivative of this (dG/dt) which represents a 'burst' of activity in the sense that it captures by definition the net change in the number of entities that become members of the cohesive unit at each moment in time. It hence has a burst shape, and it is no coincidence that it resembles a wide range of burst activities observed not just in the physical sciences, but also across living systems, and hence what can be expected in human–technology–AI systems. As a side note, it even suggests a new generic explanation for Gamma Ray Bursts (GRBs), i.e. visually different GRBs all derive from the common mechanism of time-dependent

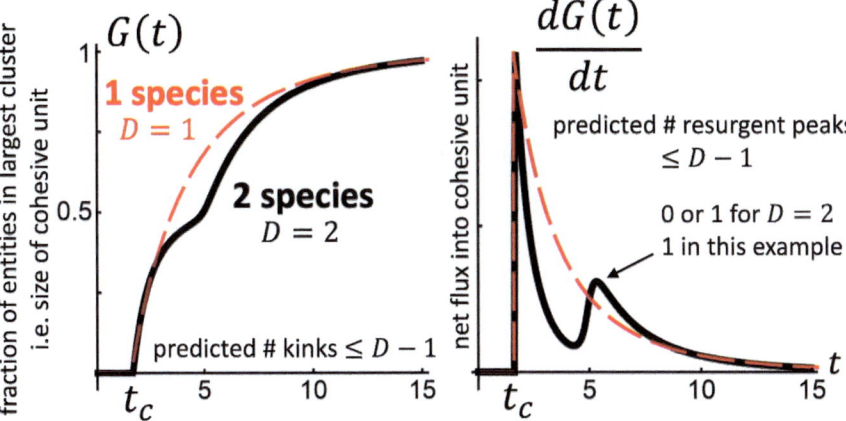

Fig. 1.6: Left panel: Size $G(t)$ of the cohesive unit that can emerge in a system of *heterogeneous* humans, technology, and AI. It appears as a surprise to anyone observing the system on the macroscale. Its kink-like structure, its growth shape, and its onset time t_c, depend on the heterogeneity of the interacting entities (see Chap. 4). The onset at time t_c becomes less abrupt as the number N of interacting entities decreases. Right panel: Derivative of $G(t)$ offers an explanation for multi-peak bursts in activity observed in many real-world systems, i.e. the burst time-series is generated by the net change in the number of entities in the cohesive unit at each successive time step (dG/dt). See Chap. 5 for the full theory which features the ability of the subsequent peaks in dG/dt to be even higher than the initial one, hence closely mimicking the bursts seen in many real-world domains, including the interesting case of Gamma Ray Bursts.

fusion of heterogeneous entities that become successively available in some region of space. [5]

1.3.2 Fusion and fission

Our discussion above ignored the presence of a reverse process, fission, whereby clusters break up. Fusion and fission compete in terms of forming large clusters. Obviously, in the case where a milk curd emerges and then just sits there in the rest of the liquid, fusion dominates. But in other settings that is not the case. Just think of a recent remote meeting you had on Zoom, or a get-together in person. Everybody 'fuses' together but then later this meeting breaks up and people leave. While the formation of this cohesive unit (i.e. the meeting) may have been organized top-down, it can also happen spontaneously, e.g. the infamous office water-cooler get-together. In the online setting discussed earlier where each cohesive unit is an in-built community (e.g. a Gab group) the fission may be enforced *en masse* by moderators when they are notified

---

[5] see Wikipedia page for Gamma Ray Burst shapes https://en.wikipedia.org/wiki/Gamma-ray_burst.

about malign activity. This fission will then likely be of the cohesive unit as a whole, which gets broken up (e.g. the in-built community gets shut down). This is akin to what is known to happen in offline riots and also human conflict, where a collection of fighters might quickly be forced to scatter. The case of such complete or near-complete fragmentation of a cluster will occupy much of our discussion in this book since it can be seen as a loss of cohesion.

In the presence of such fission, a huge cluster (cohesive unit) will either be unable to form, or if it does form, it will not survive long. Instead, we can expect a continually evolving sea of clusters as in Fig. 1.7, left panel, which comes from the empirical online data in Chap. 2. These clusters may have a broad range of sizes; hence the steady-state distribution of cluster sizes becomes of interest. If there is a well-defined steady-state distribution and that distribution has very fat tails, this means that there is a real chance of some relatively extreme size cluster appearing in the future. And if each cluster causes an observable event over time, that means that in a series of such events being observed, there is a real chance that an event of extreme size will occur in the future. Obviously, this extreme event will not be as big as one caused by the cohesive unit that forms in the absence of fission, but it will nonetheless be much larger than other events before it. Hence if future risk estimates in this system have been calculated based on prior events, this extreme event could be catastrophic, i.e. the Black Swan analogy. As we show when discussing the empirical data in Chap. 2 and the later theory chapters, an approximate power-law is of specific interest for the steady-state distribution of cluster sizes (and hence events) as sketched in Fig. 1.7, right panel. It means that above a certain minimum cluster (hence event) size s_{min}, the distribution of cluster sizes (hence events that they generate) approximately follows $n_s = As^{-\alpha}$ where A is a constant.

To explain the implications of an approximate power-law distribution as in Fig. 1.7, right panel, imagine you are sitting in a new classroom at the start of a school year and the class's new students arrive one by one. We can think of each separate student entering as representing a separate 'event', where the size of that event is the height of that student. So your observations of students entering the classroom becomes a time-series of events, where each event has its own size. Now imagine you generate a histogram of these students' height (i.e. event sizes) which you update as each student enters. It will quickly form into a bell-curve shape—or more precisely, a Normal or equivalently a Gaussian distribution. But you never give a thought to the fact that you are in a classroom with a concrete roof that is only 12 ft high. What if the next student is 60 ft tall, or 6000 ft? Common sense says that is ridiculous because a bell-curve distribution predicts that the chances of this are practically zero, not just for this class but for the entire history of humanity. But that is not the case for systems in which

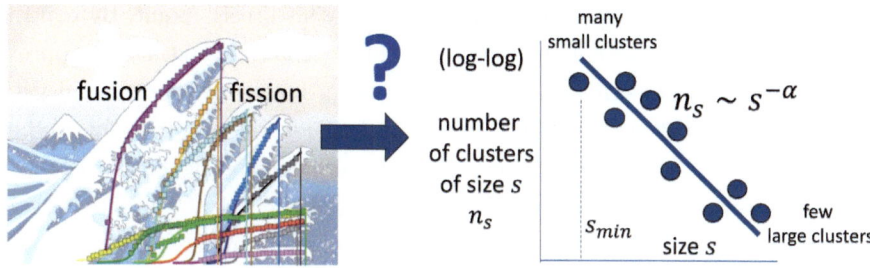

Fig. 1.7: In the presence of fission such that clusters get fragmented, a single giant cluster (cohesive unit) for the entire population N may not emerge. Instead there is a sea of clusters and the quantity of interest becomes the distribution of cluster sizes. Log-log scales are used because an approximate power-law distribution $n_s = As^{-\alpha}$ will appear as an approximate straight line $\log n_s = -\alpha \log s + \log A$. All estimates of α from empirical data in this book are performed using the recognized state-of-the-art statistical techniques, not simple log-log plots since these produce biased estimates (see https://aaronclauset.github.io/powerlaws/ for testing code). Left panel is adapted from the American Physical Society's research highlights article https://physics.aps.org/articles/v16/89 that was itself drawn from an earlier original sketch by one of us.

the size of individual entities (and hence the size of the events that they cause) feed off of each other, i.e. the systems have the attributes of a complex system such as feedback and adaptation as outlined in Section 1.2. Now the event sizes need not all be narrowly distributed around some average value. Just think of stock market crashes, where the trading processes develop feedback based on the shift to selling which then makes the system head downwards rapidly. Because such systems can have a significant number of extremely large events, the corresponding distributions will have a distribution that has fatter tails than a bell-curve distribution. A particular example of such a fat-tailed distribution is a power-law distribution, or some approximate form which is close to it. For a system with an approximate power-law distribution of event sizes, there is a real risk of an extreme event occurring in the future akin to a 6000 ft tall student turning up at the classroom door.

Much has been written about approximate power-laws in the physics literature and beyond, but particularly in physics because a power-law distribution is 'scale-free' (Newman, 2005). In other words, if we change s by some scale so that $s \to bs$ then the distribution becomes $A(bs)^{-\alpha} = A's^{-\alpha}$ which is functionally identical and just has a different constant A'. So it is free of a scale. This has important ramifications for physics since it connects to phase transitions where at a critical point—but only at a critical point—the size of the system's correlations is also scale-free and hence exhibits power-laws. The gel formation in milk for example, has an approximate power-law distribution of cluster sizes (which is akin to a size of correlations) at the exact point where $G(t)$ begins to become non-zero. But only at this precise point. By contrast, we will see in Chap. 2 that the empirical data for online–offline conflict has an approximate power-law which is not only stable over time despite non-zero fusion and fission, it is also robust to fluctuations in the fusion and fission rates. Crucially, it therefore makes no sense to try explaining this pattern in terms of the system somehow sitting magically at its transition point (critical point) for all time. In Chap. 3, we show that this pattern is instead a remarkable property of what happens when the fusion and total (or near total) fission mechanisms co-exist. Our Chap. 3 theory explanation of this power-law pattern is also robust to heterogeneity among the entities, i.e. it can hold for the generalizations in Chaps 4 and 5—and crucially it also generates values for the power-law exponent α that are in agreement with the empirical patterns in Chap. 2.

Mathematically, the standard deviation for power-law distributions of event sizes with $\alpha > 2$ is technically infinite—which is another way of saying that the standard deviation of event sizes does not properly quantify the risk (which is typically measured in industry and business by the standard deviation). In other words, unlike the distribution for heights in a classroom, there is a finite probability of a future event equivalent to a 6000 ft student suddenly turning up. An approximate power-law distribution of cluster sizes as in Fig. 1.7, right panel, means that estimates of future risk can be dictated by such extreme events. The equivalent of the 6000 ft person (event) turning up at any time becomes quite possible.

Before diving into the empirical patterns in Chap. 2, we mention that this book does not go into any depth about what the individual human, machinery, or software/AI entities are, or what they are each doing at any given time. It does not need to, because the macroscopic observed behaviour of such a multi-body interacting system is determined by what is happening at the mesoscopic level of clusters, just as the science of how the milk curd forms (Fig. 1.4) depends on the correlated pockets of molecular entities, not what each entity is doing at any one moment. Proceeding in this way, this book is able to not only provide a comprehensive theory of interacting humans, technology, and AI, it also offers a first glimpse of how AI such as an LLM may be assembling its seemingly miraculous generalization abilities over time. We refer to Chap. 2 (in particular, Fig. 2.1) and Chap. 10 for more discussion on that. We also refer to our two recent *Physical Review Letter* articles (Manrique, Huo, El Oud, Zheng, Illari, and Johnson, 2023; Huo, Manrique, and Johnson, 2024; Manrique, Huo, El Oud, and Johnson, 2024) and their online Supporting Material for additional details.

2
Empirical patterns

In this chapter, we present evidence of robust empirical patterns related to the important real-world phenomena of surprises and extremes, that emerge from current online–offline systems of interacting humans, machinery, and software. These systems will increasingly feature AI in the future. Then in Chaps 3–5, we develop successively more sophisticated theoretical descriptions of these patterns using the fission and fusion ideas mentioned in Chap. 1. The integrated human–technology–AI systems that will undoubtedly dominate our everyday lives in the near future are yet to be developed or imagined. Hence the data that we present are for current collections of humans with some machinery and software, or AI on its own—but not all together. However, the ubiquitous nature of the empirical patterns across current systems combined with the generality of our theory in Chaps 3–5 and its ability to reproduce these patterns suggests that future human–technology–AI systems will also exhibit these empirical patterns. Hence these same patterns, related to surprise and extreme events, can be used to estimate risks in future human–technology–AI systems: and they can serve in what-if scenario testing for future mitigations, interventions, and policies.

2.1 Current human–machinery–software and AI examples

Figure 2.1 shows empirical evidence of cohesive unit formation from a range of existing real-world systems involving collections of humans, machinery, software, and AI on its own. In each case, there is a sudden rise of a cohesive unit from out-of-nowhere at the macroscopic scale—but with additional structure that is not present in the simple milk example of Fig. 1.4. This additional structure (kinks), which was hinted at in Fig. 1.6, can be explained using the multi-species version of our theory that we develop in Chaps 4 and 5. There is no data available yet from a system like Fig. 1.2 with interacting humans and technology (machinery–software) *and* AI: but Figs 2.1(a), (b), and (d) all feature an activity involving humans and technology, and all show the sudden emergence of a cohesive unit and its kink-like growth which appear to an observer on the macroscopic scale as an out-of-nowhere surprise.

Figure 2.1(c) is of particular interest since the system is made up of just AI. As shown, even the simplest version of our multi-species theory (Chap. 4, $D = 2$ species)

16 *Empirical patterns*

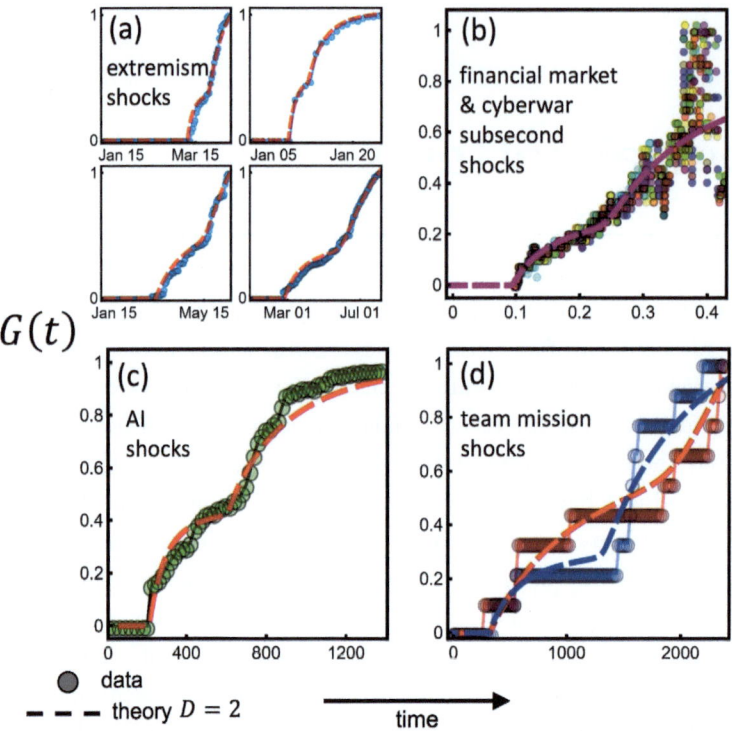

Fig. 2.1: Sudden rise and multi-kink growth of a macroscopic cohesive unit with magnitude G (vertical axis) vs time (horizontal axis) in various current real-world systems of interacting humans, machinery, software compared to the simplest version of the theory in Chap. 4 (i.e. $D = 2$). (a) Online social media anti-U.S. jihadi membership during given period; (b) subsecond human–machine trading across electronic exchanges; (c) reduction in the loss function during AI learning; (d) human–machine–software activity during two nominally identical pilot space missions. Exogenous causes can be ruled out by the nature of the systems. In (a) and (d), if the onsets and growth kinks were caused by exogenous news event(s), they should be synchronous, but they are not, so an exogenous cause is unlikely. In (b), exogenous causes are practically impossible on that subsecond timescale. Source: data from Huo, Manrique, and Johnson (2024).

seems to capture its kink-like growth, and this agreement becomes even better with more species ($D \geqslant 3$). Hence the multi-species theory in Chaps 4 and 5 offers a first understanding of how such AI actually works. Specifically, it says that AI's ability to generalize emerges in a similar way to a cohesive unit, i.e. from the fusion of its heterogeneous 'pieces'. To explain all this, we start by noting that the basic entities in AI/Machine-Learning are tokens which are then embedded—similar to the entities in our theory in Chaps 3–5—in some high dimensional space. The empirical data in Fig. 2.1(c) are obtained from a controlled investigation set up by Nanda *et al.* (Nanda et al., 2023; Nanda and Lieberum, 2024; Nanda, 2024) which involves measurement of AI performance as a function of iterations. During this process, the AI shows a sudden positive jump(s) in its ability to generalize—so-called grokking. The AI system producing the data in Fig. 2.1(c) is a Transformer trained on the problem of 5 digit addition, with inputs of the form 1 2 3 4 5 + 6 7 8 9 0 = 8 0 2 3 5 (each digit and sign are one token). Though we refer to Nanda et al.'s papers for details of the AI and the investigation performed, the main takeaway from the good agreement in Fig. 2.1(c) (Chaps 4 and 5) is that our theory explains the kink-like evolution of AI's output abilities as being caused by the formation of cohesive units among the different species of circuits that develop within the AI system. The observed sequence of cluster assembly can be accounted for using more complex temporal terms in F as described in Chap. 5. While outside the scope of the current book, we return to this AI-explanation topic again briefly in Chap. 10.

For the rest of this chapter, we focus on the robust empirical patterns that we have found to emerge from the specific human–technology example of conflict, in its broadest sense. There are several reasons why we choose conflict. In the real (i.e. offline) world, conflict mostly involves the interaction of collections of humans using technology as weapons of war—but increasingly fighters will be using AI. Conflict in the online world involves individuals using the technology of different social media platforms, and now AI (see Chap. 6), to build support for their beliefs and opinions and/or attack others beliefs and opinions, e.g. to build anti-U.S. jihadi support, or to engage in the battle of trust over vaccines (Johnson et al., 2020). Not only are offline and online conflicts urgent real-world problems, so too will be future hybrid online–offline conflicts. Moreover, the inherent struggle between opposing factions that underlie such conflict—whether online, offline or both—mean that it also qualifies as a scientific complex system as a result of feedback, adaptation, memory, etc. Since conflict is all about trying to beat an opponent, it also makes sense that future conflict will likely profit from both online and offline spaces as well as the latest technology including AI. It is therefore no accident that empirical data for both online and offline conflicts can generate robust patterns that align with the theory in Chaps 3–6. We will see the sudden surprises, bursts of activity, and extreme events alluded to in Chap. 1. For presentation purposes, we break up our conflict discussion into mostly online and mostly offline, but we emphasize that future conflicts will undoubtedly mix both, together with new AI as addressed in Chap. 6.

2.2 Examples of conflict mostly online

The online complexity of social media platform software, with its particular technological features and algorithms, provides the environment in which approximately 5 billion of us humans interact—and which will increasingly involve machinery (e.g. wearables beyond simple headsets). There are many reports of the online mix of humans, different social media platforms' technology features, and AI itself, driving conflict over issues ranging from elections through to climate change, abortion, and vaccines (Johnson et al., 2020). There are also concerns that online extremism and hate are being amplified by this combination.

But before showing the empirical patterns, we must first establish how people behave online. Prior work across the social sciences shows that people (e.g. parents) aggregate online into in-built communities around particular interests (e.g. a Facebook 'page' or VKontakte 'club' of parents) in order to share advice and experiences (Moon et al., 2019; Ammari and Schoenebeck, 2016; Laws et al., 2019). Anti-X communities are no different, where 'X' can be any particular religion, race, ethnicity, gender, etc. We stress that these in-built communities have nothing to do with community detection in networks: instead, each in-built community is a well-defined, unique entity with its own unique ID which is assigned by the particular platform that they are on. From now on, we refer to each in-built community simply as a *community*. Members of a community then develop some level of trust within that community. The same is also true for extremism and hate communities (Johnson et al., 2019a; Leahy et al., 2022; Velásquez et al., 2021).

A typical in-built community has thousands of members (i.e. users) and because of the general common focus of its members in terms of interests and opinions, each in-built community can be thought of as a cohesive unit. But each in-built community can also be seen as a node in some network, since these in-built communities tend to create hyperlinks to each other. Specifically, at any instant in time t, users in any such in-built community 1 can create a link (technically, a hyperlink) with any other in-built community 2 on the same or a different platform (Zheng et al., 2024). This exposes the linked community 1 to community 2's content and hence community 2's content can spread to community 1's members. This link also alerts community 1's members to community 2's existence, hence community 1's members can now go to community 2 to share their content including comments and replies so that community 2's members can see it.

The overall takeaway therefore, is that individuals online form into in-built communities, which can then change the size of their memberships by attracting new recruits. This fusion also happens at a higher scale. Specifically, these in-built communities can connect to each other through the hyperlinks that their members create, to form clusters, and these clusters can then grow. What constitutes an 'entity' (i.e. node in Fig. 1.4) at these two scales is therefore an individual user or an individual in-built community respectively. When moderators notice or are alerted to an individual in-built community—or a cluster of in-built communities—involved in malign activity, they can carry out complete or near-complete fragmentation (i.e. fission) by shutting down the in-built community or shutting down the links within a cluster of in-built communities. The respective entities will then scatter (fission) but may soon join with others again (fusion). An empirical example of this at the scale of an individual in-built community is shown in Fig. 2.2. Figure 2.3 shows the fusion of in-built communities (where each is a node) into clusters during the 2020 U.S. election process (Verma et al., 2024).

This fusion of different individuals into an in-built community—or different in-built communities into a cluster of in-built communities—together with their occasional abrupt fission by moderators, provides the key mechanism in our models in Chaps 3–6 in the book. Another relevant empirical finding for the modelling in Chaps 3–6, is the fact that the probability that fusion occurs between clusters varies approximately as the product of their sizes as shown in Fig. 2.4(f) inset—hence our focus is on a product kernel in Chaps 3–6. This agrees with earlier empirical findings by Palla *et al.* on human communications in cellphone settings (Palla et al., 2007). We also note the approximate power-law distributions of cluster sizes observed empirically in different online conflict settings, in Figs 2.4(c), (d), and (e), with an estimated $\alpha \approx 2.5$ in each case—exactly as predicted in Chap. 3 for fusion–fission with a product kernel.

Figure 2.4(a) illustrates the growth curves for the set of nascent cohesive units (in-built communities) that emerged online surrounding anti-U.S. jihadi support at the

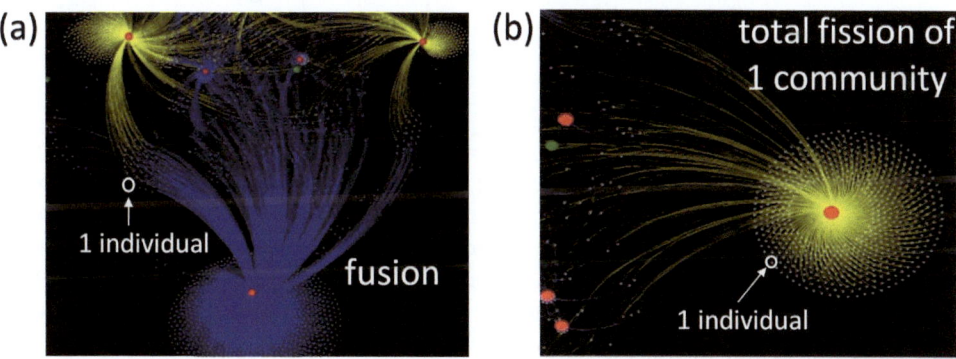

Fig. 2.2: Empirically observed (a) fusion and (b) total fission of online in-built communities featuring anti-U.S. hate on VKontakte between day t (yellow) and $t+1$ (blue). Red nodes are anti-U.S. communities that later got shut down (total fission); green nodes are those still not yet shut down; yellow links point to individuals (white dots) removed from the anti-U.S. community on day $t+1$; blue links point to individuals added to the anti-U.S. community on day $t+1$. Since the spatial layout results from (a) and (b) are closeups of a fuller network plotted using ForceAtlas2, nodes appearing closer together are more interconnected: (b) also shows that very few individuals are simultaneously also members of other communities.

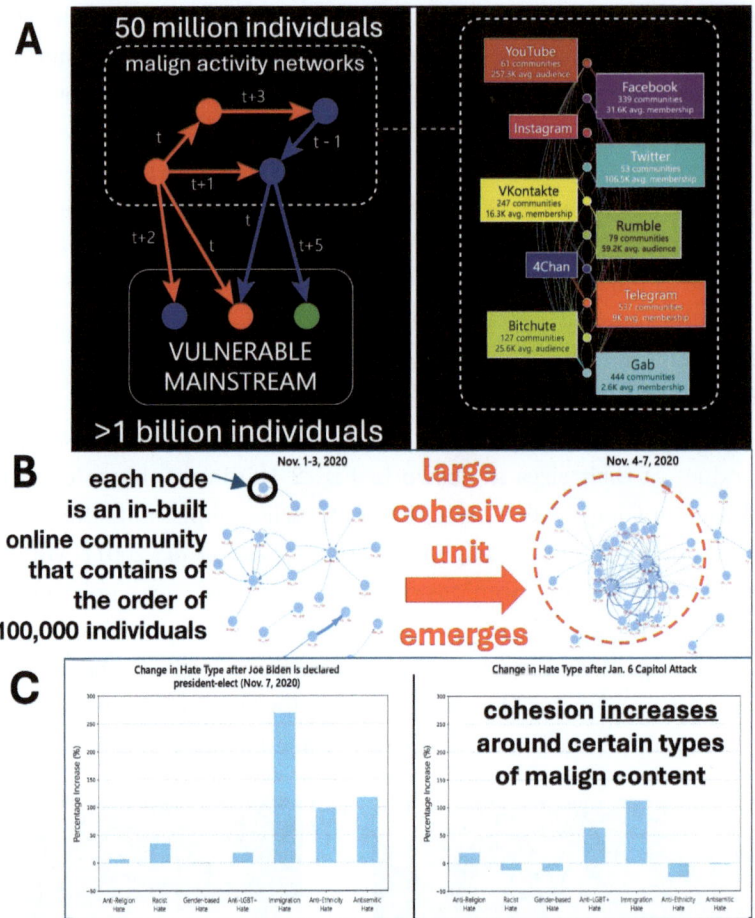

Fig. 2.3: Empirical fusion (and some fission) of clusters of anti-X hate communities across platforms, and also fusion of topics around content, surrounding the 2020 U.S. election process. Source: data from Verma *et al.* (2024). A: Hate universe infrastructure. A hate community (node) on a given social media platform (given colour) establishes a link (by sharing a URL for a piece of content) to another community (node) at a given time. Each community is a platform-provided community: a Telegram Channel, a Gab Group, a YouTube Channel, etc. Boxes show number of hate communities (nodes) in a given platform and the average number of members of each community. Edges shown are aggregated over time. Other panels show how fusion of in-built communities—and also fusion of hate topics—dominated over this time period.

time that ISIS was first appearing. This would be $G(t)$ if we divided by N, but we don't do that since N is changing in time. Figure 2.4(b) shows similar growth curves for another online conflict situation. Details of our data and the data collection process

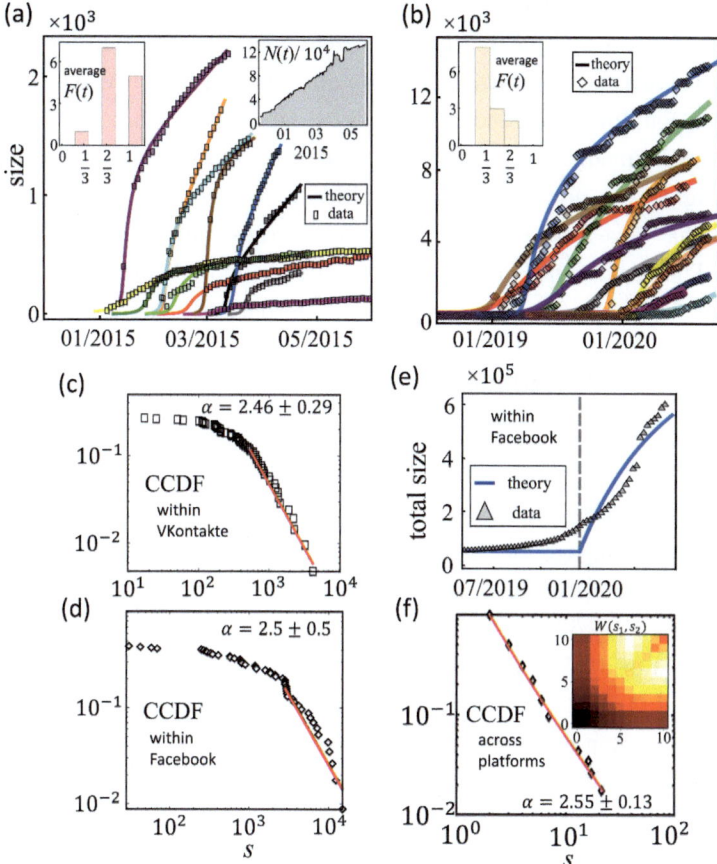

Fig. 2.4: Empirical data (symbols) and Chaps 3–4 theory predictions (lines) for in-built anti-X communities within and across platforms. (a) Size (i.e. number of members) of foreign anti-U.S. (jihadi) communities on VKontakte. (b) Size of domestic anti-U.S. government (pro-civil war) communities on Facebook. Insets: changing population size; time-averaged $F(t)$ which suggests that (b) reflects a heterophily fusion mechanism more than (a). (c–d) Complementary cumulative distribution (CCDF) of individual community sizes s from (a) and (b). (e) Evolution of total size of all communities from (b). (f) CCDF at a higher scale, i.e. sizes of clusters of interlinked communities. Inset: empirically inferred interaction kernel $W(s_1, s_2)$ obtained from data across all platforms; axes s_1 and s_2 are sizes of interacting clusters. Source: data from Manrique et al. (2023).

are provided in (Huo et al., 2024; Manrique et al., 2023; Manrique et al., 2024) and also on our lab website at https://donlab.columbian.gwu.edu.

These cohesive unit growth shapes do not typically occur for non-anti-X systems, such as a pizza-lovers community. This means that whatever fusion and fission process occurs for anti-X communities is not the same for more benign communities. This

22 *Empirical patterns*

makes sense since a pizza-lovers community is hardly likely to burst onto the scene and attract lots of members quickly, then attract moderator attention for malign activity and get shut down (i.e. total fragmentation and hence fission). Also as evidence of the inter-species coupling discussed in depth in Chap. 5 (e.g. where each species can be a different social media platform), we note that analysis of the link dynamics shows that, as compared to non-anti-X communities, anti-X communities not only generate more posts that have links in them, but also that these links are to a greater variety of platforms. This means that anti-X communities are more likely to redirect their users to fringe platforms than their non-anti-X counterparts (Zheng *et al.*, 2024).

2.3 Examples of conflict mostly offline

We now turn to humans and technology interacting offline in a conflict setting. Again, this will increasingly include AI in the future. In addition to its core academic and policy interest, the question of how to assess casualty risk in future conflicts or terrorism represents a key challenge for the trillion-dollar insurance industry. Given the sparsity of historical extreme event data, any mechanistic insight would be particularly valuable since it enables the construction of concrete what-if scenarios and counterfactuals (Woo, 2011). Indeed, the need for new academic innovation in this area is so great that it spawned the 2024 joint U.S. National Science Foundation–government–industry initiative on terrorism and catastrophic online/cyber risks. Future conflicts, including physical violence, will no doubt be fought in both the online and offline spaces using all sorts of Metaverse-like technologies that have yet to be invented, and of course humans. But because this has not yet arrived, much of what we discuss here is offline conflict activity of humans using technology in conventional ways (e.g. guns, IEDs) as machines of war. We note that the following discussion draws from an as-yet unpublished collaboration with Dr Gordon Woo and D. Restrepo who we thank for their generosity in letting us present it.

Remarkably, the robust empirical patterns of extremes and surprise that we show here are similar to those for online conflict from the previous section. The reason for this similarity is that both online and offline conflict feature similar fusion and complete (or near complete) fission processes—and hence produce similar empirical patterns (see Chap. 3 and beyond for the theoretical analysis). This online–offline similarity also means that future hybrid online–offline conflicts should also feature similar empirical patterns. The fact that our theory in Chaps 3–5 explains both the online and offline conflict patterns suggests that it will apply to hybrid conflicts in the future, whether they are mostly offline, mostly online, or anywhere in between.

Why humans fight has no easy or unique answer, e.g. the current Gaza war. However, a better understanding of *how* humans fight could provide new insights into

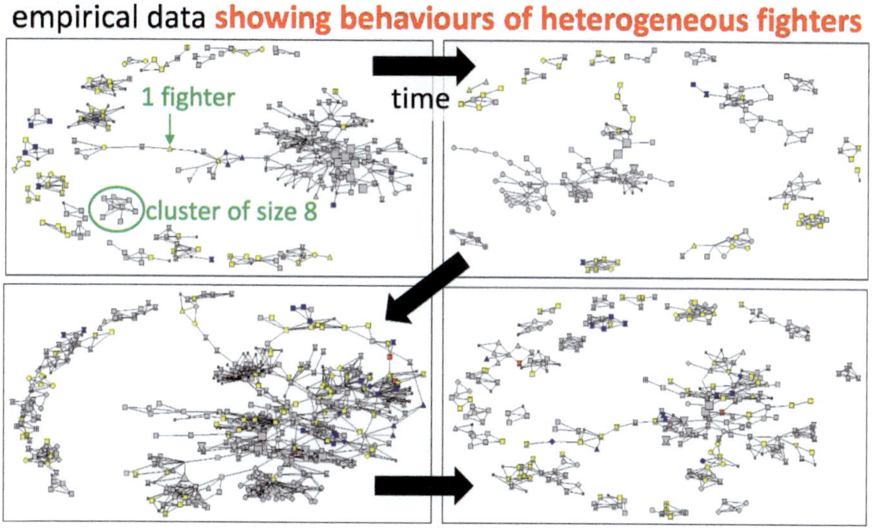

empirical data showing behaviours of heterogeneous fighters

Fig. 2.5: Fusion and fission of fighters in an offline conflict. Consecutive time windows show state-of-the-art individual level empirical data from a violent group that is a strong proxy for many others (Provisional Irish Republican Army (Horgan and Gill, 2013)). Fighters (nodes) are heterogeneous: individual attributes shown by different shape (faction allegiance), size (role) and colour (skill). A link denotes fighters (nodes) with a personal connection which indicates cohesion and hence easier coordination. We are grateful to Prof. Horgan and Prof. Gill for permission to show this data plot.

interventions, future risk, and hidden shifts. Figure 2.5 provides evidence that fusion and fission of fighters is involved in the mechanics of how fighters fight operationally over time. This data comes from the uniquely detailed study by John Horgan and Paul Gill (2013) of the fighter dynamics in PIRA (Provisional Irish Republican Army) who quite successfully fought the British for an extended period. The PIRA data are a strong proxy for many other fighting groups since PIRA actively exchanged operating details with other terrorist organizations globally, e.g. the Palestine region dating back to the PLO (Palestine Liberation Organization) and such entities are known to copy each other (Horgan and Gill, 2013). It does not matter to the theory that we develop whether they are labelled a 'terrorist organization', a faction within or across such organizations, an insurgency, a guerilla group etc., or why they are fighting.

More specifically, Fig. 2.5 reveals three key empirical features that must form part of any realistic model for such an offline fighting group. Crucially, these are *exactly the same* three features that form the basis of our model for online conflict and more generally our Chaps 3–5 theory of human–technology–AI behaviour. Written in the language of offline conflict, they are:

- *Individual heterogeneity*: Figure 2.5 shows that offline fighters (nodes) are heterogeneous in terms of individual attributes, e.g. faction allegiance, skills, role. Each different node symbol denotes a different set of individual attributes. Chapter 4 explains how we can encode a major attribute by a species label, e.g. allegiance to Hamas, Hezbollah, or Palestinian Islamic Jihad (PIJ). Each fighter's other attributes are encoded as a vector similar to existing social science studies (Centola et al., 2007) and also similar to token embedding in AI/Machine-Learning.
- *Fusion*: Figure 2.5 shows that clusters of linked nodes (fighters) form with time-dependent sizes and compositions. A link to a new node increases a cluster's size by one. A link between nodes already in clusters fuses their clusters to form a single cluster. Even though fighters may be spatially well separated, they have modern communications and so can interact over any distance. So phone/online communications make spatial separations less relevant, similar to the online situation shown in Fig. 2.2.
- *Fission*: Figure 2.5 shows that clusters can fragment. If a cluster senses imminent detection or danger, its members may scatter as in fish, bird, or animal groups—and also online (Fig. 2.2)—hence the cluster fission will likely be complete.

The same two fusion–fission limits as discussed in Chap. 1 are of interest for offline conflict:

1. **Non-negligible fission rate**: As in Fig. 1.7, the cohesive unit cannot fully form because of occasional fission. Instead, the clustering dynamics reach a steady state with an approximate power-law distribution for $s > s_{\min}$ given by $s^{-\alpha}$ which describes the number of clusters with s fighters. In the limit that fighters' interactions are independent of their spatial position and hence separations, the effective size of each cluster is s since all s fighters can always be engaged. This yields $\alpha = 5/2 \equiv 2.5$ exactly mathematically, as shown in Chap. 3. In the opposite limit of fighting on a grid-like urban battlespace and/or no long-range interactions, the effective size of a cluster is of order $s^{1/2}$ since only fighters on the cluster's perimeter are always engaged, e.g. edge of a square of size s varies as $s^{1/2}$. This yields $\alpha = 5/2 - 1/2 = 2$ exactly mathematically, which agrees with numerical simulations. For any intermediate case, our fusion–fission theory predicts that $2.0 \leq \alpha \leq 2.5$. It is reasonable that s dictates the number of casualties x that a conflict event involving that cluster produces. Hence the fusion–fission mathematics predicts that the distribution of the number of casualties per event will also be an approximate power-law distribution with exponent $2.0 \leq \alpha \leq 2.5$. This prediction is robust to many variations, including changing N.

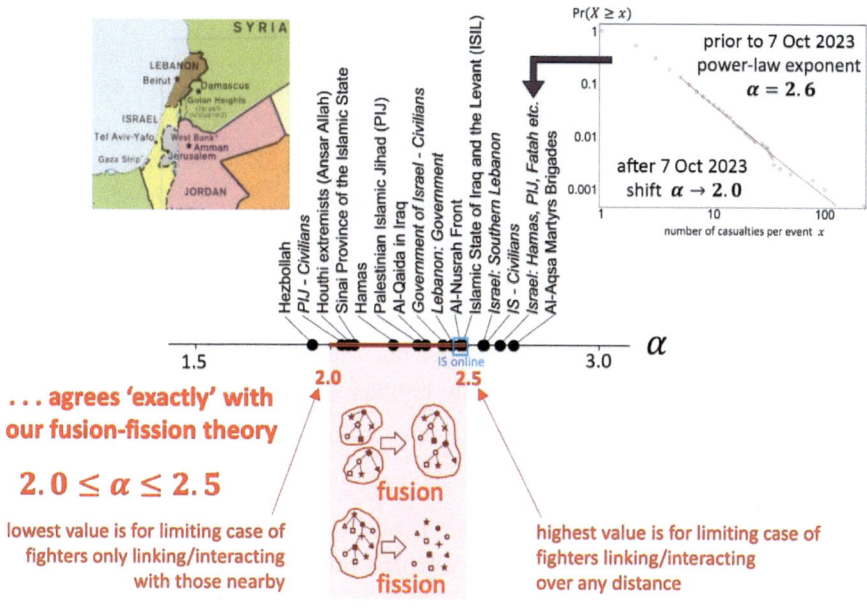

Fig. 2.6: Empirical exponent values α for approximate power-law distribution of casualties per event from offline conflicts (italic text) and terrorism (non-italic text) across Israel–Palestine region (each shown as solid black circle) (Spagat, Johnson, and Weezel, 2018). Our fusion–fission mathematics in Chaps 3–4 predicts α range 2.0–2.5 depending on distance dependence of fighter interactions: $\alpha = 2.0$ for urban grid-like battlefield and/or restricted long-range communications, $\alpha = 2.5$ for unrestricted long-range communications and/or online. As shown, this is consistent with all Israel–Palestine region conflicts and terrorism: and it explains why the data-point for Gaza and West Bank conflict shifts to $\alpha = 2.0$ after 7 October 2023 because of Gaza's grid-like battlefield and/or degradation of Hamas' etc. long-range communications. See online article https://arxiv.org/abs/2409.02816 for full details.

Our theoretical prediction $2.0 \leqslant \alpha \leqslant 2.5$ agrees remarkably well with the casualty event data for the topical example of the Israel–Palestine region (see Fig. 2.6). In addition to explaining why the empirical patterns for all Israel–Palestine region conflicts and terrorism fall broadly across $\alpha = 2.0 - 2.5$ as shown, it also explains why the data-point for the Gaza conflict shifts to $\alpha = 2.0$ after 7 October 2023, because of Gaza's grid-like battlefield and/or degradation of Hamas' etc. long-range communications. Though approximate power-laws are known to arise for conflicts and terrorism around the world (Johnson et al., 2013b), Spagat, Johnson, and Weezel (2018) showed their α values range from 1.37 to 5.21, which means that α

values for the Israel–Palestine region could in principle have fallen far outside this predicted range $2.0 \leqslant \alpha \leqslant 2.5$. But it does not.

2. **Negligible fission rate**: The coherent unit can now form and will survive for a considerable time—as is indeed observed in Fig. 2.7A akin to Fig. 1.6A. Figure 2.7B shows its rate of change which is akin to Fig. 1.6B. Chapter 4 derives formulae for t_c and the shock's onset composition for different numbers of species D. The theory determines t_c as 6 October, i.e. it predicts that large-scale cohesion emerged among the N fighters the day before the actual attack. This then left them several hours for the physical logistics of advancing into Israel. Figure 2.7A's inset shows additional empirical support for our theory, i.e. smaller-scale fusion among members of Hamas etc. momentarily bubbles up prior to macroscale emergence at t_c.

As a prelude to Chap. 4, we note that Fig. 2.7B shows good agreement between our theory for $D = 3$ species (Chap. 4) and the membership of official anti-Israel military communities. Mathematically, the minimum species number D involved is equal to the number of peaks observed (i.e. $D = 3$). Each peak has a different mixed-species (mixed-allegiance) composition. Even if the giant cluster (shock) of new members in Fig. 2.7A does not comprise entirely of initially active fighters, they likely quickly became active.

A simple—but misleading—takeaway from this fusion–fission explanation would be that the 7 October attack happened because Israel stopped generating fission events. Hence to prevent another such attack in the future, simply increase fission. However, the correct takeaway is more complex: fighters from a minimum of three different allegiances ($D = 3$ species) were all undergoing fusion in the same way at the same time. Figure 2.8 [1] allows users to explore this with a multi-species version of the cohesive unit formation in NetLogo (Fig. 1.5). Hence a multi-adversary attack force could easily assemble, i.e. clusters could easily slot into each others' activity in a planned or spontaneous way. Crucially, this multi-adversary assembly would have then been missed by any adversary-specific surveillance since no single-adversary shock ever formed. Eye-witness accounts provide additional independent support of this takeaway of multi-adversary fusion of fighters.

We note that the conflict and terrorism data underlying these results are extracted from the Georeferenced Event Dataset (GED [2]) and the Global Terrorism Database

---

[1] see https://gwdonlab.github.io/netlogo-simulator/.
[2] at https://ucdp.uu.se/downloads/.

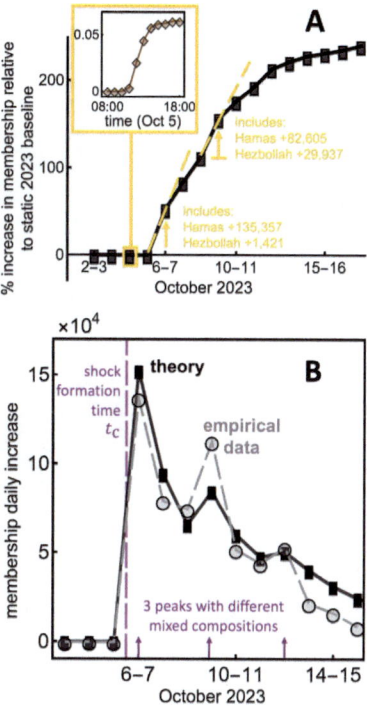

Fig. 2.7: A: Surprise emerges akin to Fig. 1.6, i.e. cohesive unit of fighters emerges. Curve shows empirical data for membership of anti-Israel official military communities on Telegram. See online article https://arxiv.org/abs/2409.02816 for full details. Only Hamas and Hezbollah communities are shown since they dominate, but these community members also include those with allegiance to PIJ etc. that lack Telegram communities of their own. Inset: glimpse of underlying mesoscale fusion. B: Rate of change of empirical data in A (dashed grey). Solid line is result from our theory for $D = 3$ species (see Chap. 4). See Chaps 3–4 for an explanation of why the surprise in panel A (i.e. cohesive unit, gel, giant connected component GCC) is also equivalent to a shock in an inviscid fluid.

(GTD [3]): see Spagat, Johnson, and Weezel (2018) for this together with full replication code and a full discussion of all the statistical testing including power-laws. The beauty of the simple fusion–fission model, in addition to its close quantitative fit to the empirical patterns, is that it does not matter whether collective violence is labelled as a conflict or terrorism, or what kind of war occurred; or whether an adversary is truly 'terrorist' or an 'organization', or a faction within or across such organizations, or an insurgency, or some informal army, or a guerilla group etc.; or why it is fighting Israel; or why particular links exist; or the amount or nature of any pre-planning.

[3] at https://www.start.umd.edu/gtd/.

28 Empirical patterns

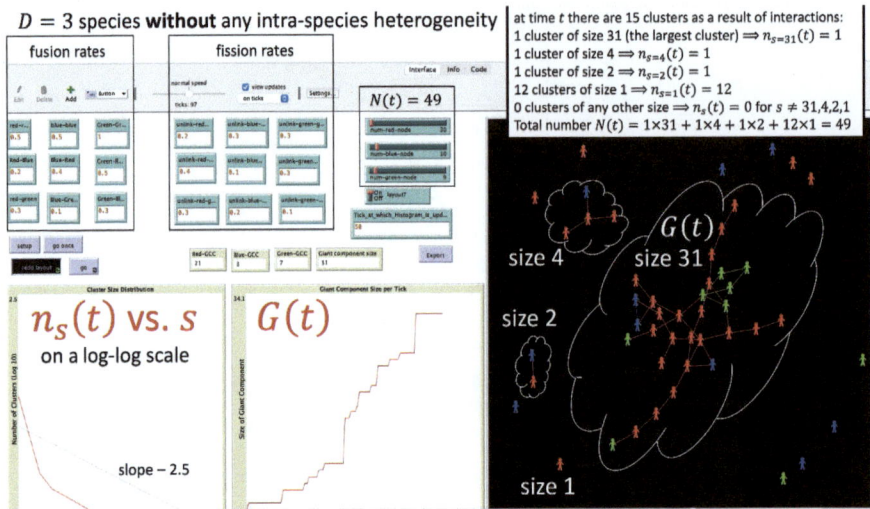

Fig. 2.8: This free online simulation allows users to explore a multi-species version of the cohesive unit formation in NetLogo (see https://gwdonlab.github.io/netlogo-simulator/). The theory for $G(t)$ etc. is given in Chaps 4–5. We are very grateful to Akshay Verma for helping build this.

To summarize, the fusion–fission theory that we develop in Chaps 3–5 provides a remarkably simple yet widely applicable quantitative description of conflict online and offline. Before ending this empirical pattern discussion, we note that several prior studies have promoted the idea of offline conflict being like a forest fire. However, the mechanism of forest fires is unrealistic for the empirical online and offline conflict situations that we have presented; moreover, obtaining similar ballpark power-law exponents to the empirical data would involve cherry-picking the parameter regime in such a forest fire model. In more detail, there is no evidence that fighters stand around in the way that trees do in a forest; or that the spark to fight passes from one to the other in a similar way to how trees burn, and long-range communications and hence cohesion among fighters cannot be ignored. A similar critique holds for other types of percolation models. Though similar exponents for the power-law can emerge, these tend to appear only at the critical point, i.e. the percolation transition. It is like the milk curd: there is a power-law in the cluster sizes at this moment, but only then. Hence any such percolation model makes little sense as a generative model of online or offline conflict given that the actual empirical exponent values α are so stable in time (Fig. 2.6).

3
Multi-body Physics 1.0: single-species fusion and fission

Our goal in Chaps 3–5 is to derive physics that can describe the collective behaviour of general human–technology–AI systems (Fig. 1.2) as well as the benchmark empirical patterns found so far in such systems (Chap. 2). In this chapter and Chaps 4–5, we develop new physics by allowing the interacting entities to have intrinsic heterogeneity, and treating this heterogeneity in an increasingly sophisticated way (i.e. successively less averaging). These chapters do not consider explicitly any internal dynamics within each entity, but this aspect is picked up in Chaps 7 and 8. But as we show here in Chaps 3–5, treating this heterogeneity at the correct level of approximation—i.e. knowing what to average over and how, which is the essence of much of theoretical physics—will enable us to reproduce the key empirical patterns in Chap. 2 without the further complications of Chaps 7 and 8.

We note before proceeding that because of the different contexts in which a **cohesive unit** (e.g. Figs 1.4-1.6) may be referred to in other areas of physics, chemistry, biology, and social science, we may refer to it in this book as a **gel**, or a **giant connected component (GCC)**, or even as a **shock** in an inviscid fluid because of the form of its governing equation, as shown in this chapter. All these terms mean the same thing at the macroscopic level: a **cohesive unit**. Also because of a similar diversity in terminology across the literature, we may use words such as **coalescence, aggregation, coagulation**, etc. interchangeably with **fusion**, and **fragmentation, disaggregation, degradation**, etc. interchangeably with **fission**.

3.1 The broader scientific picture

From the fundamental physical interactions that bind together microscopic particles within an atom, to everyday social contacts between people, to astronomical-scale bodies, interactions on different scales rule what we observe and hence the science that we are all trying to understand. We are often able to focus on a particular system scale and propose physical laws to describe it without interference from other scales. This, for example, enables us to study one-body Newtonian theory without worrying about quantum or relativistic effects, or the influence on and from planetary motion. Nevertheless, the interactions among a large collection of particles at one scale often result in some emergent phenomenon at a higher scale: the whole subject of statistical mechanics and thermodynamics is a well-known example. In fact, the whole of science is arguably just a tower of successive scales and abstraction levels, with each discipline focusing on its own preferred scale and level. Some areas of physics, however, try to study transitions across such scales. Among them, renormalization group theory systematically describes the transition between scales and hence behaviours.

This chapter does not attempt to discuss generalized theories such as the renormalization group. Instead, we focus on a particular class of interactions that is crucial to low-energy complex systems: fusion (aggregation, etc.) and fission (fragmentation, etc.). These fundamental dynamical processes can be seen as approximations to a range of behaviours observed in real-world complex systems, e.g. birds flocking, online grouping or simply formation of a physical gel as in Fig. 1.4. More importantly, they can be formulated with simple mathematics through well-established tools from statistical physics. We will look into variations of these models and, above all, their implications for scale transitions mathematically. Later on in the book we will find several surprising, if not shocking (literally) real-world manifestations of the theory discussed below.

3.2 What physics already says: fusion equations

Before diving into the theory, we first list below a few terms that we use frequently in this particular chapter, and what they mean in this context. Though the focus of the theory in Chaps 3–5 is on heterogeneous entities (which is more general than particles) that undergo fusion into clusters followed perhaps by cluster fission, we use these more physics-sounding terms below because of their common use in this area of the physics and chemistry literature.

- **Particles** refer to the constituent entities in a system. They are the smallest units in a system and we do not need to detail what they consist of and how they are built. Referring back to Fig. 1.2, they can be physical particles, birds in a flock, individual people in a society, a user of the Internet, a piece of an LLM as it evolves towards generalizing, or an LLM itself. For our purpose, they are all just treated as indivisible entities.
- **Clusters** refer to groups of *particles* that are bonded together by some certain criteria specific to the system, i.e. they have a cohesion through some correlation or coordination or relationship. For physical systems related to the Smoluchowski discussion in what follows, the bonding may simply be a physical connection due to, say, inelastic collisions, whereas for a bird ecology a cluster is naturally a bird flock defined by the highly correlated behaviours among the particles, i.e. the

birds. This book is particularly interested in online social networks, and clusters in this case are defined by linkages between users or groups of users, e.g. hyperlinks. The **size** of a cluster refers to the number of particles in the cluster. In classical systems, for historical reasons, they are also referred to as the *mass* of the cluster.
- **Aggregation** or **coalescence** (i.e. fusion) refers to creation of larger clusters (i.e. larger *size*) from smaller ones due to establishment of such bondings. Sometimes the term **merging** is also used interchangeably in the literature. When referring to the bondings microscopically, i.e. at the particle level, we say particles *connect* with one another. Its opposite is **fragmentation** (i.e. fission).

With these terms in mind, let's think about what quantities we are interested in for an aggregating system, i.e. there is fusion among the entities. Unlike a chemist who might seek the detailed aggregation mechanisms and cluster structures, a statistical physics viewpoint is to stand back from this level of microscopic detail. This concept is called *coarse-graining*. We hence coarse-grain the system to the cluster level, beneath which no microscopic details are specified. We call this scale *mesoscopic* which is a dimension between microscopic particles and potential macroscopic phenomena due to collective behaviour of the clusters. At this level of abstraction, the most detailed knowledge we can have about the system is only the size k of a cluster, and thus a crucial set of physical quantities is the cluster size distribution $\{n_k\}$ as hinted at in Chaps 1 and 2. Namely, we focus on the number of clusters n_k which have k particles. For a dynamic system that undergoes aggregation, these n_ks change in time. Now we can ask a physically interesting question: do we know anything about the dynamics of such a system?

The answer is yes, if we can condense the potentially complicated yet microscopic rate-determining aggregation mechanisms into a rate function at the cluster level. For aggregation theory, such a function is called a *kernel*. For the moment we only consider a *binary* coalescence kernel by assuming that a multi-cluster aggregation is rare and can be thought of as successive binary aggregations. However we note that in coordinated online disinformation campaigns, the merging of multiple entities and clusters may indeed occur. The binary kernel is denoted as κ_{ij} for two merging clusters with sizes i, j. Taking an aggregating-only (i.e. no fission) system with size distribution $\{n_k, k \in \mathbb{Z}_+\}$ and aggregation kernel κ_{ij}, we can write down the differential equations of $\{n_k\}$ for all k:

$$\dot{n}_k(t) = \frac{1}{2} \sum_{\substack{1 \leqslant i \leqslant k-1 \\ 1 \leqslant j \leqslant k-1}}^{i+j=k} \kappa_{ij} n_i(t) n_j(t) - \sum_{i=1}^{\infty} \kappa_{ik} n_i(t) n_k(t), \quad k \geqslant 2 \qquad (3.1)$$

$$\dot{n}_k(t) = -\sum_{i=1}^{\infty} \kappa_{ik} n_i(t) n_k(t). \quad k = 1 \qquad (3.2)$$

These are the famous *Smoluchowski coagulation equations*. These equations are, of course, much easier to write down than to solve, due to the sheer number of coupling terms within all the summations. Let's look at these sums in turn in order to fully understand each one.

1. The first summation in eqn (3.1) includes all possible pairs of clusters that can merge into one with size k, thereby contributing to the number of k-sized clusters. Here the prefactor $1/2$ eliminates the redundancy of double counting. This summation does not appear in eqn (3.2) since a size-1 cluster cannot be further divided based on our fundamental assumptions.

2. The other sum that appears in both equations accounts for loss of size-k clusters due to their aggregation with other clusters, hence the minus sign. The summation runs throughout all possible cluster sizes in the system. This is shown in the upper summation limit, which, being infinite, indicates that the total number of particles is so large that it is effectively another level of magnitude, and consequently the largest possible cluster is also of macroscopic scale (this will be central to our later discussions).

We can summarize the two cases in one equation by expanding the range of summation to include $k = 0$ ($k \in \mathbb{N}$) and defining $n_0(t) = 0$. Equations (3.1) and (3.2) together then become

$$\dot{n}_k(t) = \frac{1}{2} \sum_{\substack{0 \leq i \leq k \\ 0 \leq j \leq k}}^{i+j=k} \kappa_{ij} n_i(t) n_j(t) - \sum_{i=1}^{\infty} \kappa_{ik} n_i(t) n_k(t), \tag{3.3}$$

where $k \in \mathbb{Z}_+$. This ensures that the first term goes to zero when $k = 1$, thus recovering eqn (3.2). Since eqn (3.3) incorporates all mesoscopic information of its underlying aggregation system, we often call them the *master equations*.

Up to now, we have found the key set of dynamic functions of an aggregating system $\{n_k\}$, and constructed a set of ordinary differential equations in terms of these functions. This is in fact rather surprising: our equations contain no information about the physical locations or topological structures of any clusters because of their intrinsic mesoscopic scale. Those who are familiar with describing an evolving system with *partial* differential equations such as Newton's equations of motion may be looking for spatial terms: however, the system is at this stage approximated as being *homogeneous* in space so that a cluster is able to interact with all other clusters with the same ease—this is why we sum up $n_i n_j$ on the RHS (Right-Hand Side) in order to include all possible pairs of size-i, j clusters, which all contribute to formation of size-k clusters. This assumption turns out to be very relevant when studying some specific complex systems such as online interactions, since physical locations or distances become irrelevant in cyberspace. We will actually obtain some very important PDEs (Partial Differential

Equations) later in the chapter, but the variables besides time will still not be spatial. As for the geometry of the clusters, our discussion goes beyond the scope of such microscopic details as stressed above. In other words, in order to faithfully describe a system, we encapsulate important system-wide information about its cohesion into parameters such as the kernels, but ignore microscopic details. For example, if we study an aggregation system where particles only form linear clusters, i.e. one particle only connects to at most two others in a cluster, there is a specific kernel for this system (see later), but we do not care if the cluster is straight, curved, or spiral.

Mathematically, the complexity of the master equations lies not only in the number of terms, but also in the nature of these terms. They are all of second order due to the binary aggregation. The non-linearity would be of even higher order if multi-cluster aggregations were allowed. Non-linearity is a prominent property of any aggregation system. It leads to complicated mathematics but also fascinating phenomena, as we will come to later in the book.

3.2.1 Generalization of Smoluchowski equations: continuous size

In the above form of the master eqn (3.3), the index of variable n_k is taken to be discrete since the cluster size k is an integer. However, it doesn't have to be the case if we take k to be some form of continuous mass. In the continuum limit, we can turn the suffix k into a continuous variable x. Equation (3.3) then becomes:

$$\dot{n}(x,t) = \frac{1}{2}\int_0^x \kappa(x-y,y)n(x-y,t)n(y,t)\,dy - \int_0^\infty \kappa(x,y)n(x,t)n(y,t)\,dy, \quad (3.4)$$

whereby we simply transform the discrete sums into continuous integrals. We won't dive into details of this form for this book since our main interest still lies in discrete clusters. However, one insight from this formalism is that the first integration looks like a *convolution* operation. This insight helps with some key mathematical transformations coming up later.

3.3 Kernels

In this section, we look into various forms of the (binary) kernel function, some of which are found in the real world or serve as accurate approximations while others are mathematically valid but have not yet been observed. There are two straightforward properties that κ_{ij} follows: it is non-negative (i.e. it denotes fusion) and symmetric (that is, we reasonably assume that cluster i merging with cluster j is identical to the reversed merging process). Based on these, we are able to write down a generic form of the kernel, but first let's start from discussing a few simple ones. These kernels all lead to analytic solutions of the master equations. Moreover, they may at some stage be empirically justified for certain types of human–technology–AI systems.

3.3.1 Special kernels

Constant kernel $\kappa_{ij} = 1$. This is mathematically the simplest possible kernel, and the simplest constant 1 suffices here instead of an arbitrary constant c since we can always rescale the time t to absorb the arbitrary constant, by substituting $t \to ct$. This argument carries over to the other kernels discussed in this section. With a constant kernel eqn (3.3) becomes

$$\dot{n}_k(t) = \frac{1}{2} \sum_{\substack{0 \leqslant i \leqslant k \\ 0 \leqslant j \leqslant k}}^{i+j=k} n_i n_j - n_k \sum_{i=1}^{\infty} n_i. \tag{3.5}$$

Note that the sum

$$M_0(t) = \sum_{i=1}^{\infty} n_i(t) \tag{3.6}$$

is the total number of clusters in the system. Here we perform a trick to extract some valuable information from eqn (3.5) without lengthy calculation: we sum up both sides for all k to obtain

$$\frac{d}{dt}\left(\sum_{k=1}^{\infty} n_k(t)\right) = \frac{dM_0(t)}{dt} = \frac{1}{2} \sum_{k=1}^{\infty} \sum_{\substack{0 \leqslant i \leqslant k \\ 0 \leqslant j \leqslant k}}^{i+j=k} n_i n_j - M_0^2. \tag{3.7}$$

However, the double summation on the RHS is actually equivalent to M_0^2 with a regrouping of the sums by associativity. Hence, we can reduce the RHS to

$$\frac{dM_0}{dt} = -\frac{1}{2} M_0^2. \tag{3.8}$$

This is straightforward to solve. Assuming that the initial condition is $M_0(0) = N$, then

$$M_0(t) = \frac{2N}{2 + Nt}. \tag{3.9}$$

In obtaining eqn (3.7), we have actually completed a calculation of a *moment function*, which is a key concept that will be used frequently. $M_0(t)$ is the zeroth moment of the system. The dynamics of $M_0(t)$ is as expected: it decreases with time as more and more clusters merge together into larger ones. Now we can substitute it back into eqn (3.5) and solve the master equations one by one. Since they are recursive, we start from the smallest cluster and calculate bottom-up. For convenience, we assume an all-monomer initial condition, i.e.

$$n_k(0) = \delta_{k,1} N, \tag{3.10}$$

so that we don't have to carry too many constants in our calculations. This initial condition will come up frequently in the book: it describes a system that aggregates all the way up from monomers (size-1 clusters). Here we only provide solutions to the smallest two clusters.

1. Monomer. This is directly solvable with a separation of variables:

$$\dot{n}_1(t) = -\frac{2N}{2+Nt}n_1$$

$$\Rightarrow n_1(t) = N\left(\frac{2}{2+Nt}\right)^2.$$
(3.11)

2. Size-2 cluster.

$$\dot{n}_2(t) = \frac{1}{2}n_1^2 - \frac{2N}{2+Nt}n_2 = \frac{1}{2}N^2\left(\frac{2}{2+Nt}\right)^4 - \frac{2N}{2+Nt}n_2.$$

We reorganize the equation and use the integrating factor method.

$$\dot{n}_2(t) + 2\left(t+\frac{2}{N}\right)^{-1}n_2 = \frac{8N^2}{N^4}\left(t+\frac{2}{N}\right)^{-4}$$

$$\Rightarrow \frac{d}{dt}\left[n_2(t)\left(t+\frac{2}{N}\right)^2\right] = \frac{8}{N^2}\left(t+\frac{2}{N}\right)^{-2}$$

$$\Rightarrow n_2(t)\left(t+\frac{2}{N}\right)^2 = -\frac{8}{N^2}\left(t+\frac{2}{N}\right)^{-1} + \frac{4}{N}$$

$$= \frac{4t}{N}\left(t+\frac{2}{N}\right)^{-1},$$

where in the second line we multiply on both sides the integrating factor

$$\left(t+\frac{2}{N}\right)^2,$$

and in the third line we integrate from $t' = 0$ to t. The final result is therefore

$$n_2(t) = \frac{4t}{N}\left(t+\frac{2}{N}\right)^{-3} = \frac{4N^2 t}{(Nt+2)^3}.$$
(3.12)

By inspecting the forms of these two solutions, we can readily assume a general formula for $n_k(t)$ and prove it through induction. We leave this as an exercise.

Having found the solutions, now we can think about what kind of systems result in a constant kernel. Having a constant kernel means that the aggregation rate in the system is independent of cluster sizes, so there are only a *fixed* number of particles within a cluster that are open to interaction regardless of its size. One such example is a system of linear clusters, where one particle can only interact and connect with at most two particles, such that each cluster can only merge with others from its head or tail.[1]

Sum kernel $\kappa = i+j$. The corresponding master equations are

$$\dot{n}_k(t) = \frac{k}{2} \sum_{\substack{0 \leqslant i \leqslant k \\ 0 \leqslant j \leqslant k}}^{i+j=k} n_i n_j - \sum_{i=1}^{\infty} (i+k) n_i n_k. \tag{3.13}$$

This kernel is proportional to the size of the merged cluster, and is thus suitable for systems whose aggregation rate depends on the combined mass of the two merging clusters. Equation (3.5) can be solved with a similar approach as above by induction, although the solutions are more complicated with a double exponent structure.

Product kernel $\kappa = ij$. One of the most commonly found kernels, its physical intuition is clear. As 2 clusters, A of size i, and B of size j, are merging, the product kernel results from the interaction between each of all i particles in A and each of all j particles in B: hence the multiplicative rule, since the summed product ij provides the number of ways and hence the chance that the two clusters will merge (fuse). The equation reads

$$\dot{n}_k(t) = \frac{1}{2} \sum_{\substack{0 \leqslant i \leqslant k \\ 0 \leqslant j \leqslant k}}^{i+j=k} ij n_i n_j - \sum_{i=1}^{\infty} ik n_i n_k. \tag{3.14}$$

This kernel is found in a host of complex system settings, and will be a main focus of some upcoming sections for two reasons: (1) it is the kernel form that is observed empirically in Chap. 2 and also arises in human electronic communications (Palla et al., 2007), and (2) it correctly predicts the empirically observed patterns in Chap. 2. Similar inductive methods exist to solve the equations but we do not elaborate them here. As in the sum scenario, the solutions are not simple functions. We will, however, show important properties corresponding to this kernel by a simpler and more insightful approach later in this chapter.

3.3.2 Generic kernels

After discussing the three simple, analytically solvable examples of kernel functions, we now aim to find the generic form of all kernel functions based on non-negativity

---

[1] One must take care of the distinguishability of particles and directionality of clusters when dealing with a specific linear system. For a system of indistinguishable particles and accordingly linear clusters without directionality, the kernel is strictly constant, but it gets more complicated otherwise.

and symmetry. The most general symmetric and discrete two-variable function takes the following form:

$$\kappa_{ij} = \left| \sum_{\alpha,\beta \in \mathbb{Z}} A_{ij}^{\alpha\beta} (i^\alpha j^\beta + i^\beta j^\alpha) \right| \quad (3.15)$$

—an infinite series. Note that it is sufficient to only include polynomials with integer exponents α, β, since part of this infinite series can always be the series expansions of some more complicated functions, a couple of which will be seen in the examples below. There really isn't much useful information in such a general expression: we seek for symmetry, and we obtain only symmetry, of the most general form. The sheer number of degrees of freedom (infinite, from the double summation of all integers), however, mean that there should always exist a particular kernel, however complex it may be, that corresponds to a specific type of aggregation with its unique rules and characteristics. We will now provide a few typical kernels, because such kernels may prove useful in some online–offline human–technology–AI settings, e.g. where the spatial separation of entities becomes important.

Brownian kernel. Brownian motion describes the diffusion of a particle due to random thermal collisions with its surrounding particles. Its mechanism has been successfully interpreted by Einstein using random walk and diffusion models. Now let's consider a coagulation model based on the Brownian motion: if two particles coalesce right after collision instead of reflecting off (in other words, the collisions are perfectly inelastic), we then acquire a coagulation model with a Brownian coalescence rule, i.e. a *Brownian kernel*. Such a situation could certainly occur among certain entities in an online–offline human–technology–AI setting, so we will spend some time analysing it. In particular, we will derive the form from first principles.

We first consider a simpler model: instead of possible collisions and coalescence between every pair of particles, we assume that there is one special cluster (i.e. a nucleus) in the system that coalesces with other particles upon collision, while all the other particles are non-interacting. Though a toy model, this could be realistic in an online–offline human–technology–AI setting in which there is one dominant cluster (e.g. some dominant electronic communications or news-sharing system that mops up users). This special cluster will grow to a massive size (i.e. much larger than a single particle) and with an arbitrary shape. We consider its stationary frame of reference. From the setup assumptions, all the non-interacting particles undergo random walks, and the concentration of the particles satisfies the *diffusion equation*

$$\frac{\partial c(\boldsymbol{x}, t)}{\partial t} = D \nabla^2 c(\boldsymbol{x}, t), \quad (3.16)$$

where D is the diffusivity, with the boundary condition

38 Multi-body Physics 1.0: single-species fusion and fission

$$c(\boldsymbol{x},t)\big|_{\boldsymbol{x}\in\partial S} = 0, \tag{3.17}$$

with S being the boundary of the cluster. This is due to the instantaneous coalescence assumption in our model. In addition, we assume for simplicity a uniform initial condition

$$c(\boldsymbol{x},0) = 1. \tag{3.18}$$

Our goal is to calculate the coalescence rate. From basic calculus we know that this is simply the flux of $\nabla c(\boldsymbol{x},t)$

$$r(t) = D \int_{\partial S} \nabla c(\boldsymbol{x},t) \cdot \mathrm{d}\boldsymbol{\sigma}, \tag{3.19}$$

where σ is the infinitesimal area element. The next step might seem to be straightforwardly solving the diffusion equation for $c(\boldsymbol{x},t)$. But the usual Gaussian solution does not fit our boundary and initial conditions. So we can only settle for an approximate approach. This is OK provided that we choose the right condition and that the system is in the correct regime. Inspecting the system, we see that there is only one cluster for the particles to collide with. Hence if the system is so large that the frequency of collision is very small compared with the rate of diffusion, then considering the uniform distribution initially, $c(\boldsymbol{x},t)$ changes very slowly in time and we can therefore take a steady state approximation

$$D\nabla^2 c(\boldsymbol{x},t) = 0, \text{ with } c(\boldsymbol{x})\big|_{\boldsymbol{x}\in\partial S} = 0,\ c(\boldsymbol{x})\big|_{|\boldsymbol{x}|\to\infty} = 1. \tag{3.20}$$

Note that the last boundary condition is translated from the initial condition eqn (3.18). This is now reduced to a Laplace equation, but solving this still requires a bit of effort. However, we can take a shortcut. If we define a *complementary function*

$$\phi(\boldsymbol{x}) = 1 - c(\boldsymbol{x}), \tag{3.21}$$

which apparently satisfies the Laplace equation also, we can simplify the boundary conditions considerably into something familiar:

$$\phi(\boldsymbol{x})\big|_{\boldsymbol{x}\in\partial S} = 1,\ \phi(\boldsymbol{x})\big|_{|\boldsymbol{x}|\to\infty} = 0. \tag{3.22}$$

There is now a nice connection to the domain of electrostatics, since $\phi(\boldsymbol{x})$ acts like an electrostatic potential. Our rate function written in terms of ϕ then becomes

$$r(t) = -D \int_{\partial S} \nabla \phi(\boldsymbol{x},t) \cdot \mathrm{d}\boldsymbol{\sigma}. \tag{3.23}$$

However, we know that Gauss' law states that the flux equals the total electrostatic charge enclosed in the boundary ∂S:

$$-\frac{1}{4\pi}\int_{\partial S}\boldsymbol{\nabla}\phi(\boldsymbol{x},t)\cdot\mathrm{d}\boldsymbol{\sigma}=Q. \quad (3.24)$$

Note also that we have taken the vacuum permittivity constant ϵ_0 to be 1 since it is simply a trivial selection of a unit system. Hence

$$r(t)=4\pi DQ, \quad (3.25)$$

with this 'charge' Q being the only unknown to calculate. What we do know is the potential on the boundary, so if we have knowledge of the geometry of this boundary (i.e. the cluster), we can write down its capacitance C and thereby obtain $Q=C\phi(\boldsymbol{x}\in\partial S)$. Now consider that we establish this steady state at large time $t\to\infty$ and the clustering has been homogeneous, i.e. approximately spherical, then the capacitance $C=R$; and hence finally

$$r(t)=4\pi DR. \quad (3.26)$$

So far, we have obtained a concise result for our toy model, and the remaining work is to migrate the derivations onto the real Brownian coalescence model, where all pairs of particles are able to coalesce by collision. For the above model to still hold, we need to insist on the *low collision frequency* limit. Again, we are comparing the collision frequency to a diffusion timescale τ_0. The former can be measured by the average time between two collisions τ_c, whereas the latter can be evaluated from the Langevin theory to be of the following form,

$$\tau_0=\frac{mD}{k_BT}. \quad (3.27)$$

The Brownian coalescence regime then occurs under the condition

$$\tau_c\gg\tau_0. \quad (3.28)$$

Since we no longer have a central cluster that is much bigger and heavier than all the other particles due to its monopoly on coalescence, taking the rest frame of a specific cluster is not useful any more. As we return to the lab frame, *relative motion* requires us to modify eqn (3.26) accordingly. The first modification is the diffusivity D:

$$D_{12}=\frac{\langle(\boldsymbol{x}_1-\boldsymbol{x}_2)^2\rangle}{6t}=\frac{\langle\boldsymbol{x}_1^2\rangle}{6t}-\frac{\langle\boldsymbol{x}_1\cdot\boldsymbol{x}_2\rangle}{3t}+\frac{\langle\boldsymbol{x}_2^2\rangle}{6t}=D_1+D_2 \quad (3.29)$$

since $\langle\boldsymbol{x}_1\cdot\boldsymbol{x}_2\rangle=0$. Furthermore, the size scale R also needs changing. In our toy model, the central cluster is much bigger than the particles so the collision cross-section is effectively just that of the cluster. However, for two clusters 1 and 2 with comparable sizes it has to be modified into $(R_1 + R_2)$, the distance of their centres when colliding together. Therefore, the Brownian coalescence kernel is

$$\kappa_{ij} = 4\pi(D_i + D_j)(R_i + R_j). \tag{3.30}$$

Can we express it with cluster sizes only, just like the earlier examples? The answer is affirmative if we assume that the masses and densities of all particles are identical. We can then write the masses of a cluster of size k as

$$m_k = \rho R_k^3 = km_1,$$

where m_1 is the mass of a monomer. It follows that $R_k \propto k^{\frac{1}{3}}$. Moreover, the Langevin theory provides the relation between diffusivity and cluster size

$$D = \frac{T}{6\pi\eta R}, \tag{3.31}$$

where T is the temperature and η the viscosity. Hence, $D_k \propto k^{-\frac{1}{3}}$, and ultimately,

$$\kappa_{ij} = \left(i^{-\frac{1}{3}} + j^{-\frac{1}{3}}\right)\left(i^{\frac{1}{3}} + j^{\frac{1}{3}}\right), \tag{3.32}$$

where again we omit all constant coefficients.

Exponential kernel. As previously mentioned, the generic kernel in eqn (3.15), or at least part of it, can be an infinite series. This means that theoretically, kernels are not limited to polynomials. Below is one example

$$\kappa_{ij} = ije^{-\alpha(i+j)}, \tag{3.33}$$

a simple product kernel modified by an exponential function, where α is constant. Its effect is quite obvious: it suppresses coalescence between large clusters. There is not much to introduce beyond that at this point, but we will see it again later in the book as well as the exercises.

3.4 Inclusion of heterogeneity

We have so far discussed the basic constructions of Smoluchowski's coagulation theory. However, if we are to apply this theory reliably to fusion of entities in a human–technology–AI system such as Fig. 1.2, we really need to introduce the *heterogeneity* of the entities. The above discussions are built upon an assumption that particles (entities) are akin to homogeneous spheres, especially in deriving the Brownian kernel. While this identical particle assumption can often be applied to classical systems under physical interactions, it most certainly does not for complex systems of our interest in

Fig. 1.2. The particles are inherently different and the difference will affect coalescence rates. Our task is now to propose a means to include the heterogeneity. We will do this at successively higher levels of sophistication—which mathematically means with successively lower amounts of approximation—in this and the following two chapters.

The strategy that we adopt is to seek some kind of mean-field theory. Retaining microscopic details of heterogeneous attributes that particles (entities) possess, is not of interest—we care instead about the collective effects of this collective heterogeneity. In other words, the challenge is to average the heterogeneous details in some way, so as to produce a coarse-grained measure of the heterogeneity. But first we need to be able to quantify this heterogeneity. To do this, we appeal to an idea which lies at the heart of how information of all sorts is encoded in Machine-Learning and AI, i.e. we assume that it is possible to encapsulate all the relevant microscopic information about a particle (entity) into a d-dimensional vector $\vec{y}$, each of whose d dimensions define a heterogeneous attribute. We call it the *character* of the particle. In this way, each entity in Fig. 1.2 will have its own vector and hence its own character. From this definition, we can now refer to a particle p by its character $\vec{y}_p$. The study of what vectors are the most accurate characters is itself worth a PhD thesis and not pursued here: but we often find that the simplest ones suffice for our purpose. A couple of examples will be shown later.

Since we are studying aggregation (fusion), we need to determine the effect of these characters and hence the underlying entity heterogeneity, on the coalescence rates. Mathematically, this means defining a function $S(\vec{y}_1, \vec{y}_2, t)$, which we could name a similarity measure. It could, for example, modify the aggregation rate of particles $\vec{y}_1$ and $\vec{y}_2$ according to the difference between the two particles' characters, though in general, more complicated functional dependencies may occur (Johnson et al., 2013a; Manrique et al., 2015; Manrique and Johnson, 2018). Moreover, the precise form of the modification will depend on the human–technology–AI system of interest: *homophily* (similar entities are more likely to aggregate), *heterophily* (similar entities are less likely to aggregate), or no influence, in which case S is constant, among other possibilities. We are now one step away from obtaining a mean-field measurement of heterogeneity. From the definition above, the quantity to average is just $S(\vec{y}_1, \vec{y}_2, t)$. This requires the knowledge of the system-wide character distribution $q(\vec{y}, t)$, which can be time-dependent in principle. Averaging over the system then produces a function $F(t)$ defined below,

$$F(t) = \int_{\vec{y}_1 \in \mathcal{V}} \int_{\vec{y}_2 \in \mathcal{V}} S(\vec{y}_1, \vec{y}_2, t) q(\vec{y}_1, t) q(\vec{y}_2, t) \, d\vec{y}_1 \, d\vec{y}_2 \,, \qquad (3.34)$$

where $\mathcal{V}$ is the vector space of the character vectors $\vec{y}_{1,2}$, i.e. the range of all possible characters. This resultant function is free from microscopic dependence on any particle, and is thus a valid macroscopic variable suitable for our theory. It follows that the kernel function κ_{ij} should be modified to $F(t)\kappa_{ij}$, and so the modified Smoluchowski's equations for a heterogeneous system at this level of mean-field approximation become:

$$\dot{n}_k(t) = \frac{F(t)}{N^2} \sum_{\substack{0 \leq i \leq k \\ 0 \leq j \leq k}}^{i+j=k} \kappa_{ij} n_i(t) n_j(t) - \frac{2F(t)}{N^2} \sum_{i=1}^{\infty} \kappa_{ik} n_i(t) n_k(t), \qquad (3.35)$$

where N is the total population. Compared with eqn (3.3), we note that here a factor $2/N^2$ is multiplied on the RHS. This is meant to take into consideration the probability of random selection of particle pairs to connect. It does not contradict eqn (3.3) since we are free to rescale the time t and the system and hence cancel the factor. For convenience, we define the rescaled time $\tau = 2t/N$ and normalized size distribution $\nu_k(t) = n_k(t)/N$. Equation (3.35) then reduces to

$$\dot{\nu}_k(\tau) = \frac{F(\tau)}{2} \sum_{\substack{0 \leq i \leq k \\ 0 \leq j \leq k}}^{i+j=k} \kappa_{ij} \nu_i \nu_j - F(\tau) \sum_{i=1}^{\infty} \kappa_{ik} \nu_i \nu_k. \qquad (3.36)$$

Note that we omitted the time dependence notation of $\nu_i(\tau)$ on the RHS for conciseness. From now on, by *master equations* we refer to this set of equations.

Now we duly provide a few simple examples of the minimal models we use to calculate $F(t)$. For simplicity, we take a 1-D character $y \in [0, 1]$, and a time-independent uniform character distribution $q(y) = 1$. For aggregation under homophily, we can define $S(y_1, y_2) = 1 - |y_1 - y_2|$, and hence the mean-field aggregation probability becomes

$$F = \int_0^1 dy_1 \, dy_2 \, (1 - |y_1 - y_2|) = \frac{2}{3}. \qquad (3.37)$$

By contrast, in the case of heterophily, we can define $S(y_1, y_2) = |y_1 - y_2|$, and the result is naturally $F = 1 - 2/3 = 1/3$. These toy models can work surprisingly well in describing real-world systems, proving the capability and robustness of such a minimal model.

3.5 Emergent phenomena from fusion

Our goal in this chapter is to construct a theory which faithfully describes collective behaviours in complex systems that exhibit aggregation. Up to now we have obtained the master equations (eqn (3.35)) in terms of the cluster size distribution $\{n_k\}$, generalized from Smoluchowski's coagulation theory. While each of these mesoscopic equations operates at the cluster level, we hope to use them collectively in order to deduce a system-level (i.e. macroscopic) description of the system's collective behaviour. This turns out to be possible, and in this section we will discuss some simple mathematics that lead to intriguing macroscopic phenomena.

3.5.1 Moment functions

First of all, we need to define some macroscopic quantities obtainable from our model, i.e. quantities that can be expressed with $\{n_k\}$. In the case of system-level functions, two obvious candidates are a couple of infinite sums:

1. sum of all n_ks, or the total number of clusters

$$M_0 = \sum_{k=0}^{\infty} n_k, \qquad (3.38)$$

2. sum of all kn_ks, or the total number of particles

$$N = \sum_{k=0}^{\infty} kn_k. \qquad (3.39)$$

Note that the lower limits in the summations are both 0 instead of 1, since following the definition in Section 3.2, we are simply adding to these sums a redundant $n_0 = 0$. Extending the range of k simplifies the calculations that follow. Equation (3.38) should look familiar as we have used the zeroth moment when calculating the size distribution in the case of the constant kernel in Section 3.3. Mathematically, these belong to a family of functions called *moments*. Generally, the pth moment of the system is defined as

$$M_p(t) = \sum_{k=0}^{\infty} k^p n_k(t). \qquad (3.40)$$

We quickly recognize that the total number of particles N is just the first moment of the system

$$N = M_1. \qquad (3.41)$$

The moment is not an unfamiliar concept in physics; one can easily invoke such definitions as moments of inertia, moments of force (torque), etc. Many of the physical moments appear in rotational systems precisely due to the '$\sum$ distribution × index' structure of moments. In our context, moments of the lowest three orders are the most important: we have shown the physical significance of the zeroth and first moments, and the meaning of M_2 will shortly be revealed.

3.5.2 Dynamics of the moments and generating functions

To study the properties of these moment functions, it is straightforward to look into their dynamics. By definition, these can be calculated from the master equations, since

$$\frac{dM_p(t)}{dt} = \sum_{k=0}^{\infty} k^p \dot{n}_k, \qquad (3.42)$$

where $\dot{n}_k$ can be substituted by eqn (3.35):

$$\frac{dM_p(t)}{dt} = \sum_{k=0}^{\infty} k^p \left[\frac{F(t)}{N^2} \sum_{\substack{0 \leqslant i \leqslant k \\ 0 \leqslant j \leqslant k}}^{i+j=k} \kappa_{ij} n_i(t) n_j(t) - \frac{2F(t)}{N^2} \sum_{i=1}^{\infty} \kappa_{ik} n_i(t) n_k(t) \right]$$

$$= \frac{F(t)}{N^2} \sum_{i=0}^{\infty} \sum_{j=0}^{\infty} \kappa_{ij} (i+j)^p n_i(t) n_j(t) - \frac{2F(t)}{N^2} \sum_{k=1}^{\infty} \sum_{i=1}^{\infty} \kappa_{ik} n_i(t) k^p n_k(t). \quad (3.43)$$

Note that in the first term after the second equality, we are able to regroup the double summations simply by the addition of a $k = 0$ term, and then separate k into i, j. This approach had come up earlier when we were explicitly calculating $\{n_k\}$ under the constant kernel. Now with κ_{ij} known and ideally being one of the special kernels previously discussed, this can be further reduced into ODEs of $M_p(t)$, albeit non-linear in most cases (see exercises). However, this becomes impossible for more generic kernels, and we need a more universal approach to the dynamics.

Hence we introduce a *generating function* $\mathcal{D}(x,t)$ defined as follows,

$$\mathcal{D}(x,t) = \sum_{k=0}^{\infty} n_k(t) e^{-kx}, \quad (3.44)$$

where $x \in [0, \infty)$ is a continuous variable—its non-negative domain guarantees that $\mathcal{D}(x,t)$ converges. It is so called because it can generate all moment functions via partial differentiations:

$$M_p(t) = (-1)^p \left[\frac{\partial^p}{\partial x^p} \mathcal{D}(x,t) \right]_{x=0}. \quad (3.45)$$

A closer look at the definition reveals that eqn (3.44) is actually a discrete *Laplace transform* from $n_k(t)$ to $\mathcal{D}(x,t)$. The Laplace transform is one of the integral transformations of functions, alongside the familiar *Fourier transform*.[2] While the latter uses trigonometric basis functions, the Laplace transform utilizes exponential ones. Heuristically, many of the Laplace transform's properties bear a similarity to those of the Fourier transform, including the linearity of the transformation and a similar transform table.

Equation (3.44) establishes a duality between the mesoscopic quantity $n_k(t)$ and the macroscopic function $\mathcal{D}(x,t)$. Its significance goes beyond a mere definition and mathematical tool to calculate moments. The abstract variable x lives in the reciprocal space of the size variable k. Moreover, it is this definition that bridges the two scales of

---

[2]We can now further appreciate the addition of the $k = 0$ term: mathematically it fulfils the definition of the transform.

our interest. Therefore, whenever we study macroscopic emergent phenomena, including but also beyond coalescence, we should come back to this definition.

We can draw another interesting comparison to eqn (3.44): it looks like the *partition function* familiar to us from equilibrium statistical mechanics. Here x occupies a similar position to the *inverse temperature* β. The partition function of a statistical ensemble contains all information of the system, and we find that the same is true for our generating function, assuming we ignore microscopic behaviours beneath the cluster level.

Recall that we have defined a normalized cluster size distribution $\{\nu_k(\tau)\}$ above. Correspondingly we can also define a normalized generating function

$$\delta(x,\tau) = \sum_{k=0}^{\infty} \nu_k(\tau) e^{-kx} = \frac{n_k(\tau)}{N} e^{-kx} \tag{3.46}$$

and normalized moment functions

$$m_p(t) = (-1)^p \left[\frac{\partial^p}{\partial x^p} \delta(x,\tau) \right]_{x=0}. \tag{3.47}$$

In the following sections, we will focus on the generating function in order to complete the mathematical construction of the aggregation model based upon the master equations.

3.5.3 Generative PDEs

Our motivation for introducing the generating function was the dynamics of the moment functions. This requires us to first find the differential equation that governs the dynamics of $\mathcal{D}(x,t)$. As a result of the transformation relation established in eqn (3.44), such an equation can be obtained directly by transforming the master equations (eqn (3.35)). This is of course equivalent to direct substitution as shown in eqn (3.43), but being able to apply well-known transformations massively simplifies the calculation. In addition, as will be shown soon, the PDE of $\mathcal{D}(x,t)$ implies much more than just the dynamics of the moments.

Now we demonstrate how to transform the master equations eqn (3.35) into x-space. To begin, note that the linearity of the Laplace transform enables us to transform each term in the master equations separately. To illustrate the generality of this approach, we choose the generic kernel: but taking the whole infinite series (eqn (3.15)) is rather redundant due exactly to the linearity; hence without loss of generality, we can simply take the kernel to be one generic term $A_{\alpha\beta} \left(i^\alpha j^\beta + i^\beta j^\alpha \right)$, $\alpha, \beta \in \mathbb{Z}$. The master equations with the above kernel read

$$\dot{\nu}_k(\tau) = \frac{F(\tau)}{2} A_{\alpha\beta} \sum_{\substack{0 \leqslant i \leqslant k \\ 0 \leqslant j \leqslant k}}^{i+j=k} \left(i^\alpha j^\beta + i^\beta j^\alpha \right) \nu_i \nu_j - F(\tau) A_{\alpha\beta} \sum_{i=1}^{\infty} \left(i^\alpha k^\beta + i^\beta k^\alpha \right) \nu_i \nu_k. \quad (3.48)$$

First, note that i, j in the first term on the RHS under summation are symmetric, so the equation can be simplified into

$$\dot{\nu}_k(\tau) = \frac{F(\tau)}{2} A_{\alpha\beta} \sum_{\substack{0 \leqslant i \leqslant k \\ 0 \leqslant j \leqslant k}}^{i+j=k} i^\alpha \nu_i j^\beta \nu_j - F(\tau) A_{\alpha\beta} \left[k^\beta \nu_k \sum_{i=1}^{\infty} i^\alpha \nu_i + k^\alpha \nu_k \sum_{i=1}^{\infty} i^\beta \nu_i \right]$$

$$= \frac{F(\tau)}{2} A_{\alpha\beta} \sum_{\substack{0 \leqslant i \leqslant k \\ 0 \leqslant j \leqslant k}}^{i+j=k} i^\alpha \nu_i j^\beta \nu_j - F(t) A_{\alpha\beta} \left[k^\beta \nu_k m_\alpha(\tau) + k^\alpha \nu_k m_\beta(\tau) \right]. \quad (3.49)$$

Now we can apply well-known transform table results:

1. $\nu_i(\tau)$ is simply transformed into $\delta(x, \tau)$.
2. $(-i)^\alpha \nu_i(\tau)$ is transformed into the αth derivative of $\delta(x, \tau)$ for $\alpha > 0$, or the αth anti-derivative for $\alpha < 0$; likewise the transform of $j^\beta \nu_j(\tau)$.
3. Motivated by the continuous Smoluchowski's equation (eqn (3.4)), the first term on the RHS is a discrete convolution between $i^\alpha n_i(t)$ and $j^\beta \nu_j(\tau)$ and is transformed into products of the respective transformed functions.
4. Time dependence is not touched by the Laplace transform, since it occurs between k-space and x-space only.

The final result hence follows:

$$\frac{\partial \delta(x, \tau)}{\partial \tau} = F(\tau) A_{\alpha\beta} \left[(-1)^{\alpha+\beta} \frac{\partial^\alpha \delta}{\partial x^\alpha} \frac{\partial^\beta \delta}{\partial x^\beta} - (-1)^\beta \frac{\partial^\beta \delta}{\partial x^\beta} m_\alpha(\tau) - (-1)^\alpha \frac{\partial^\alpha \delta}{\partial x^\alpha} m_\beta(\tau) \right]. \quad (3.50)$$

The resulting equation is a non-linear PDE in terms of the generating function. We call it the *generative equation*. We note for later that a typical physical system characterized by non-linear PDEs is fluid dynamics, in which famous equations include the Navier–Stokes equation for example.

Following the general derivation, we can then look into the three special kernels, especially the product and sum kernels, for which solutions are known to exist but involve too lengthy algebra. We will find out if the generative equations tell us anything about their behaviours with less effort.

Constant kernel revisited. For the constant kernel, we take $\alpha = \beta = 0$ and $A_{\alpha\beta} = 1/2$, and eqn (3.50) becomes

$$\frac{\partial \delta(x, \tau)}{\partial \tau} = \frac{F(\tau)}{2} \delta^2 - F(\tau) m_0(\tau) \delta. \quad (3.51)$$

$m_0(t)$ has been calculated in eqn (3.9) when we were solving the Smoluchowski's equations directly for the same kernel. Now with the addition of the heterogeneity factor $F(t)$ and the new notation, we solve for $m_0(t)$ from eqn (3.51). Simply taking $x = 0$ as there are no x-derivatives in the equation to evaluate, we obtain

$$\frac{dm_0(\tau)}{d\tau} = -\frac{F(\tau)}{2} m_0(\tau)^2, \tag{3.52}$$

by substituting eqn (3.45). We can see that this only differs from eqn (3.8) by the additional factors in the master equations, hence proving that the generative equation approach is working. For a simple calculation, we assume that the heterogeneity factor F is constant in time. Solving this as before with the all-monomer initial condition (the canonical initial condition for the aggregating system)[3]

$$m_0(0) = 1, \tag{3.53}$$

we then have

$$m_0(\tau) = \frac{1}{1 + F\tau/2}. \tag{3.54}$$

The (un-normalized) first moment is by definition N and constant (hence $m_1 = 1$). To calculate $m_2(\tau)$, we start by taking two partial derivatives with respect to x in eqn (3.51):

$$\frac{\partial \delta''(x,t)}{\partial \tau} = F\left[\delta'^2 + \delta\delta'' - m_0(\tau)\delta''\right], \tag{3.55}$$

where δ' simply denotes a partial derivative in x. Next, we take $x = 0$ again to obtain

$$\frac{dm_2(\tau)}{d\tau} = F, \tag{3.56}$$

where the second and third terms cancel. The solution then becomes straightforwardly

$$m_2(\tau) = 1 + F\tau. \tag{3.57}$$

This linear behaviour of $m_2(\tau)$ will prove to be crucial later in this chapter when we discuss the significance of the second moment. For now we just leave the simple result as it is.

Product kernel revisited. The sum kernel is left as an exercise, as we jump straight into the most important kernel in this chapter. We return to eqn (3.50) and take $\alpha = \beta = 1$ and $A_{\alpha\beta} = 1/2$ to obtain

$$\frac{\partial \delta(x,\tau)}{\partial \tau} = \frac{F(\tau)}{2}\left(\frac{\partial \delta}{\partial x}\right)^2 + F(\tau)\frac{\partial \delta}{\partial x}. \tag{3.58}$$

---

[3] In the canonical initial condition, all moments are equal to the total number of particles.

To further simplify this equation, we define another generating function

$$\mathcal{E}(x,t) = -\frac{\partial \mathcal{D}(x,t)}{\partial x} = \sum_{k=0}^{\infty} k n_k e^{-kx} \qquad (3.59)$$

and its normalized form

$$\varepsilon(x,\tau) = -\frac{\partial \delta(x,\tau)}{\partial x} = \sum_{k=0}^{\infty} k \nu_k e^{-kx}, \qquad (3.60)$$

and then substitute it into eqn (3.58):

$$\frac{\partial \delta(x,\tau)}{\partial \tau} = \frac{F(\tau)}{2}\varepsilon^2 - F(\tau)\varepsilon. \qquad (3.61)$$

Taking one further derivative in x gives

$$-\frac{\partial \varepsilon(x,\tau)}{\partial \tau} = F(\tau)\varepsilon \frac{\partial \varepsilon(x,\tau)}{\partial x} - F(\tau)\frac{\partial \varepsilon(x,\tau)}{\partial x}$$

$$\Rightarrow \frac{\partial \varepsilon(x,\tau)}{\partial \tau} + F(\tau)(\varepsilon - 1)\frac{\partial \varepsilon(x,\tau)}{\partial x} = 0. \qquad (3.62)$$

To see its structure more clearly, we make the following substitution

$$\mathcal{G}(x,\tau) = \varepsilon - 1, \qquad (3.63)$$

and it follows that (since N is constant)

$$\frac{\partial \mathcal{G}(x,\tau)}{\partial \tau} + F(\tau)\mathcal{G}\frac{\partial \mathcal{G}(x,\tau)}{\partial x} = 0. \qquad (3.64)$$

Those who are familiar with fluid dynamics will recognize this equation as the *inviscid Burgers' equation*. This is an intriguing discovery: our generative equations not only share the non-linear property with PDEs in fluid dynamics, one of them *is* actually a famous hydrodynamic equation. We will discuss this important equation later, but now let's first calculate the moments.

Zeroth moment. We again take F to be constant for simplicity. Take $x = 0$ in eqn (3.58):

$$\frac{dm_0(\tau)}{dt} = -\frac{F}{2} \Rightarrow m_0(\tau) = 1 - F\frac{\tau}{2}. \qquad (3.65)$$

Second moment. Taking one more derivative in x from eqn (3.62) gives

$$\frac{\partial \varepsilon'(x,\tau)}{\partial \tau} + F\left[\varepsilon'^2 - (\varepsilon - 1)\varepsilon''\right] = 0. \qquad (3.66)$$

We then follow the same procedure that x is taken to be 0 and get

$$\frac{dm_2(\tau)}{d\tau} = Fm_2(\tau)^2$$

$$\Rightarrow m_2(t) = \frac{1}{1 - F\tau}. \tag{3.67}$$

Different from the constant kernel case, here this second moment diverges at a specific time given by

$$\tau = \frac{2t}{N} = \frac{1}{F} \tag{3.68}$$

which as we discuss next, becomes the onset time t_c of the cohesive unit from Chap. 1.

3.5.4 Gelation and second moment

We have calculated the second moments corresponding to two aggregation kernels: the constant kernel yields a second moment that grows linearly in time, whereas the product kernel yields a second moment that diverges at some finite time. Why do they behave differently? What does divergence mean in this context? Do generative equations have connections to such behaviours?

Let us recall the previous discussions on the infinite sums of n_k, kn_k, etc., which have appeared numerous times in this chapter already. We mentioned the reason why the upper limit of the sums is taken to be infinite: since $N \gg 1$, any cluster with size comparable to N would be considered a macroscopic entity that is essentially in a different phase from the rest of the clusters. Such a giant cluster is called a *gel* in traditional physics terminology (i.e. a cohesive unit in this book), and the process of gel formation is called *gelation*. In this case the value of N is 'almost infinite' at the cluster level. Realistically, for a size distribution $\{n_k(t)\}, k = (0,)1, 2, \cdots, \infty$, the largest possible k for which $n_k(t)$ can be non-zero is $k = N$, and

$$n_k(t) = 0, \ \forall k > N. \tag{3.69}$$

Note that in this original model we have developed so far, N is always conserved, but generalizations of the aggregation model can make it time-dependent as well. We will actually discuss those systems later. Now back to gelation and moments: let's assume that at some later time t a giant cluster of size $g \sim N$ has formed and is much larger than all the other clusters. In other words, it is on the brink of becoming a gel. To be more specific, we further assume that the largest cluster among the rest is of size $k_m \ll g$. Then we can write down the following constraints on the size distribution:

$$n_k(t) = \begin{cases} 0 & \forall k > g \\ 1 & k = g \end{cases}.$$

Furthermore, we can deduce some relations that the moment functions (of the three lowest orders) must satisfy.

1. The zeroth moment

$$M_0(t) = \sum_{k=1}^{g} n_k(t) = \sum_{k=1}^{k_m} n_k(t) + 1 < \sum_{k=1}^{g-1} n_k(t) + g \leqslant N, \qquad (3.70)$$

where in the last inequality we note that the equality is reached iff (if and only if) all the clusters except the gel are monomers, in which case $\sum_{k=1}^{g-1} n_k(t)$ takes the maximum value $(N-g)$. The result shows that $M_0(t)$ is always bounded by the first moment N, and so can never diverge.

2. The first moment

$$M_1(t) = N \qquad (3.71)$$

trivially.

3. The second moment

$$M_2(t) = \sum_{k=1}^{g} k^2 n_k(t) = \sum_{k=1}^{k_m} k^2 n_k(t) + g^2.$$

We claim that its value is dominated by g^2, i.e. $M_2(t) \approx g^2$. Below we prove that

$$g^2 \gg \sum_{k=1}^{k_m} k^2 n_k(t). \qquad (3.72)$$

Proof We know with certainty that

$$\sum_{k=1}^{k_m} k n_k(t) = N - g. \qquad (3.73)$$

Then,

$$\sum_{k=1}^{k_m} k^2 n_k(t) = \sum_{k=1}^{k_m} k \cdot k n_k(t)$$

$$\leqslant \sum_{k=1}^{k_m} k_m \cdot k n_k(t) = k_m (N - g), \qquad (3.74)$$

where we note that equality can be reached iff

$$n_k(t) = \begin{cases} 0 & 1 \leqslant k \leqslant k_m \\ (N-g)/k & k = k_m \end{cases}. \qquad (3.75)$$

However, from our assumption, the gel size g is of the same scale as N, i.e. $g \gg k_m$ while $g \sim N - g$. Therefore, $g^2 \gg k_m(N-g) \geqslant \sum_{k=1}^{k_m} k^2 n_k(t)$ and

$$M_2(t) \approx g^2 \to \infty, \qquad (3.76)$$

since $M_2(t)$ is a second-order term of a close-to-infinite value. □

The above behaviours of the first three moments in gelation show that the divergence of the *second moment* corresponds to the occurrence of gelation (i.e. formation of a cohesive unit) while the other two moments remain bounded. We have finally derived the significance of the second moment after much teasing. Now we can review the previous results from eqns (3.57) and (3.67). The linear second moment under the constant kernel means that the system does *not* gel, a result that the reader will verify in the exercises by finding the general expression of the size distribution functions $\{n_k\}$. By contrast, the diverging moment under the product kernel indicates that gelation occurs at time $t_g = N/2F$ from eqn (3.68). This is the *gelation time*, which in the language of this book is the onset time of the cohesive unit.

3.6 Gelation and shock wave

After addressing one of the most important emergent phenomena that can arise in the system, we come back to study the generative equations—especially the inviscid Burgers' equation corresponding to the product kernel. In fluid dynamics, this equation describes a shock wave, which is, incidentally, also an instance of diverging phenomena. This implies that the generative equation is much more than a mere precursor of the ODEs of moment functions: rather, the generative equation ought to be an intrinsic law of the emergent dynamics of such an aggregation-based system. To confirm this, we are going to solve the Burgers' equation as in eqn (3.64) and compare the result with the one from the second moment. Most generative equations, being non-linear PDEs, have no analytical solutions, but fortunately the inviscid Burgers' equation is simple enough to solve.

The standard approach to solve an inviscid Burgers' equation is via the method of characteristics. For that the normalized variables come in handy. Recall that eqn (3.62) takes the following form:

$$\frac{\partial \varepsilon}{\partial \tau} + F(\varepsilon - 1)\frac{\partial \varepsilon}{\partial x} = 0.$$

From this, we can write down the ODE satisfied by the characteristic curves of $\varepsilon(x, \tau)$, i.e. their gradients,

$$\frac{\mathrm{d}x}{\mathrm{d}\tau} = F(\varepsilon - 1). \tag{3.77}$$

By definition, ε remains constant along each characteristic curve, so eqn (3.77) can be integrated directly to obtain

$$x = (\varepsilon - 1)F\tau + f_0(\varepsilon), \tag{3.78}$$

where the function $f_0(\varepsilon)$ is independent of x, τ and depends on the initial condition. Following the canonical all-monomer initial condition

$$\varepsilon(x, \tau = 0) = e^{-x} \Rightarrow x(\tau = 0) = -\ln\varepsilon. \tag{3.79}$$

Comparing eqns (3.78) and (3.79), we find that

$$f_0(\varepsilon) = -\ln\varepsilon, \tag{3.80}$$

and the characteristics follow

$$x = (\varepsilon - 1)F\tau - \ln\varepsilon. \tag{3.81}$$

This is the solution, albeit implicit, to eqn (3.62). Unfortunately, it does not come in an explicit form,[4] but it is enough for our analysis.

3.6.1 Characteristic curves

To begin, we emphasize the range of these three variables. Naturally, the (rescaled) time $\tau \in [0, \infty)$, and the range of x has been defined earlier in the chapter: $x \in [0, \infty)$. Meanwhile,

$$\varepsilon = \sum_{k=0}^{\infty} k\nu_k e^{-kx} \equiv \frac{1}{N}\sum_{k=0}^{\infty} kn_k e^{-kx} \in (0, 1], \tag{3.82}$$

since $e^{-kx} \in (0, 1]$ for $k, x \geq 0$. Now we focus on the characteristic curves. From eqn (3.81), it is clear that the characteristics belong to a set of straight lines l_ε defined by parameter $\varepsilon \in (0, 1]$, with gradient

$$F(\varepsilon - 1) \in (-F, 0], \tag{3.83}$$

where, as before, F is assumed to be constant, and x-interception

$$-\ln\varepsilon \in [0, \infty). \tag{3.84}$$

In addition, the τ-interception is given by

$$\frac{\ln\varepsilon}{F(\varepsilon - 1)} \in (1/F, \infty). \tag{3.85}$$

When $\varepsilon = 1$, the specific characteristic l_1 coincides with the τ-axis. As ε decreases to approach 0, the gradient gradually drops to $-F$ asymptotically. In the meantime, the x-interception moves upwards, while the τ-interception increases from $\tau = 1/F$. The contour plot of ε in terms of x, τ is shown in Fig. 3.1.

There is something highly disconcerting in the plot: l_1 crosses with *every* other characteristic at $\tau > 1/F$. Since each characteristic corresponds to a distinct value of ε, something wrong or wild must occur at $\tau > 1/F$. In this case it's the latter, since we

---
[4]This is an example of the *Lambert W-function*.

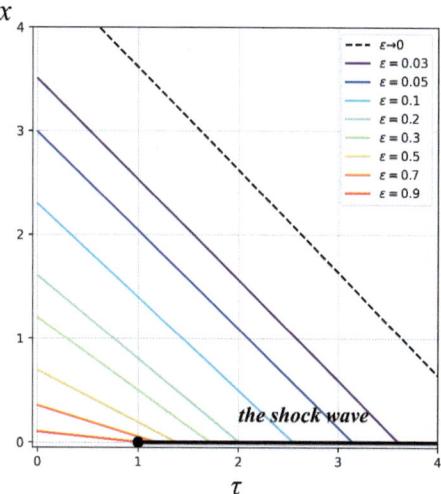

Fig. 3.1: Contour plot of $\varepsilon(x, \tau)$.

have encountered a divergence at $x = 0, \tau = 1/F$. As we will see shortly, this is precisely the *shock wave* behaviour that was promised when the inviscid Burgers' equation was first introduced. The intersection between characteristics is mathematically explained by the doubly valued property of the Lambert W-function.

Before moving on, we comment on the divergence point $x = 0, \tau = 1/F$, where the shock wave emerges as claimed above. Translated into the original variable, the time of shock wave emergence is

$$t_{\text{sw}} = \frac{N}{2F} = t_{\text{g}} \qquad (3.86)$$

which is exactly the gelation time derived with the second moment, and hence the cohesive unit onset time. Furthermore, moment calculation also takes x to be zero. This means that the emergence of a cohesive unit is another name for the emergence of a gel *and* that both are equivalent to the emergence of a shock wave in the reciprocal space of the cluster size k.

3.6.2 The waveform of the shock wave

To study the shock wave (and equivalently the cohesive unit) we need to plot the curves of ε with respect to x, as a one-parameter set in terms of $\tau \in [0, \infty)$. The relation can be obtained by slight transformation of eqn (3.81) into

$$x = F\tau\varepsilon - \ln\varepsilon - F\tau. \qquad (3.87)$$

We can first plot x against ε and then invert the axes. Keeping in mind that $x \geqslant 0$ by definition, the plot then follows as shown in Fig. 3.2. From the plot we see that

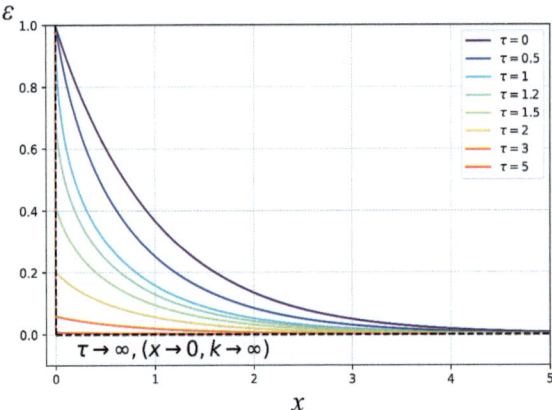

Fig. 3.2: ε–x curves with τ from 0 to ∞, namely the evolution of the wave with time.

for $\tau \leqslant 1/F$, the waveform is smooth and continuous, starting from the fixed point at $x = 0, \varepsilon = 1$, and converges to 0 as $x \to \infty$. However after the shock wave emergence time $\tau_{sw} = 1/F$, the fixed point remains part of the solution but becomes detached from the waveform which intersects the ε-axis at a point $\varepsilon_0 < 1$. Technically they are still connected at $x < 0$ but that is beyond the domain. The wave thus becomes discontinuous and contains an abrupt jump at $x = 0$. The jump is exactly the manifestation of a shock wave: one wavefront in the front towards the direction of propagation travels slower than one at the back, such that the latter catches the former and causes the discontinuity. Our wave here is non-oscillatory, and hence does not have a 'wavefront' like a travelling wave, but the cause of the shock follows the same principle.

3.6.3 Shock wave = gel curve = cohesive unit, and the phase transition

Finally, we examine the last pair combination from the three variables. Setting x to be the parameter, transformation of eqn (3.81) gives

$$\tau = \frac{x + \ln \varepsilon}{F(\varepsilon - 1)}, \qquad (3.88)$$

where $0 \leqslant \varepsilon \leqslant 1$. Special treatment is required for $\varepsilon = 1$ to avoid division by 0, where from eqn (3.81) it is trivially the solution at $x = 0$ for all $\tau \in [0, \infty)$. The plot is shown in Fig. 3.3.

We see that at $x = 0$, besides the trivial solution there also exists a non-trivial one when $\tau \geqslant 1/F$. This is another manifestation of the divergence at $\tau = 1/F$. But out of

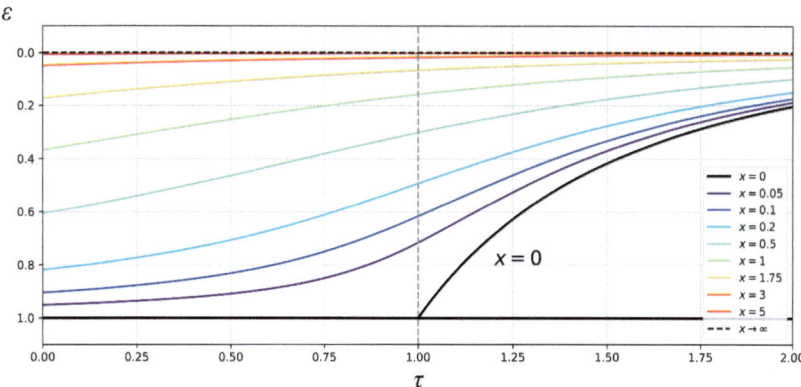

Fig. 3.3: ε–τ curves with x as a parameter. Notice the double solutions of $x = 0$ at $\tau > 1/F$. It can be seen that the non-trivial branch of the solution is not smooth at the branch point.

the two branches of this curve, which one is unphysical? From the asymptotic behaviours of the curves, it seems reasonable to keep the non-trivial solution. However there is a more fundamental reason: this bendy curve, with a discontinuous first derivative at $\tau = 1/F$, has been observed in both real-world gelation systems and ideal simulations to be the *gel curve*. The discontinuity implies that a *phase transition* occurs together with gelation, and the gel is indeed a new phase of existence for the particles. This is suggested also from the conflict between the definitions of coalescence and the gel: a gel is the largest possible aggregate of particles, but the coalescence rule denies the existence of the *largest* cluster. Therefore, the gel can only be a condensate beyond the phase of coalescing clusters. This also explains the decrease of $\varepsilon(x = 0, \tau > 1/F)$, which is equivalent to the first moment, as portrayed in the plot. We know that the first moment is simply the number of particles and does not change when the particle number is conserved. When the gel is treated as a different phase, however, the number of particles in the cluster phase does reduce, and our theory remains consistent and self-contained. As in all this analysis, the equivalence of the various terminologies of cohesive unit, gel curve, and shock wave is clear. The ε–τ graph turns out to very closely fit the growth of online extremist groups—see Fig. 3.4

3.6.4 Shock wave emergence under general initial conditions

Our discussions have so far focused on one specific initial condition—the all-monomer initial condition. Here we discuss further the solutions with general initial conditions and the effect of the x-dependence.

56 *Multi-body Physics 1.0: single-species fusion and fission*

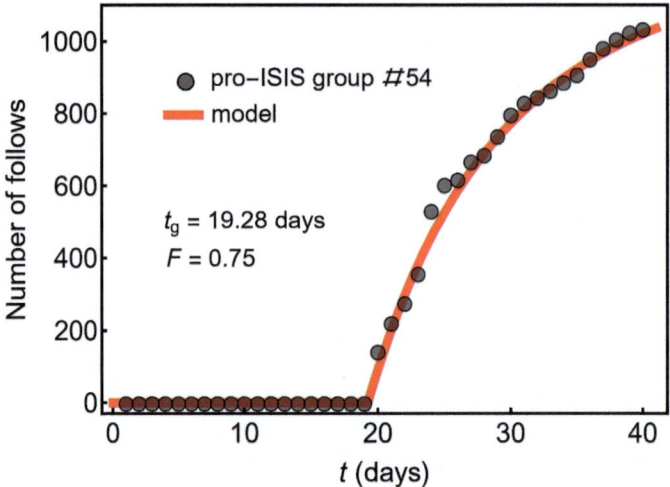

Fig. 3.4: A snapshot of a giant pro-ISIS cluster of users on a social network platform and its evolution over time, from the empirical data discussed in Chap. 2. This is the emerging cohesive unit as discussed in Chap. 1 and shown empirically in Chap. 2. Its growth agrees reasonably well with the current Chap. 3 gelation theory, but the fit to the kink(s) can be improved using the more sophisticated theories for individual users' heterogeneity as presented in Chaps 4 and 5. The equivalence of the various terminologies of cohesive unit, gel curve, and shock wave is clear.

To begin, we denote the initial condition as

$$f(x) = \varepsilon(x, 0). \tag{3.89}$$

We know that the characteristic equation reads

$$x = (\varepsilon - 1)F\tau + f^{-1}(\varepsilon), \tag{3.90}$$

where $f^{-1}(\varepsilon)$ is the inverse function of the initial condition. As discussed above, we regard τ as a parameter when studying the relation between x and ε, which gives the waveforms by taking the inverse of eqn (3.90) at different time instants τ. Since an explicit expression of ε generally does not exist, we just look into eqn (3.90) and the waveform is simply the inverse ε–x graph. Equation (3.90) consists on the RHS of a linear function $\tau(\varepsilon - 1)$ added with $f^{-1}(\varepsilon)$, and the x–ε plot is accordingly given by superposition of the two. Since it is the inverse function that we are after, we want eqn (3.90) to be strictly monotonic such that a well-defined, single-valued ε with respect to x indeed exists. Whenever this fails, we have a shock wave. However, shock waves are inevitable in a clustering system, since a) we *know* that it occurs, and

b) mathematically $f(x)$ must be monotonically decreasing with x (due to its series definition from a mesoscopic perspective[5]). So there *exists* a time τ when

$$\tau = -\frac{1}{F}\left[\frac{\mathrm{d}f^{-1}(\varepsilon)}{\mathrm{d}\varepsilon}\right]_{\max}, \qquad (3.91)$$

which is the gel time, i.e. the onset or formation time of the cohesive unit. It is the first time instant when ε is multi-valued in terms of x. This is not surprising since a shock wave is a basic property of the inviscid Burgers' equation. Assume that $\left[\mathrm{d}f^{-1}(\varepsilon)/\mathrm{d}\varepsilon\right]_{\max}$ is reached at $\varepsilon_0 = f(x_0)$, then we can simplify the result into

$$\tau_g = -\frac{1}{F}\left[\frac{\mathrm{d}f^{-1}(\varepsilon)}{\mathrm{d}\varepsilon}\right]_{\max} = -\frac{1}{F}\left[\frac{\mathrm{d}f^{-1}(\varepsilon)}{\mathrm{d}\varepsilon}\right]_{\varepsilon=\varepsilon_0} = -\frac{1}{Ff'(x_0)} = -\frac{1}{Ff'(x)|_{\min}}. \qquad (3.92)$$

This is the general expression for the gel time (shock wave onset time) with a general initial condition of $\varepsilon(x,0) = f(x)$. In the exercises, the reader is asked to check the gel time under the canonical initial condition, which is given by eqn (3.68), against the general expression, and to calculate it for some other specific initial conditions.

From eqn (3.92), to delay the gel time we need to maximize $f'(x)|_{\min}$. Note that for the generating function $\varepsilon(x,t)$, its definition states that

$$f(0) = \varepsilon(0,0) = \sum_{k=1}^{\infty} k\nu_k(0)e^{-kx} = 1. \qquad (3.93)$$

Moreover, $f(x)$ is a sum of exponential functions e^{-kx} with $k = 1, 2, \ldots$. This guarantees that $f(x)$ monotonically decreases, and therefore

$$f(x) \in (0,1]. \qquad (3.94)$$

From what we know about exponential functions, at $x \geqslant 0$, the larger k is, the steeper e^{-kx} becomes as it decreases with x. So to maximize the gradient we must take $f(x) = e^{-x}$, i.e. the canonical initial condition. In other words, the all-monomer initial condition has the latest possible gelation among all initial conditions. This is also as expected, since any other initial condition must contain clusters of size larger than 1, and it must be closer in time to gelation. For a mathematical proof though, first we calculate the derivative

$$f'(x) = -\sum_{k=1}^{\infty} k^2 \nu_k(0) e^{-kx} < 0. \qquad (3.95)$$

This function monotonically increases with x. Since $x \in [0, \infty)$,

---

[5]From a macroscopic point of view, this can be interpreted by the fact that a wave has to be localized.

58 Multi-body Physics 1.0: single-species fusion and fission

$$f'(x)\big|_{\min} = f'(0). \tag{3.96}$$

Therefore, we need to maximize $f'(0)$ among all possible initial conditions. Note that $\sum_{k=1}^{\infty} k\nu_k(0) = 1$, so

$$\begin{aligned} f'(0) &= -\sum_{k=1}^{\infty} k^2 \nu_k(0) \\ &= -\sum_{k=1}^{\infty} k\big(k\nu_k(0)\big) \\ &\leqslant -\sum_{k=1}^{\infty} k\nu_k(0) = 1, \end{aligned} \tag{3.97}$$

where equality is reached iff $\nu_k(0) = \delta_{1,k}$, which represents exactly the all-monomer 'canonical' condition.

In conclusion, the waveform (i.e. the x-dependence of the generating function) determines the gel time through its minimal gradient. In principle, we can delay the gel time by maximizing this minimum when choosing initial conditions, but the latest possible is merely an all-monomer 'canonical' initial condition since by definition the size distribution n_ks are non-negative, and subsequently more exotic forms of $f(x)$ are forbidden. To provide a heuristic interpretation of this result, we return to Fig. 3.2 which is basically snapshots of a wave propagating towards $x = 0$ as it decays more steeply in x-space, until it becomes tangential to the vertical axis at the gel time. The need for a gently decaying initial waveform in order to delay gelation is then manifested here as it takes longer to reach the shock wave emergence.

3.7 Generalizing to monomer addition and fission

The previous section has demonstrated the capability of our aggregation model in treating coalescence and gelation, and equivalently the emergence of the cohesive unit from Chaps 1 and 2. However, in the real world, systems that exhibit coalescence often contain some other cluster-level interactions at the same time. Due to the vast heterogeneity within and among these systems, generalizability hence becomes a crucial requirement for a working model. In this section we will explore how our theory can be generalized into a theoretical framework that can be applied to realistic as well as theoretically plausible extensions of aggregation-based systems.

3.7.1 Aggregation with monomer injection

Our first generalization has an obvious motivation. The total number of entities (particles) N is conserved in the original Smoluchowski coalescence system, and yet completely closed systems are rare. Real human–technology–AI systems (Fig. 1.2) will definitely come with dynamic $N(t)$. Sometimes $N(t)$ may just fluctuate around a mean value and so can be assumed to be a constant, but at other times, as in Fig. 2.4 insets where there is steady stream of new recruits, $N(t)$ will change in time in some way. To keep calculations and discussions free from lengthy algebra and dense formulae, we assume that the particles added to (taken from) the system are all monomers, hence the *monomer injection* model reference in the title. The injection rate is simply

$$Q(t) = \frac{\mathrm{d}N(t)}{\mathrm{d}t}. \tag{3.98}$$

It follows that when $Q(t) > 0$ monomers are injected into the system, while in the opposite case, where $Q(t) < 0$, they are extracted. Hence this makes sense for a steady stream of incoming recruits as in Fig. 2.4 insets, or for a steady stream leaving.

Changing the number of monomers results in a modification of the master equations. For the sake of closed-form solutions as well as its real-world ubiquity, we still apply the product kernel. The modified master equations read

$$\dot{n}_k(t) = \frac{F(t)}{N(t)^2} \sum_{\substack{0 \leqslant i \leqslant k \\ 0 \leqslant j \leqslant k}}^{i+j=k} ij n_i(t) n_j(t) - \frac{2F(t)}{N(t)} k n_k(t) + Q(t) \delta_{k,1}, \tag{3.99}$$

where the added term at the end of the equation contains a Kronecker delta function. Here we use the original set of variables instead of the rescaled ones, as they are more convenient when expressing the time-varying $N(t)$. Then we can transform eqn (3.99) into

$$\frac{\partial \mathcal{D}(x,t)}{\partial t} = \frac{F(t)}{N(t)^2} \mathcal{E}^2 - \frac{2F(t)}{N(t)} \mathcal{E} + Q(t) e^{-x}. \tag{3.100}$$

Taking one further partial derivative with respect to x leads to a Burgers'-like PDE,

$$\frac{\partial \mathcal{E}(x,t)}{\partial t} + \frac{2F(t)}{N(t)^2} [\mathcal{E}(x,t) - N(t)] \frac{\partial \mathcal{E}}{\partial x} - Q(t) e^{-x} = 0. \tag{3.101}$$

The addition of the inhomogeneous term spoils the method of characteristics. However, we can still look into the second moment as before. Taking yet another x-derivative yields

$$\frac{\partial \mathcal{E}'}{\partial t} + \frac{2F(t)}{N(t)^2} \left[\mathcal{E}'^2 + (\mathcal{E} - N) \mathcal{E}'' \right] + Q(t) e^{-x} = 0. \tag{3.102}$$

Finally, we take $x = 0$:

$$\frac{\mathrm{d}M_2}{\mathrm{d}t} = \frac{2F(t)}{N(t)^2} M_2^2 + Q(t). \tag{3.103}$$

ODEs of this form, whose RHS is a quadratic function of the unknown function itself, are called *Riccati equations*. For the constant-N case, eqn (3.67) is a trivial example of

60 Multi-body Physics 1.0: single-species fusion and fission

the Riccati equation and can be solved directly. The equation here, however, requires more effort. For simplicity, we only solve here for the case of constant F and steady injection, where $Q(t) = q$ is constant. The reduced ODE for M_2 now becomes

$$\frac{dM_2}{dt} = \frac{2F}{(N+qt)^2} M_2^2 + q, \qquad (3.104)$$

where N denotes the initial number of particles, i.e. $N(0) = N$. Since what we really care about is the divergence of M_2 and time of this divergence (i.e. t_{sw} or, equivalently, the gel time, or the onset or time of formation of the cohesive unit) it is more straightforward to study the dynamics of M_2^{-1} and look for its zeros. After some reorganization, eqn (3.104) gives

$$\frac{dM_2^{-1}}{dt} + qM_2^{-2} + \frac{2F}{(N+qt)^2} = 0. \qquad (3.105)$$

Now is a good time to reintroduce the rescaled variables and define some new ones for simplification. We first denote a rescaled time

$$\sigma = \frac{q}{2}\tau + 1 \in [1, \infty), \qquad (3.106)$$

and

$$\eta(t) = m_2(\tau)^{-1} = NM_2^{-1}(t).$$

Note that together with

$$\tau = \frac{2t}{N}$$

we find that σ is simply

$$\sigma = \frac{N(t)}{N}, \qquad (3.107)$$

i.e. the scaled time-varying total population. Substituting in the variables above, eqn (3.105) becomes

$$\frac{d\eta}{d\sigma} + \eta^2 + \frac{2F}{q\sigma^2} = 0 \qquad (3.108)$$

with initial condition

$$\eta(\sigma = 1) = 1. \qquad (3.109)$$

The standard method of solving a Riccati equation is to transform it into a second-order ODE by substitution. Here we substitute with a variable u such that

$$\eta = \frac{1}{u}\frac{du}{d\sigma} = \frac{d\ln u}{d\sigma}. \qquad (3.110)$$

Then we have

$$\frac{d^2 u}{d\sigma^2} + \frac{2F}{q\sigma^2} u = 0. \tag{3.111}$$

The solution is clearly in a polynomial form σ^β for some β. If we substitute u with σ^β, we can obtain the *characteristic equation*

$$\beta^2 - \beta + \frac{2F}{q} = 0. \tag{3.112}$$

According to the values of

$$\Delta = 1 - \frac{8F}{q}, \tag{3.113}$$

the solutions can be divided into three regimes depending on the sign of Δ. Correspondingly, $q = 8F$ marks the boundary of these regimes. The solutions are given below.

1. $0 < q < 8F$,

$$M_2 = 2N(t) \frac{a + \tan\left(\frac{\alpha}{2} \ln \sigma\right)}{2a + (1 - a^2) \tan\left(\frac{\alpha}{2} \ln \sigma\right)} \tag{3.114}$$

2. $q = 8F$,

$$M_2 = 2N(t) \frac{2 + \ln \sigma}{4 + \ln \sigma} \tag{3.115}$$

3. $q > 8F$,

$$M_2 = 2N(t) \frac{\alpha + \tanh\left(\frac{\alpha}{2} \ln \sigma\right)}{2\alpha + (1 + \alpha^2) \tanh\left(\frac{\alpha}{2} \ln \sigma\right)}, \tag{3.116}$$

where we denote

$$\alpha = \sqrt{1 - \frac{8F}{q}}, \quad a = \sqrt{\frac{8F}{q} - 1} = -i\alpha. \tag{3.117}$$

Here we derive the solutions for $q < 8F$ as an example, and leave the other two as an exercise at the end of the chapter. This is the non-trivial case, with $\Delta < 0$ and hence two complex roots of eqn (3.112)

$$\beta = \frac{1}{2}\left(1 \pm i\sqrt{\frac{4F}{q} - 1}\right). \tag{3.118}$$

From eqn (3.117),

$$a = \sqrt{\frac{8F}{q} - 1} \in (0, +\infty) \tag{3.119}$$

and the general solution of $u(\sigma)$ is

$$u = \sigma^{1/2} \left[A\sigma^{\frac{a}{2}i} + B\sigma^{-\frac{a}{2}i} \right] \tag{3.120}$$

for some constants A, B. The result has a somewhat uncanny appearance: the solution for a physical system should clearly be real, but the function u is written in terms of

complex functions. Hence we should change its functional basis and make it appear real. By noting that
$$\sigma^{\frac{a}{2}\mathrm{i}} = \exp\left(\mathrm{i}\frac{a}{2}\ln\sigma\right), \tag{3.121}$$
the general solution of $u(\sigma)$ can be converted with trigonometric bases into
$$u = \sigma^{1/2}\left[C\cos\left(\frac{a}{2}\ln\sigma\right) + D\sin\left(\frac{a}{2}\ln\sigma\right)\right], \tag{3.122}$$
for some constants $C, D \in \mathbb{C}$. We know that physicality requires the constants to be real, but for mathematical completeness they are still allowed to be complex. It follows that
$$\frac{du}{d\sigma} = \frac{u}{2\sigma} + \sigma^{-1/2}\frac{a}{2}\left[-C\sin\left(\frac{a}{2}\ln\sigma\right) + D\cos\left(\frac{a}{2}\ln\sigma\right)\right], \tag{3.123}$$
and the solution of η is therefore
$$\eta(\sigma) = \frac{1}{u}\frac{du}{d\sigma} = \frac{1}{2\sigma}\left[1 + a\frac{-C\sin\left(\frac{a}{2}\ln\sigma\right) + D\cos\left(\frac{a}{2}\ln\sigma\right)}{C\cos\left(\frac{a}{2}\ln\sigma\right) + D\sin\left(\frac{a}{2}\ln\sigma\right)}\right]. \tag{3.124}$$

We can see that although eqn (3.122) comes with two unknown constants, here it is reduced into one, the ratio C/D—the correct number since we are simply solving a first-order ODE. Applying the initial condition
$$\eta(1) = \frac{1}{2} + \frac{a}{2}\frac{D}{C} = 1 \Rightarrow \frac{C}{D} = a, \tag{3.125}$$
the solution turns out to be
$$\eta(\sigma) = \frac{1}{2\sigma}\left[1 + a\frac{1 - a\tan\left(\frac{a}{2}\ln\sigma\right)}{a + \tan\left(\frac{a}{2}\ln\sigma\right)}\right] = \frac{1}{2\sigma}\frac{2a + (1-a^2)\tan\left(\frac{a}{2}\ln\sigma\right)}{a + \tan\left(\frac{a}{2}\ln\sigma\right)}. \tag{3.126}$$

This function has a zero at
$$\tan\left(\frac{a}{2}\ln\sigma_{\mathrm{sw}}\right) = \frac{2a}{a^2 - 1}. \tag{3.127}$$

Since the tangent function comes with multiple branches, it is crucial that we choose the correct range for the argument $(a/2)\ln\sigma_{\mathrm{sw}}$ such that

1. $\sigma_{\mathrm{sw}} > 1$ as required by its definition,
2. σ_{sw} is a continuous function in terms of a.

Note that when $a = 1$, eqn (3.127) goes to infinity, and that when $0 < a < 1$ it becomes negative. The above requirements thus determine that
$$\frac{a}{2}\ln\sigma_{\mathrm{sw}} \in [0, \pi) \tag{3.128}$$
and that $a = 1$ simply corresponds to $\ln\sigma_{\mathrm{sw}} = \pi/a$, whereas $0 < a < 1$ corresponds to $\pi/a < \ln\sigma_{\mathrm{sw}} < 2\pi/a$. The determination of this range enables us to take the inverse tangent function to obtain the shock wave onset time:

$$\ln \sigma_{\text{sw}} = \frac{2}{a} \arctan\left(\frac{2a}{a^2 - 1}\right) \Rightarrow \tau_{\text{sw}} = \frac{2}{q}\left[\exp\left(\frac{2}{a} \arctan \frac{2a}{a^2 - 1}\right) - 1\right]. \qquad (3.129)$$

Now we heuristically analyse the behaviour of τ as a decreases from $+\infty$ to 0, corresponding to increasing q from 0 to $8F$. We know that

$$\frac{2a}{a^2 - 1} = 2\left(a - \frac{1}{a}\right)^{-1} \qquad (3.130)$$

monotonically decreases with respect to a in the ranges $a \in (1, \infty)$ and $(0, 1)$, respectively. Furthermore, we also know that the inverse tangent function monotonically increases, and that σ_{sw} is continuous in terms of a. Therefore σ_{sw} monotonically decreases with a. It also follows that τ_{sw} decreases with a, and hence *increases* with respect to q despite the prefactor $2/q$. This is because its rate of change cannot beat that of the exponential function. This aligns with our result that as the injection rate q increases, the shock wave emergence time gets delayed until it becomes infinite when $q = 8F$. Finally, we can rewrite eqn (3.129) in terms of the original parameters as

$$t_{\text{sw}} = \frac{N}{q}\left[\exp\left(\frac{2}{a} \arctan \frac{2a}{a^2 - 1}\right) - 1\right], \text{ where } a = \sqrt{\frac{8F}{q} - 1}, \qquad (3.131)$$

when $q < 8F$.

Having understood the delay of the gel time as q increases in the range $(0, 8F)$, we go back to study gelation under the other two schemes. However, from the expressions in eqns (3.115) and (3.116), M_2 does not diverge in finite time when $q \geqslant 8F$ (see exercises). Hence $q = 8F$ becomes an important boundary for monomer injection: when the injection rate is equal to or beyond this boundary, no gel is formed in the system in finite time. By contrast, a similar proof can be presented to show that gelation does indeed occur with monomer extraction, i.e. for $q < 0$. The conclusion is intriguing: if we inject into the system monomers with a large enough and constant rate (or a slow-varying one compared to other timescales of the system, e.g. coalescence rate), gelation can be prevented, and no shock wave will develop in the system. However, this is not surprising after some further thought. What we really achieve by injecting monomers is to dilute the system and thereby decrease the probability of merging between clusters. For a graphical explanation, we return to Fig. 3.2 and note that adding monomers is equivalent to pulling the waveform to the right towards $x \to \infty$ direction, while the wave spontaneously propagates towards $x \to 0$ to form the shock wave. If we maintain the 'pull', we can indefinitely delay the shock wave's emergence, hence avoiding gelation, i.e. no cohesive unit emerges and hence there is no 'surprise' (Chaps 1 and 2).

3.7.2 Adding fission

So far, we have been discussing various types of aggregation. Now we add the process in the reverse direction: *fragmentation* (i.e. fission, Chaps 1 and 2). The name suggests its nature: large clusters break up into several smaller ones. When discussing coalescence, we only modelled binary aggregation, since we assumed for simplicity that the system is dilute enough that aggregations that involve more than two clusters are highly unlikely, and the rare instances where it happens could be treated as successive binary aggregations. For fragmentation, there are many possible mechanisms. To start with, we can consider a toy model where each fragmentation event only removes one monomer from the cluster—we call it the *chipping* model, and the reader can play around with it in the exercises. Depending on the human–technology–AI system of interest, we may also want to look into binary fragmentations, which are exactly the reversed process of binary aggregation. For the moment, we consider complete fragmentation since this has been observed empirically in online conflict (Fig. 2.2) and offline conflict, and it also aligns with what is known about how non-human living systems (e.g. birds, fish) respond when facing danger. Specifically, we consider total fragmentation (fission) which breaks up a cluster completely into monomers—though we note that mathematically similar results hold if the fission is not complete (e.g. approximate power-law distribution, Fig. 1.7, right panel) as long as the size of the remnant clusters is smaller than some small value $s_{\min}$. It can arise when a cluster experiences external threats from, for example, moderators in the online case, or from the environment, or other censorship or attacks depending on the system. Apart from its realism for online and offline conflict (Chap. 2), it will also turn out that the resulting model is very tractable mathematically.

In this section we add this process of complete fragmentation to what was presented. We call it the *EZ model* after the scientists who first explored it (Eguíluz and Zimmermann, 2000). As for the earlier coalescence (fusion) discussion, we need a heterogeneity parameter $\varphi_{\text{frag}}(t)$ and a kernel to measure fragmentation rates: since fragmentation is a unitary process, unlike binary aggregation, we simply take a single-variable discrete function λ_k in terms of the cluster size k.[6] The modified master equations including complete fragmentation then take the following form

$$\dot{n}_k(t) = \frac{F}{N^2} \sum_{\substack{1 \leqslant i \leqslant k \\ 1 \leqslant j \leqslant k}}^{i+j=k} \kappa_{ij} n_i(t) \nu_j(\tau) - \frac{2F}{N^2} \sum_{i=1}^{\infty} \kappa_{ik} n_i(t) n_k(t) - \frac{\varphi_{\text{frag}} \lambda_k n_k}{N}, \quad k \geqslant 2 \quad (3.132)$$

---

[6] Depending on what exactly we want to encapsulate in the model, we *could* utilize higher-order kernels, e.g. a double-variable kernel for binary fragmentation if we assume that it is the new clusters created after a fragmentation process that determine the reaction rate.

$$\dot{n}_1(t) = \frac{\varphi_{\text{frag}}}{N}\sum_{j=2}^{\infty} j\lambda_j n_j - \frac{2Fn_1}{N^2}\sum_{j=1}^{\infty}\kappa_{1j}n_j,\quad k=1 \tag{3.133}$$

where the first term on the RHS of eqn (3.133) represents monomer gains from the total fragmentations of clusters of all sizes. As before, we can again compress the two expressions into one with the Kronecker delta and in rescaled units:

$$\dot{\nu}_k(\tau) = \frac{F}{2}\sum_{\substack{0\leqslant i\leqslant k \\ 0\leqslant j\leqslant k}}^{i+j=k}\kappa_{ij}\nu_i(\tau)\nu_j(\tau) - F\sum_{i=1}^{\infty}\kappa_{ik}\nu_i(\tau)\nu_k(\tau) - \frac{\varphi_{\text{frag}}}{2}\lambda_k\nu_k + \delta_{1k}\frac{\varphi_{\text{frag}}}{2}\sum_{j=1}^{\infty}j\lambda_j\nu_j. \tag{3.134}$$

There is one subtlety in eqn (3.134): the summation limit in the last term is changed from $j \geqslant 2$ into $j \geqslant 1$. This is because monomers cannot further fragment, and we have to compensate for the permanent fragmentation term $-\varphi_{\text{frag}}\lambda_k\nu_k/2$, i.e. the second term from the right on the RHS, when $k=1$. This actually facilitates the transformation, as we will see shortly.

The canonical kernel $\lambda_k = k$. This kernel is arguably the simplest, apart from a constant kernel. In short, it is to fragmentation what the product kernel is to aggregation. We call it the canonical kernel because it is natural to select and break up clusters according to their sizes, especially for an external 'destroyer' from whose perspective larger clusters may be more obvious targets. The following derivation assumes a product aggregation kernel $\kappa_{ij} = ij$ and a canonical fragmentation kernel $\lambda_k = k$. We first substitute in these values to eqn (3.134):

$$\dot{\nu}_k(\tau) = \frac{F}{2}\sum_{\substack{0\leqslant i\leqslant k \\ 0\leqslant j\leqslant k}}^{i+j=k}ij\nu_i\nu_j - \left(F + \frac{\varphi_{\text{frag}}}{2}\right)k\nu_k + \delta_{1k}\frac{\varphi_{\text{frag}}}{2}m_2. \tag{3.135}$$

The standard transformation procedure follows:

$$\frac{\partial\delta(\tau)}{\partial\tau} = \frac{F}{2}\varepsilon^2 - \left(F + \frac{\varphi_{\text{frag}}}{2}\right)\varepsilon + \frac{\varphi_{\text{frag}}}{2}m_2 e^{-x} \tag{3.136}$$

$$\Rightarrow \frac{\partial\varepsilon}{\partial\tau} = -F\varepsilon\frac{\partial\varepsilon}{\partial x} + \left(F + \frac{\varphi_{\text{frag}}}{2}\right)\frac{\partial\varepsilon}{\partial x} + \frac{\varphi_{\text{frag}}}{2}m_2 e^{-x}. \tag{3.137}$$

No shock wave emerges in this scenario due to the negative feedback of complete fragmentation—exactly as stated in Chap. 1. This means that a steady state in some long time limit is possible, possibly akin to Fig. 1.7, right panel. Taking the steady state limit and setting $x=0$ for eqn (3.136), we obtain that

$$\frac{M_2}{N} = m_2 = \frac{F + \varphi_{\text{frag}}}{\varphi_{\text{frag}}}. \tag{3.138}$$

This well-behaved and non-trivial relation confirms that a steady state is reachable in the long time limit. Throughout the derivation, no assumption is needed to be made about the time dependence of F and φ_{frag}: but eqn (3.138) has to be evaluated in the long time limit. In eqn (3.138) if we take the no-fragmentation limit $\varphi_{\text{frag}} \to 0$ (i.e. negligible fission), M_2 diverges as found earlier and hence the cohesive unit would emerge.

The quantity M_2/N is interesting: if we consider the average probability P that two particles are in the same cluster, it follows that

$$P = \frac{\sum_{k=1}^{\infty} \binom{k}{2} n_k}{\binom{N}{2}} = \frac{M_2 - N}{N^2 - N} \approx \frac{1}{N}\left(\frac{M_2}{N} - 1\right). \tag{3.139}$$

We call this the *connection probability*. Its application will be seen later in the book, but for now it is clear that eqn (3.138) determines the connection probability.

This is the first system we have studied where an equilibrium is reached. The next question would then be: what is the size distribution at steady state?

In order to calculate that, we can solve eqn (3.136) under steady-state conditions and then inverse transform the solution to obtain n_k. In eqn (3.136), taking $\dot{D} = 0$ and hence steady state, we get a quadratic equation

$$F\varepsilon^2 - (2F + \varphi_{\text{frag}})\varepsilon + \varphi_{\text{frag}} m_2 e^{-x} = 0 \tag{3.140}$$

where the factor $1/2$ cancels. Its solution is then

$$\varepsilon(x,\tau) \equiv \frac{\mathcal{E}(x,\tau)}{N} = \frac{1}{2F}\left[(2F + \varphi_{\text{frag}}) \pm \sqrt{(2F + \varphi_{\text{frag}})^2 - 4F\varphi_{\text{frag}} m_2 e^{-x}}\right]$$

$$= \frac{1}{2F}\left[(2F + \varphi_{\text{frag}}) \pm \sqrt{(2F + \varphi_{\text{frag}})^2 - 4F(F + \varphi_{\text{frag}})e^{-x}}\right]$$

$$= \frac{2F + \varphi_{\text{frag}}}{2F}\left[1 \pm \sqrt{1 - \frac{4F(F + \varphi_{\text{frag}})}{(2F + \varphi_{\text{frag}})^2}e^{-x}}\right], \tag{3.141}$$

where in the second equality we substitute in eqn (3.138). Now we must decide which one of the two roots to keep, which can be worked out by taking $x = 0$. Since $\varepsilon(0,\tau) = 1$ by definition, we have

$$1 = \frac{2F + \varphi_{\text{frag}}}{2F}\left[1 \pm \sqrt{1 - \frac{4F(F + \varphi_{\text{frag}})}{(2F + \varphi_{\text{frag}})^2}}\right]$$

$$= \frac{2F + \varphi_{\text{frag}}}{2F}\left(1 \pm \frac{\varphi_{\text{frag}}}{2F + \varphi_{\text{frag}}}\right), \tag{3.142}$$

and this requires that we take the minus sign and discard the other solution. Finally as $\tau \to \infty$, we get

$$\varepsilon(x,\tau) = \frac{2F + \varphi_{\text{frag}}}{2F}\left[1 - \sqrt{1 - \frac{4F(F + \varphi_{\text{frag}})}{(2F + \varphi_{\text{frag}})^2}e^{-x}}\right]. \tag{3.143}$$

What is left is the inverse transformation. We do not have to adhere to the exact transformation dogma, which would require a contour integral in the complex plane. Instead, we can achieve the same goal by expanding the square root into a power series as $x \to \infty$. The binomial expansion leads to

$$\lim_{x\to\infty} \varepsilon(x,\tau) = \frac{2F + \varphi_{\text{frag}}}{2F} \sum_{k=1}^{\infty} (-1)^{k-1} \binom{1/2}{k} \left[\frac{4F(F + \varphi_{\text{frag}})}{(2F + \varphi_{\text{frag}})^2}\right]^k e^{-kx} = \sum_{k=1}^{\infty} k\nu_k e^{-kx}. \tag{3.144}$$

Hence, we have

$$\nu_1 = \frac{F + \varphi_{\text{frag}}}{2F + \varphi_{\text{frag}}}, \tag{3.145}$$

$$\nu_k = \frac{(2k-3)!!}{(2k)!!k} 2^{2k-1} \left[\frac{F(F + \varphi_{\text{frag}})}{(2F + \varphi_{\text{frag}})^2}\right]^k \frac{2F + \varphi_{\text{frag}}}{F}$$

$$= \frac{(2k-2)!}{(k!)^2}\left[\frac{F(F + \varphi_{\text{frag}})}{(2F + \varphi_{\text{frag}})^2}\right]^k \frac{2F + \varphi_{\text{frag}}}{F} \qquad k \geqslant 2. \tag{3.146}$$

In particular, eqn (3.145) can be examined by setting $k = 1$ and $\dot\nu_k = 0$ in eqn (3.135). For large k, we employ Stirling's approximation

$$k! \approx \sqrt{2\pi k}\left(\frac{k}{e}\right)^k \tag{3.147}$$

to obtain

$$\nu_k \approx \frac{\sqrt{2\pi(2k-2)}}{2\pi k} \frac{e^2(2k-2)^{2k-2}}{k^{2k}} \left[\frac{F(F + \varphi_{\text{frag}})}{(2F + \varphi_{\text{frag}})^2}\right]^k \frac{2F + \varphi_{\text{frag}}}{F}$$

$$= \frac{e^2}{4\sqrt{\pi}}\left(1 - \frac{1}{k}\right)^{2k} k^{-1}(k-1)^{-3/2}\left[\frac{4F(F + \varphi_{\text{frag}})}{(2F + \varphi_{\text{frag}})^2}\right]^k \frac{2F + \varphi_{\text{frag}}}{F}$$

$$\approx k^{-5/2}\left[\frac{4F(F + \varphi_{\text{frag}})}{(2F + \varphi_{\text{frag}})^2}\right]^k \frac{2F + \varphi_{\text{frag}}}{4\sqrt{\pi}F},$$

where in the last equality we make the approximation $k - 1 \approx k$ for $k \to \infty$, and identify

$$\lim_{k\to\infty}\left(1 - \frac{1}{k}\right)^{2k} = e^{-2} \tag{3.148}$$

in the same limit. Also we note that

$$\frac{4F(F + \varphi_{\text{frag}})}{(2F + \varphi_{\text{frag}})^2} = \frac{4F^2 + 4F\varphi_{\text{frag}}}{(2F + \varphi_{\text{frag}})^2} = 1 - \left(\frac{\varphi_{\text{frag}}}{2F + \varphi_{\text{frag}}}\right)^2 < 1, \tag{3.149}$$

so the exponential function does converge to 0 as $k \to \infty$. If we write

$$e^{-\gamma} = 1 - \left(\frac{\varphi_{\text{frag}}}{2F + \varphi_{\text{frag}}}\right)^2 \tag{3.150}$$

for some constant

$$\gamma = -\ln\left[1 - \left(\frac{\varphi_{\text{frag}}}{2F + \varphi_{\text{frag}}}\right)^2\right] > 0, \tag{3.151}$$

then the size distribution for large k behaves like

$$n_k \propto k^{-5/2} e^{-\gamma k}. \tag{3.152}$$

If the fragmentation rate φ_{frag} is very large, the exponential behaviour dominates and there turn out to be very few clusters compared to monomers, since they all get broken up. For small $\varphi_{\text{frag}} \ll 2F$, however, the power-law dominates, which yields **the result of a $\alpha = 5/2 \equiv 2.5$ power-law as reported empirically for the online conflict scenario in Chap. 2**. It is also the upper limit predicted for offline conflict, as explained in Chap. 2.

In the case that the clusters only interact at their perimeters, and hence the effective size of a cluster is given by the square root of k, the same derivation follows but we end up with $\alpha = 4/2 = 2.0$ as claimed in Chap. 2. We leave this as an exercise. Anything in between means that the effective size of a cluster lies between the square root power of k and k itself, which means that α lies in the range between $2/2 \equiv 2.0$ and $5/2 \equiv 2.5$. Given that clusters produce events as explained in Chap. 2, this theoretical prediction is in close agreement with the empirical results in Fig. 2.6 for offline conflict.

Existence of steady state and critical parameters for gel emergence. In the discussions above, we omit a critical question: is the steady state always reachable? A guess would be that the system may never reach steady state when the fragmentation rate φ_{frag} is too small compared to F. The answer, it turns out, is close to this guess, but with the parameter N for the number of entities now coming into play, as we now explain.

The critical condition can be evaluated using the moment functions. To begin, it is clear that if the whole system of particles are gelled together, we have

$$M_2 = N^2. \tag{3.153}$$

Furthermore, it is not hard to prove that

$$M_2 \leqslant N^2 \tag{3.154}$$

for all times, with equality only reached in the above completely gelled scenario (see exercises). Recall that the relation between M_2 and N under steady state is given by eqn (3.138)

$$\frac{M_2}{N} = \frac{F + \varphi_{\text{frag}}}{\varphi_{\text{frag}}}.$$

Equations (3.138) and (3.154) together imply that steady state is only reachable for

$$\frac{F + \varphi_{\text{frag}}}{\varphi_{\text{frag}}} \ll N \Rightarrow \frac{F}{\varphi_{\text{frag}}} \ll N - 1 \approx N. \qquad (3.155)$$

This boundary condition defined by the '$\ll$' sign seems a bit vague. This vagueness appears due to our vague definition of gelation for a finite-population system. Recall that we defined gelation to be 'a giant cohesive unit that consists of a non-negligible fraction of the total population', and this $\ll$ sign is simply saying that. If instead there is a criterion which specifies that a cluster of size $1/\alpha$ of the whole population is big enough to be a gel, then the boundary will be $F/\alpha N$, and the gel appears when

$$\varphi_{\text{frag}} < \frac{\alpha^2 F}{N}. \qquad (3.156)$$

But what about an ideal aggregating system with an infinite number of particles? How can one find the gelation condition in those circumstances given that $m_2 < 1$ is not a strict enough boundary condition? Gelation of such a deterministic system can be said to be determined by its size distribution at gelation onset. In other words, although it is impossible to gauge a fraction of the population for an infinite system, it is nonetheless possible to find a dependence between the cluster size distribution ν_k and size k, which is a power-law for binary aggregation under the product kernel. We can see a transition in this size distribution exactly at the gelation onset: a phase transition. Then instead of comparing M_2 and N, or m_2 and 1, we compare the EZ-size distribution from eqn (3.152) with the gelation distribution at onset, such that we can obtain a condition on factor γ, and hence φ_{frag} and F. We do not pursue a detailed analysis here, because expanding out $\varepsilon(x, \tau)$ for the aggregation-only model in order to obtain the gelation condition requires a number of steps.

3.7.3 Potential generalizations

The linearity of the Laplace transform guarantees that we can simply add new terms (e.g. cluster-based interactions) in the master equations and get back the corresponding macroscopic terms in the generative equation in the abstract x-space. Hence in future human–technology–AI systems, any form of the interactions that arise empirically can be used to modify the master equations and hence generate interesting new PDEs for the generative equation. Likewise, there are a wealth of interesting hydrodynamic PDEs that resemble modified versions of the Burgers' equation that could be treated as generative equations in order to explore interesting new cluster mechanisms at the level of the master equations. A list of such equations is provided in the Appendix. The formalism discussed in this chapter provides the key to transitioning between the master equation and the generative equation, or vice versa. Similarly, if we can find a PDE to describe a particular emergent phenomenon in a multi-entity system, we can use this as the generative equation and employ the inverse transform to find the form of the interactions at the cluster level. A few examples are provided in the exercises, where the reader is asked to transform some famous non-linear PDEs to k-space and obtain various terms with respect to n_k. With this in mind, our theory has the potential to incorporate a range of new cluster-based interactions that characterize particular future human–technology–AI systems. More broadly, it provides a general theoretical framework for describing and connecting new human–technology–AI collective phenomena across the mesoscopic and macroscopic scales.

Exercises

3.1 In Section 3.3.1 we calculated explicitly n_k for $k = 1, 2$ with constant kernel. By inspecting the form of $n_1(t)$ and $n_2(t)$ in eqns (3.11) and (3.12), propose a general expression of n_k for arbitrary k.

   (a) Verify your expression against the master equations to prove its correctness.
   (b) We have proved from the second moment that the constant kernel does not lead to gelation in finite time. Explain this property from the expression you obtained.

3.2 Derive the generative equations corresponding to the sum kernel $\kappa_{ij} = i + j$ with respect to $\mathcal{D}$ and $\mathcal{E}$ respectively, and then obtain the ODEs of the zeroth and the second moments.

3.3 This chapter only considers binary aggregations. In this exercise, however, write down the master equations for an aggregation-only system that admits only tertiary aggregation, and transform it into the generative equation corresponding to this system. What is the main feature of this equation?

3.4 This question reviews the shock wave emergence time under the product kernel.

   (a) Using eqn (3.92), calculate the gel time under the canonical initial condition.
   (b) Calculate the respective gel times for i) an all-dimer initial condition (with $f(x) = Ne^{-2x}$), and ii) a mixed initial condition ($f(x) = (N/2)e^{-x} + (N/4)e^{-2x} + (N/4)e^{-3x}$). You should find that they are all smaller than the all-monomer condition, as discussed previously.

3.5 For the exponential kernel as mentioned in Section 3.3.2, derive its corresponding generative equation. What is the effect of the exponential suppression on the generative equation?

3.6 This question reviews aggregation with constant monomer injection.

(a) Solve the Riccati equation (eqn (3.108)) for $q/F \geqslant 8$ and prove the results from eqns (3.115) and (3.116).

(b) Prove that the gel times under these two scenarios are both infinite. This means that we can prevent gelation by injecting monomers into the system with a high enough rate.

3.7 Following the previous question, prove that for constant monomer *extraction*, i.e. $q < 0$, the gel time is finite, and find the gel time.

3.8 This question focuses on a simple aggregation–fragmentation model—the *chipping model*, whose mechanism has been described in Section 3.7.2.

(a) Write down its corresponding master equations according to the definition.

(b) Transform the master equations to obtain the generative equation. From there determine if a steady state is reachable. If so, calculate the ratio M_2/N under steady state. If not explain why.

3.9 In Section 3.7.2 we derived the $-5/2$ power-law for size distribution $\{n_k\}$ under product aggregation kernel and the canonical fragmentation kernel. In fact, similar power-laws exist widely in 'product-like systems', where the aggregation kernel $\kappa(i,j)$ can be written as a product of fragmentation kernels $f(k)$:

$$\kappa(i,j) = f(i)f(j). \qquad (3.157)$$

(a) Prove that in such a system, the size distribution in steady state is of the form

$$n_k \propto \frac{k^{-3/2}}{f(k)} e^{-\gamma k} \qquad (3.158)$$

for some constant γ. (Hint: use a modified generating equation $\mathcal{H} = \sum_{k=1}^{\infty} f(k) n_k e^{-kx}$.)

(b) For $f(k) = k^{1/2}$, explain its physical significance in terms of the interaction patterns between clusters.

(c) More generally, for clusters that interact effectively in a d-dimensional lattice (instead of interacting freely with arbitrary clusters), find its power-law of cluster size distribution.

3.10 Prove that $M_2 \leqslant N^2$ and specify the equality condition. (Hint: is equality reached right at gelation onset?)

3.11 Transform the following hypothetical generative equations to obtain their corresponding master equations.

(a) Viscid Burgers' equation for $\mathcal{D}$

$$\dot{\mathcal{D}} + \mathcal{D}\mathcal{D}' = \nu \mathcal{D}'', \qquad (3.159)$$

(b) viscid Burgers' equation for $\mathcal{E}$

$$\dot{\mathcal{E}} + \mathcal{E}\mathcal{E}' = \nu \mathcal{E}'', \qquad (3.160)$$

(c) KdV equation for $\mathcal{E}$

$$\dot{\mathcal{E}} + \mathcal{E}''' - 6\mathcal{E}\mathcal{E}' = 0. \qquad (3.161)$$

4
Multi-body Physics 2.0: multi-species fusion and fission

We continue here with our goal of deriving physics that can describe the collective behaviour of general human–technology–AI systems (Fig. 1.2). In the previous chapter, we managed to derive some intriguing results with the simplest version of our fusion–fission model in which we treated the heterogeneity of entities in a basic way: we assigned characters from a vector space to every entity and applied a mean-field theory across the entire multi-entity population. The resulting physics did a reasonable job of explaining benchmark empirical patterns found so far in such systems (Chap. 2).

While Chap. 3 took into consideration the heterogeneity of individual entities, it can be too crude an approach. In particular, the complexity of Fig. 1.2 still seems quite far from such a simple theory. Humans, machinery, software/AI have countless different types of heterogeneity within their own type (i.e. species) and across species. Any member of the human species is definitely different to any type of robot, but humans are also quite different to each other, even after allowing for the fact that we form like-minded communities. So while some of the differences between entities indeed fall on a continuous spectrum (e.g. different heights and weights) and hence can be faithfully represented as in Chap. 3, other differences can be quite discrete with a small number of possible labels. For example, an online in-built community will belong to a specific platform of which there are only a countable number. Another example concerns opinions about vaccines which are typically divided into three species: pro-vaccination, anti-vaccination, or neutral (Johnson et al., 2020). For such finite and limited heterogeneity spaces, a mean-field averaging over a continuum as in Chap. 3 is too crude—and as we will find out in this chapter, it is also quite unnecessary. In fact, we will show in this chapter's more sophisticated approach, that the model we developed in Chap. 3 is easily generalizable to include multiple *species* of particles—Fig. 4.1 provides an overview, and gives rise to some nice analytic results.

As noted at the start of Chap. 3, we may continue to refer to a cohesive unit as a **gel**, or a **giant connected component (GCC)**, or even as a **shock** in an inviscid-like fluid because of the form of its generative equation. All these terms mean

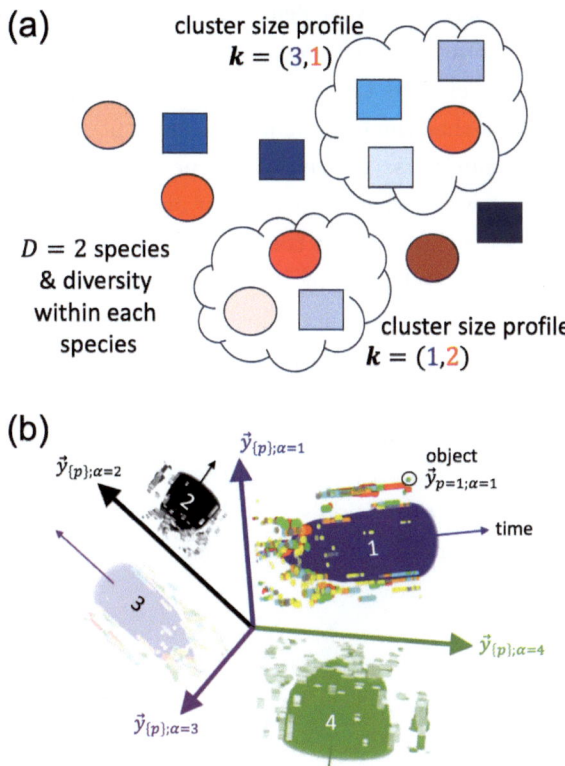

Fig. 4.1: (a) Schematic of the multi-species version of our theory at a given time step for $N = 13$ interacting entities from $D = 2$ species (circle, square). Spectrum of given colour represents intra-species traits for entities of that species. (b) Our theory's output in absence of cross-species fusion (i.e. off-diagonal F elements all zero). Adding cross-species fusion (i.e. some or all F elements non-zero), our theory predicts $1, 2, \ldots D$ cohesive units will emerge along different (mixed) directions, each with its own onset time and mixed-species composition.

the same thing at the macroscopic level: a **cohesive unit**. Also because of a similar diversity in terminology across the literature, we may use words such as **coalescence**, **aggregation**, **coagulation**, etc. interchangeably with **fusion**, and **fragmentation**, **disaggregation**, **degradation**, etc. interchangeably with **fission**.

4.1 Multi-dimensional fusion model

4.1.1 Representation of multiple species

Representing the different species in a system seems to be straightforward, at least microscopically: we just need each particle to carry an attribute that denotes its species.

This is exactly the same idea as how we define electric charges and colours of quarks. In other words, the representation is vectorial. But what about the mesoscale description? How shall we denote the clusters that are now made up of these different species?

Recall from Chap. 3 that the only quantities that interest us mesoscopically are cluster sizes and their numbers. This remains true for however many species, so we first need to figure out what clusters are like in an D-species system. One thing is clear: we should allow for clusters that consist of multiple species, otherwise the new system would simply be a collection of D decoupled single-species systems, and we already have the Chap. 3 theory for that. This also has realistic significance: mesoscale heterogeneity emerges from these inter-species interactions, as shown schematically in Fig. 1.2.

Hence the allowed clusters should be categorized not just by their sizes (the total number of particles), but also their composition (the number of particles of each species). So our cluster representation should be a D-dimensional vector $\boldsymbol{k}$, in which the number of species-α particles is k_α and its size

$$k = \sum_{\alpha=1}^{D} k_\alpha. \tag{4.1}$$

The other quantity then follows straightforwardly: the number of clusters $\boldsymbol{k}$ is $n_{\boldsymbol{k}}$, which is still a scalar.

4.1.2 Master equations and kernels

By introducing more than one species, we seem to have massively increased the complexity. However, in terms of the formalism, we have so far only extended the subscript k, i.e. the cluster size, into a vector $\boldsymbol{k}$. By doing this, we hope to retain most of the mathematical structure when extending the dimensionality, and below we will find out whether this is possible.

We begin by looking at the form of aggregation terms. For a single species, aggregation is realized by a simple scalar addition of sizes

$$k = i + j. \tag{4.2}$$

The vectorial representation of the multi-species system smoothly generalizes aggregation into a vector addition

$$\boldsymbol{k} = \boldsymbol{i} + \boldsymbol{j} \tag{4.3}$$

to allow particles in different species to connect together. It follows that now we can write down the aggregation term in the master equations almost exactly the same as before:

$$\sum_{k=i+j} \kappa(i,j) n_i n_j. \tag{4.4}$$

The summation here accounts for all possible combinations of i and j. It is a shorthand notation of a multiple summation on each of the D species:

$$\sum_{\substack{0 \leqslant i_1 \leqslant k_1 \\ 0 \leqslant j_1 \leqslant k_1}}^{k_1=i_1+j_1} \sum_{\substack{0 \leqslant i_2 \leqslant k_2 \\ 0 \leqslant j_2 \leqslant k_2}}^{k_2=i_2+j_2} \cdots \sum_{\substack{0 \leqslant i_D \leqslant k_D \\ 0 \leqslant j_D \leqslant k_D}}^{k_D=i_D+j_D} \kappa(i,j) n_i n_j. \tag{4.5}$$

Comparing this with the aggregation term in eqn (3.3), this extension in dimensionality becomes apparent. Now the master equations for multiple species readily follow:

$$\dot{n}_k(t) = \frac{1}{2} \sum_{k=i+j} \kappa(i,j) n_i n_j - \sum_i \kappa(i,k) n_i n_k, \tag{4.6}$$

where the last summation sums over all possible i vectors—again it's the shorthand notation for D-dimensional summation over all its components, this time within $\mathbb{N}$, all natural numbers.

From eqn (4.6), the only requirement for kernels $\kappa(i,j)$ in D-species aggregation, besides the inherent non-negativity and symmetry from 1-species model, is that they should be scalar in nature. As before, the possibilities are endless. Here we discuss a few with interesting mathematics or interpretations.

Extended product kernel. In 1-D, the product kernel represents the aggregation rate from particle-wise interactions. Its natural extension in multi-species models is a dot product:

$$\kappa(i,j) = i \cdot j. \tag{4.7}$$

However, the problem with a simple dot product is that it only admits same species aggregations: for example, a cluster $(2,0)$ can never merge with another cluster $(0,3)$, and so the system is, at the end of the day, still a decoupled system of multiple species, and the high-dimensional construction is unnecessary. Furthermore, if we introduce heterogeneity $F(t)$ under the 'dot product kernel', since it is just a scalar multiplier, this means that all the species have to adhere to the same underlying interaction mechanism—in other words, there is no heterogeneity between species.

Fixing these issues only requires some simple tweaks. Since we want cross-species interactions to be possible and that each pair of species should be able to have different heterogeneous parameters, we simply need to insert a metric tensor between the two vectors:

$$\kappa(i,j) = i^\mathrm{T} \mathsf{F} j. \tag{4.8}$$

Due to the symmetric assumption of aggregation, the tensor F is naturally symmetric. Each non-zero entry $F_{\alpha\beta}$ indicates an existing pair of interactive species α and β, and its value is simply the heterogeneous parameter (hence the symbol F is retained). We can then call F the heterogeneous matrix.

Extended sum kernels. In 1-D, the sum kernel measures the size of the resulting cluster due to aggregation. There are two simple extensions of this into multi-species aggregation systems.

1. We can look at the total cluster size regardless of species. Applying the definition in eqn (4.1), we have
$$\kappa(\boldsymbol{i},\boldsymbol{j}) = \boldsymbol{i} + \boldsymbol{j}. \tag{4.9}$$
Note that this kernel does nothing to distinguish different species, so it could be useful for a coarse-grained system where aggregation rate depends on size only and irrelevant to species.

2. Another possibility is to look at each species and weigh them by their corresponding heterogeneous parameters respectively. We obtain the following expression:
$$\kappa(\boldsymbol{i},\boldsymbol{j}) = \sqrt{(\boldsymbol{i}+\boldsymbol{j})^\mathrm{T} \mathsf{F} (\boldsymbol{i}+\boldsymbol{j})}, \tag{4.10}$$
where F here should be a diagonal matrix. This formalism is more mathematically tractable than the previous one as it utilizes the fundamental vectorial operations instead of an extrinsic definition as eqn (4.1).

4.1.3 Generating functions and moment tensors

After extending the two important variables $\boldsymbol{k}$ and $n_{\boldsymbol{k}}$ for multi-species aggregation, we can naturally define the multi-species equivalence of the generating functions. The first generating function $\mathcal{D}$ is defined by
$$\mathcal{D}(\boldsymbol{x},t) \equiv \sum_{\boldsymbol{k}} n_{\boldsymbol{k}} e^{-\boldsymbol{k}\cdot\boldsymbol{x}}. \tag{4.11}$$

Note that the reciprocal space of size profile $\boldsymbol{k}$ where $\boldsymbol{x}$ lives, is now a D-dimensional vector space as well, just like $\boldsymbol{k}$-space. The duality between the generating function and $n_{\boldsymbol{k}}(t)$ still holds. We can soon notice, however, that the second generating function is now vectorial:
$$\mathcal{E}(\boldsymbol{x},t) \equiv -\nabla \mathcal{D}(\boldsymbol{x},t) = \sum_{\boldsymbol{k}} \boldsymbol{k} n_{\boldsymbol{k}} e^{-\boldsymbol{k}\cdot\boldsymbol{x}}. \tag{4.12}$$

By taking $\boldsymbol{x} = \boldsymbol{0}$ we can obtain the first moment, also a vector:
$$\boldsymbol{M}_1 = \mathcal{E}(\boldsymbol{0},t) = \sum_{\boldsymbol{k}} \boldsymbol{k} n_{\boldsymbol{k}}. \tag{4.13}$$

It still measures the number of particles—only now its components correspond to different species. We retain our previous notation and denote this as

$$N \equiv M_1. \tag{4.14}$$

Generally for higher orders of moment functions, they all become tensorial, as we can see below under suffix notation:

$$(M_n)_{\alpha_1\ldots\alpha_n} = (-1)^n \partial_{\alpha_1} \ldots \partial_{\alpha_n} \mathcal{D}(x_\alpha, t), \quad \alpha_{1,2,\ldots,n} = 1, 2, \ldots, D. \tag{4.15}$$

4.1.4 Multi-dimensional generative PDEs

Now we look into the generating functions and their corresponding generative PDEs. In the remaining discussions we will be mainly using the extended product kernel due to its significance and intriguing dynamics. Rewriting eqn (4.6) with this kernel, we have

$$\dot{n}_{\boldsymbol{k}}(t) = \frac{1}{N^2} \sum_{\boldsymbol{k}=\boldsymbol{i}+\boldsymbol{j}} \boldsymbol{i}^{\mathrm{T}} \mathsf{F} \boldsymbol{j} n_{\boldsymbol{i}} n_{\boldsymbol{j}} - \frac{2}{N^2} \boldsymbol{k}^{\mathrm{T}} \mathsf{F} \sum_{\boldsymbol{i}} \boldsymbol{i} n_{\boldsymbol{i}} n_{\boldsymbol{k}}, \tag{4.16}$$

where N is defined as

$$N = \sum_{\alpha=1}^{D} N_\alpha, \tag{4.17}$$

the total number of all D species of particles. We are here using the same normalization convention as before.

Before we proceed with transformation, we first simplify the equations by rescaling to dimensionless form. Define the normalized cluster distribution

$$\nu_{\boldsymbol{k}} = \frac{n_{\boldsymbol{k}}}{N}, \tag{4.18}$$

which is normalized against the total number of particles so that it now represents the density of clusters instead of the number of clusters. As in Chap. 3 we also normalize the rescaled time

$$\tau = \frac{2t}{N}. \tag{4.19}$$

The rescaled eqn (4.16) reads

$$\frac{\mathrm{d}\nu_{\boldsymbol{k}}}{\mathrm{d}\tau} = \frac{1}{2} \sum_{\boldsymbol{k}=\boldsymbol{i}+\boldsymbol{j}} \boldsymbol{i}^{\mathrm{T}} \mathsf{F} \boldsymbol{j} \nu_{\boldsymbol{i}} \nu_{\boldsymbol{j}} - \boldsymbol{k}^{\mathrm{T}} \mathsf{F} \left(\frac{\boldsymbol{N}}{N}\right) \nu_{\boldsymbol{k}}. \tag{4.20}$$

Though we are now fully aware of how to transform eqn (4.16) into generative equations from Chap. 3, we begin with simply solving eqn (4.16) for monomers $\boldsymbol{k} = (1, 0)$ in a 2-species system, just to have an intuitive understanding of monomer dynamics. It is left as an exercise to show that for the all-monomer initial condition, the dynamics of monomers is given by

$$\nu_{1,0}(\tau) = \frac{N_1}{N} \exp\left[-\left(F_{11}\frac{N_1}{N} + F_{12}\frac{N_2}{N}\right)\tau\right] \quad (4.21)$$

and

$$\nu_{0,1}(\tau) = \frac{N_2}{N} \exp\left[-\left(F_{21}\frac{N_1}{N} + F_{22}\frac{N_2}{N}\right)\tau\right]. \quad (4.22)$$

A simple inspection shows that when the size of one of the species (N_1 or N_2) drops to zero, the evolution of monomers of the surviving species reduces to the expression found in the 1-dimensional case (eqn (3.11)). On the other hand, when both species are present but they do not interact with each other (i.e. $F_{12} = F_{21} = 0$), the monomer dynamics of one species is affected by the size of the other via the denominator $N = N_1 + N_2$. This can be understood as a dilution effect, that the second species adds to the system and hence slows down the aggregation process of the first species, and vice versa.

This analysis can be easily generalized for D species. Assuming that all objects are initially disaggregated (i.e. $n_{\hat{e}_\alpha}(0) = N_\alpha$, $\alpha = 1, 2, .., D$, where $\hat{e}_\alpha$ is the unit vector in the αth direction, and hence $n_{\hat{e}_\alpha}(\tau)$ the number of species-α monomers at time τ), it can be shown that the dynamics of monomers is given by

$$\nu_{\hat{e}_\alpha}(\tau) = \iota_\alpha \exp\left(-\iota^\mathrm{T} \mathsf{F} \hat{e}_\alpha \tau\right), \quad \alpha = 1, 2, \ldots, D, \quad (4.23)$$

where we define vector

$$\iota_\alpha = \frac{N_\alpha}{N} \quad \forall \alpha = 1, \ldots, D, \quad (4.24)$$

representing the proportion of all D species of the particles among the whole population. It naturally satisfies the condition $\sum_{\alpha=1}^{D} \iota_\alpha = 1$.

Now back to the transformation. We follow the normalization scheme above and further define the rescaled generating functions

$$\delta(\boldsymbol{x}, \tau) = \frac{\mathcal{D}(\boldsymbol{x}, \tau)}{N}, \quad \varepsilon_\alpha(\boldsymbol{x}, \tau) = \frac{\mathcal{E}_\alpha(\boldsymbol{x}, \tau)}{N} \quad (4.25)$$

in accordance with the dimensionless functions $\{\nu_k\}$, such that

$$\delta(\boldsymbol{x}, t) = \sum_k \nu_k e^{-\boldsymbol{k}\cdot\boldsymbol{x}}, \quad \varepsilon_\alpha(\boldsymbol{x}, t) = \sum_k k_\alpha \nu_k e^{-\boldsymbol{k}\cdot\boldsymbol{x}}. \quad (4.26)$$

The generative equation follows:

$$\frac{\partial \delta}{\partial \tau} = \frac{1}{2} F_{\alpha\beta} \varepsilon_\alpha \varepsilon_\beta - F_{\alpha\gamma} \iota_\alpha \varepsilon_\gamma. \quad (4.27)$$

One more partial derivative in x_α yields

$$\frac{\partial \varepsilon_\alpha}{\partial \tau} + F_{\beta\gamma}(\varepsilon_\beta - \iota_\beta)\partial_\alpha \varepsilon_\gamma = 0. \tag{4.28}$$

This is a **direct D-dimensional generalization of the inviscid Burgers' equation with shear interactions added.**

Theoretically since $N_\alpha \gg 1$, i.e. the numbers of each species are of a different scale of magnitude and hence can be seen as being infinitely large, we may as well assume that all N_α values are the same in order to further simplify our calculations, i.e. $N = DN_\alpha$, $\forall \alpha = 1, 2, \ldots, D$, and hence

$$\iota_\alpha = \frac{1}{D}, \quad \forall \alpha = 1, 2, \ldots, D. \tag{4.29}$$

Indeed, this 'identical population' assumption does not always hold in real systems, and in the following discussions we will alternate between the simpler results that obey the assumption and the generalizations that hold for any combination of species sizes.

4.1.5 Dynamics of the second moment tensor

In Chap. 3, we have seen that divergence of the second moment M_2 corresponds to the onset of gelation. However, for multi-species systems, the second moment is elevated into a rank-2 tensor. Fortunately, its dynamics is still rather straightforward, as we will see below. We first take one more partial derivative *wrt* (i.e. with respect to) ε_α on eqn (4.28) to obtain

$$\partial_\tau \partial_\beta \varepsilon_\alpha + F_{\gamma\delta}\left[\partial_\beta \varepsilon_\gamma \partial_\delta \varepsilon_\alpha + (\varepsilon_\gamma - \iota_\gamma)\partial_\beta \partial_\delta \varepsilon_\alpha\right] = 0. \tag{4.30}$$

Then we set $x_\alpha = 0$, $\forall \alpha = 1, \ldots, D$, and by rescaling M_2 the same way as rescaling the generating functions

$$(m_2)_{\alpha\beta} = -\left[\partial_\alpha \varepsilon_\beta\right]_{x=0} = -\left[\partial_\beta \varepsilon_\alpha\right]_{x=0}, \tag{4.31}$$

we can obtain its governing ODE:

$$\frac{dm_2}{d\tau} = m_2 F m_2, \tag{4.32}$$

This is a matrix Riccati equation. It is analytically solvable—we just need to treat with care the non-commutativity for matrices. We start by left-multiplying the inverse matrix m_2^{-1} to obtain

$$m_2^{-1}\frac{dm_2}{d\tau} = Fm_2. \tag{4.33}$$

Since

$$\frac{d}{d\tau}(m_2^{-1}m_2) = m_2^{-1}\frac{dm_2}{d\tau} + \frac{dm_2^{-1}}{d\tau}m_2 = 0, \tag{4.34}$$

we have

$$-\frac{dm_2^{-1}}{d\tau}m_2 = Fm_2, \qquad (4.35)$$

and thus

$$m_2^{-1} = m_2^{-1}(0) - F\tau. \qquad (4.36)$$

As usual we consider the all-monomer initial condition. Now under multiple species, this means that there are no mixed clusters, and all diagonal entries of $m_2(0)$ are exactly ι_α. Hence,

$$m_2^{-1}(0) = \mathrm{diag}(\iota_\alpha^{-1}). \qquad (4.37)$$

It follows that

$$m_2^{-1} = \mathrm{diag}(\iota_\alpha^{-1}) - F\tau. \qquad (4.38)$$

Under the 'identical population' assumption this becomes

$$m_2^{-1} = D\mathbb{I} - F\tau. \qquad (4.39)$$

What follows immediately is the gel time τ_g, at which point the determinant of m_2 diverges, i.e.

$$\det\left(m_2^{-1}(\tau_g)\right) = 0, \qquad (4.40)$$

a Dth order polynomial equation. Solutions for $D = 2$ are straightforward and will be presented shortly. Analytic solutions for $D = 3$ exist but are highly complicated, and for $D > 3$ solutions are generally numerical only. In the next couple of sections we will present detailed discussions for the 2-species and 3-species scenarios due to their analyticity. Then we will extend the discussions into arbitrary values of D and reveal nice physical interpretations for the mathematical constructions.

4.1.6 Multi-species gel curves

So far, we have seen that our D-dimensional extension leads to a working description of a D-species aggregation system and all the nice maths in 1-species remains. Even better, elevation in dimensionality does not strip eqn (4.28) of analyticity; it remains solvable with the characteristics method. Note that by definition,

$$\partial_\alpha \varepsilon_\gamma = \sum_s \partial_\alpha s_\gamma \nu_s e^{-s_\beta x_\beta} = -\sum_s s_\alpha s_\gamma \nu_s e^{-s_\beta x_\beta} = \partial_k \varepsilon_\alpha; \qquad (4.41)$$

eqn (4.28) then becomes

$$\frac{\partial \varepsilon_\alpha}{\partial \tau} + F_{\beta\gamma}(\varepsilon_\beta - \iota_\beta)\partial_\gamma \varepsilon_\alpha = 0. \qquad (4.42)$$

This structure is familiar from Chap. 3, and we can apply the method of characteristics to all components ε_α. Recall the characteristic equation in 1-D and we can write down the multi-dimensional equivalent:

$$\frac{dx_\alpha}{d\tau} = F_{\alpha\beta}(\varepsilon_\beta - \iota_\beta), \tag{4.43}$$

where we use the fact that $F_{\alpha\beta}$ is symmetric. Solving it the same way as the 1-D case by assuming $F_{\alpha\beta}$ to be independent of time, we have

$$x_\alpha(\tau) = F_{\alpha\beta}(\varepsilon_\beta - \iota_\beta)\tau + f_\alpha(\varepsilon), \tag{4.44}$$

where f_α is to be determined by initial condition. With an all-monomer initial condition

$$\varepsilon_\alpha(\boldsymbol{x}, \tau = 0) = \iota_\alpha e^{-x_\alpha}, \quad \text{(no sum in } \alpha\text{)} \tag{4.45}$$

we have

$$x_\alpha(0) = -\ln(\varepsilon_\alpha/\iota_\alpha) = f_\alpha, \tag{4.46}$$

and the implicit yet exact solution to eqn (4.28) is

$$\ln(\varepsilon_\alpha/\iota_\alpha) = F_{\alpha\beta}(\varepsilon_\beta - \iota_\beta)\tau - x_\alpha \tag{4.47}$$

Finally, we take $x_\alpha = 0$ to look at gelation behaviours. It follows that

$$F_{\alpha\beta}(\varepsilon_\beta - \iota_\beta)\tau - \ln(\varepsilon_\alpha/\iota_\alpha) = 0$$
$$\Rightarrow \varepsilon_\alpha = \iota_\alpha e^{-\sum_\beta F_{\alpha\beta}(\iota_\beta - \varepsilon_\beta)\tau}, \tag{4.48}$$

where in the last equality we no longer use the summation convention. Equivalently we can also write the exponent in matrix multiplication form:

$$\varepsilon_\alpha = \iota_\alpha \exp\left[-\mathsf{F}(\boldsymbol{\iota} - \boldsymbol{\varepsilon}) \cdot \hat{\mathbf{e}}_\alpha \tau\right], \quad \alpha = 1, 2, \ldots, D. \tag{4.49}$$

In the previous chapter, eqn (3.81) gives the solution for 1-species systems. Let's recast it into the same form as above:

$$\varepsilon = e^{-F(1-\varepsilon)\tau}. \tag{4.50}$$

We can see that eqn (4.48) is a direct generalization of eqn (4.50)—each cross-species interaction simply corresponds to an additional multiplicative factor $e^{-F_{\alpha\beta}(1-\varepsilon_\beta)\tau}$ that indicates coupling between species α and β. The prefactor ι_α accounts for the fact that the population of each species is now a fraction ι_α of the total number of particles.

4.2 Scenario 1: 2-species fusion and gelation

4.2.1 Gelation onset time

We begin with some notational convention: we write the diagonal entries of F as f_α and the off-diagonal entries $\epsilon_{\alpha\beta}$. For $D = 2$, since the only two off-diagonal elements are identical, we further denote that $\epsilon_{12} = \epsilon_{21} = \epsilon$. For simplicity we take the identical-population assumption during the following exact derivations. Then, from eqn (4.39), we have

$$\mathsf{m}_2^{-1} = 2 \begin{pmatrix} 1 - f_1\tau/2 & -\epsilon\tau/2 \\ -\epsilon\tau/2 & 1 - f_2\tau/2 \end{pmatrix} \qquad (4.51)$$

and

$$\begin{aligned}\Delta \equiv \det\big((2\mathsf{m}_2)^{-1}\big) &= (1 - f_1\tau/2)(1 - f_2\tau/2) - \epsilon^2\tau^2/4 \\ &= (f_1 f_2 - \epsilon^2)(\tau/2)^2 - (f_1 + f_2)(\tau/2) + 1.\end{aligned} \qquad (4.52)$$

Hence, τ_g satisfies

$$(f_1 f_2 - \epsilon^2)(\tau_g/2)^2 - (f_1 + f_2)(\tau_g/2) + 1 = 0, \qquad (4.53)$$

and its two roots are therefore

$$\tau_g^{(1,2)} = \frac{(f_1 + f_2) \mp \sqrt{(f_1 - f_2)^2 + 4\epsilon^2}}{(f_1 f_2 - \epsilon^2)}. \qquad (4.54)$$

Now let's look into the properties of τ_g and m_2, especially their dependence on the parameters $f_{1,2}$ and ϵ and their underlying physical implications. First, we note that the two roots of eqn (4.53), as shown in eqn (4.54), are always real. Within this 3-D parameter space, it is physically intriguing to inspect the relative value of cross-species interaction ϵ against the intra-species interactions $f_{1,2}$. For this matter we

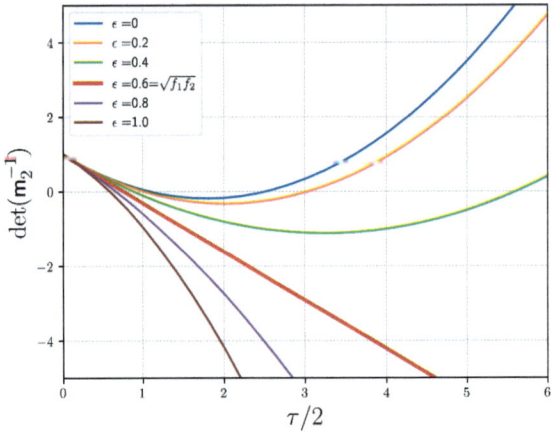

Fig. 4.2: $\Delta(\tau)$ for different ϵ values. f_1, f_2 are taken to be $0.9, 0.4$, respectively.

fix $f_{1,2}$. A changing ϵ only alters the quadratic coefficient of $\Delta(\tau)$—more specifically, $\tau_g^{(1,2)}$ is determined by ϵ compared with $\sqrt{f_1 f_2}$, i.e. the geometric average of f_1 and f_2. For small $\epsilon < \sqrt{f_1 f_2}$, both roots are positive. As ϵ increases, the smaller root $\tau_g^{(1)}$ contracts while the larger one $\tau_g^{(2)}$ diverges. Just at the geometric average $\epsilon = \sqrt{f_1 f_2}$, the quadratic coefficient cancels, $\Delta(\tau)$ becomes linear in τ, and there is one sole root, positive. For relatively stronger inter-species interaction $\epsilon > \sqrt{f_1 f_2}$ only the larger root $\tau_g^{(2)}$ is now positive, i.e. the physical solution, and it decreases further as ϵ continues increasing. As $\epsilon \to \infty$, the gel time $\tau_g \to 0$. This is shown with a concrete example in Fig. 4.2.

We are able to obtain the big picture, at least qualitatively, simply with some elementary mathematics. Of course we can go one step further and plot the two roots exactly as a function of ϵ: Fig. 4.3 shows this with the same set of parameters $f_{1,2}$ as in Fig. 4.2. Now that the mathematics is clear, let's turn to the physics: how do we interpret the two roots? Are they both physical at all? Why does one disappear into infinity at $\epsilon = \sqrt{f_1 f_2}$?

To answer those questions, we start by looking at the simplest scenario, at $\epsilon = 0$. This corresponds to two species being decoupled and essentially two independent 1-D systems. In this case, the two roots of τ_g are both physical, corresponding to the two separate gelation onsets. We can see this more clearly from the expression of m_2. By taking the inverse of $m_2{}^{-1}$, it follows that

$$m_2 = \frac{1}{2\Delta}\begin{pmatrix} 1 - f_2\tau/2 & \epsilon\tau/2 \\ \epsilon\tau/2 & 1 - f_1\tau/2 \end{pmatrix}. \tag{4.55}$$

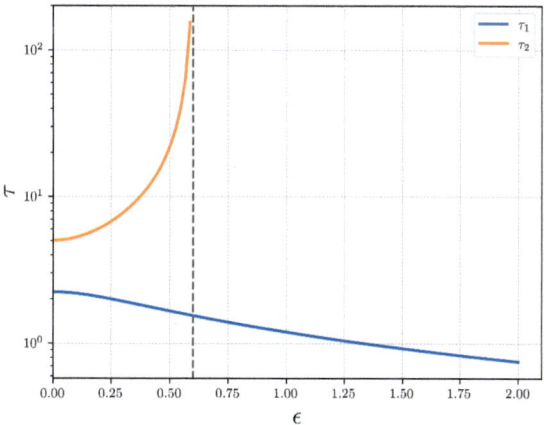

Fig. 4.3: The two roots $\tau_g^{(1,2)}$ against ϵ for $f_1 = 0.9, f_2 = 0.4$.

When $\epsilon = 0$, this becomes

$$\frac{1}{2}\begin{pmatrix} (1-f_1\tau/2)^{-1} & 0 \\ 0 & (1-f_2\tau/2)^{-1} \end{pmatrix}, \qquad (4.56)$$

and we note that the divergence of the two entries occur separately. This verifies the decoupling mathematically. However, once we turn on $\epsilon > 0$, the system no longer diverges separately by species; instead, all the entries of m_2 diverge simultaneously the first instant when Δ becomes 0, i.e. at the smaller root $\tau_g^{(1)}$. This means that a mixed gel of both species must start forming at time $\tau_g^{(1)}$. To see this from the perspective of aggregation, when $\epsilon > 0$, however small it is, cross-species aggregation must be occurring, so the species that gels more slowly should now instead be attached to the bigger clusters of the other species during its gelation, hence leading to a mixed gel. From this we can see that the exact proportion of two species within the gel should depend on the relative strength of f_1, f_2, and ϵ, and we will soon find an analytic result for this proportion in the next section. The decrease of $\tau_g^{(1)}$ as ϵ grows is also clear—the overall aggregation rate increases, hence resulting in earlier gelation onset.

There is one question yet to be answered: how to explain the fact that $\tau_g^{(2)}$ becomes unphysical the very instance ϵ is turned on? We leave it unanswered for now, and will provide a nice and detailed explanation later in this chapter.

4.2.2 Gel composition at gelation onset

After looking at the determinant of m_2, we now turn to its components. Qualitatively, the second moment near the gel time should provide a measurement of the gel composition around the onset of gelation, i.e. the proportions of two species in the gel. This is not hard to see qualitatively—due to its quadratic nature, the contribution from the largest cluster (the 'quasi-gel') becomes dominant near the gel time. Consequently, it should be a reasonable approximation that, near gel time,

$$m_2 \sim A \begin{pmatrix} K_1^2 & K_1 K_2 \\ K_1 K_2 & K_2^2 \end{pmatrix}, \qquad (4.57)$$

where K_1, K_2 are the respective proportions of type-1,2 particles in the 'quasi-gel', and A is a very large factor that accounts for the almost-divergent magnitude of m_2.

It turns out that eqn (4.57) is more than an approximation. It is possible to derive it directly from the matrix components at gel time. But at the gel time they are all singular: how can we extract any other useful information than divergence? Our way around this is to extract out a common factor $1/\Delta$. The denominator Δ is the determinant of m_2, i.e. *the* cause of divergence. We can then define a divergence-free matrix

Scenario 1: 2-species fusion and gelation

$$\mu = \begin{pmatrix} 1 - f_2\tau/2 & \epsilon\tau/2 \\ \epsilon\tau/2 & 1 - f_1\tau/2 \end{pmatrix}, \quad (4.58)$$

such that

$$m_2 = \frac{1}{2\Delta}\mu. \quad (4.59)$$

Now let's examine the entries of μ at gelation onset and compare them with eqn (4.57). For $f_1 f_2 \neq \epsilon^2$, at gel time, we plug in $\tau_g^{(1)}$ found in eqn (4.54) and obtain

$$\mu_g^{(11)} = 1 - f_2 \frac{(f_1 + f_2) - \sqrt{(f_1 - f_2)^2 + 4\epsilon^2}}{2(f_1 f_2 - \epsilon^2)}$$

$$\mu_g^{(12)} = \epsilon \frac{(f_1 + f_2) - \sqrt{(f_1 - f_2)^2 + 4\epsilon^2}}{2(f_1 f_2 - \epsilon^2)}$$

$$\mu_g^{(22)} = 1 - f_1 \frac{(f_1 + f_2) - \sqrt{(f_1 - f_2)^2 + 4\epsilon^2}}{2(f_1 f_2 - \epsilon^2)}.$$

Note that these expressions are homogeneous in terms of $f_{1,2}$ and ϵ, so we can actually write them in terms of two variables only:

$$\phi_1 = \frac{f_1}{\epsilon}, \quad \phi_2 = \frac{f_2}{\epsilon}. \quad (4.60)$$

This indicates that when it comes to gel composition, only the relative values of these three parameters make a difference. As discussed earlier, their exact values only determine how soon the system gels. The three entries are reduced into

$$\mu_g^{(11)} = \frac{\phi_1\phi_2 - 2 - \phi_2^2 + \phi_2\sqrt{(\phi_1 - \phi_2)^2 + 4}}{2(\phi_1\phi_2 - 1)}, \quad (4.61)$$

$$\mu_g^{(12)} = \frac{(\phi_1 + \phi_2) - \sqrt{(\phi_1 - \phi_2)^2 + 4}}{2(\phi_1\phi_2 - 1)}, \quad (4.62)$$

$$\mu_g^{(22)} = \frac{\phi_1\phi_2 - 2 - \phi_1^2 + \phi_1\sqrt{(\phi_1 - \phi_2)^2 + 4}}{2(\phi_1\phi_2 - 1)}. \quad (4.63)$$

In order to find their relations, instead of using brute force algebra, we turn to eqn (4.51) and note that the inverse of the second moment at gel time can simply be written as

$$m_2^{-1}(\tau_g) = \begin{pmatrix} \mu_g^{(22)} & -\mu_g^{(12)} \\ -\mu_g^{(12)} & \mu_g^{(11)} \end{pmatrix}, \quad (4.64)$$

and since $\det(m_2^{-1}(\tau_g)) = 0$, we must have

$$\mu_g^{(22)}\mu_g^{(11)} = \mu_g^{(12)^2}. \quad (4.65)$$

However, if we return to the definition of the second moment, we have

$$\mu_g^{(11)}\mu_g^{(22)} = \left(\sum_{i=0}^{\infty}\sum_{j=0}^{\infty} i^2 n_{(i,j)}\right)\left(\sum_{i=0}^{\infty}\sum_{j=0}^{\infty} j^2 n_{(i,j)}\right) = \sum_{i=0}^{\infty}\sum_{j=0}^{\infty} i^2 j^2 n_{(i,j)}^2$$

$$\leqslant \left(\sum_{i=0}^{\infty}\sum_{j=0}^{\infty} ij n_{(i,j)}\right)^2 = \mu_g^{(12)^2}, \qquad (4.66)$$

where equality can only be reached if there exists only one cluster profile $\mathbf{k}_0 = (i_0, j_0)$ throughout the system, such that there are no cross terms in the perfect square after the inequality. In the aggregation system, this occurs only at or near gelation, where a giant cluster, i.e. the gel, dominates the system. In reality, at this point there may well be other smaller clusters in the system, but their contribution to the second moment is negligible compared to the gel or quasi-gel, due to the vast difference in size. Thus, eqn (4.65) leads to the following:

$$\mu_g^{(11)} \approx K_1^2, \qquad (4.67)$$
$$\mu_g^{(12)} \approx K_1 K_2, \qquad (4.68)$$
$$\mu_g^{(22)} \approx K_2^2. \qquad (4.69)$$

These expressions confirm our earlier guess, and further specify that $\sqrt{\mu_g^{(11)}}$ and $\sqrt{\mu_g^{(22)}}$ indeed measure the gel composition at onset. The off-diagonal component $\mu_g^{(12)}$ does not provide extra information apart from being the product $K_1 K_2$. This naturally makes sense, since our mesoscopic model does not distinguish connections by types of aggregating species (intra or inter).

So far we have understood that the proportions of both species, $\eta_{1,2}$, within the gel around onset time are given by

$$\eta_1 \equiv \frac{\sqrt{\mu_g^{(11)}}}{\sqrt{\mu_g^{(11)}} + \sqrt{\mu_g^{(22)}}}, \qquad (4.70)$$

$$\eta_2 \equiv \frac{\sqrt{\mu_g^{(22)}}}{\sqrt{\mu_g^{(11)}} + \sqrt{\mu_g^{(22)}}} = 1 - \eta_1. \qquad (4.71)$$

Substituting in eqns (4.67–4.69), we can now express this pair of quantities in terms of parameters $\phi_{1,2}$:

$$\eta_1 = \left(1 + \sqrt{\frac{\mu_g^{(22)}}{\mu_g^{(11)}}}\right)^{-1}$$

$$= \left(1 + \sqrt{\frac{\phi_1\phi_2 - 2 - \phi_1^2 + \phi_1\sqrt{(\phi_1 - \phi_2)^2 + 4}}{\phi_1\phi_2 - 2 - \phi_2^2 + \phi_2\sqrt{(\phi_1 - \phi_2)^2 + 4}}}\right)^{-1}.$$

We can yet simplify this further. Substituting once more with $x = \phi_1 - \phi_2$, we have

$$\eta_1 = \left[1 + \sqrt{\frac{\phi_1\left(\sqrt{x^2+4}-x\right)-2}{\phi_2\left(\sqrt{x^2+4}+x\right)-2}}\right]^{-1}$$

$$= \left[1 + \sqrt{\frac{(\phi_2+x)\left(\sqrt{x^2+4}-x\right)-2}{\phi_2\left(\sqrt{x^2+4}+x\right)-2}}\right]^{-1}$$

$$= \left[1 + \sqrt{1 + x\frac{\sqrt{x^2+4}-x-2\phi_2}{\phi_2\left(\sqrt{x^2+4}+x\right)-2}}\right]^{-1}$$

$$= \left[1 + \sqrt{1 + x\left(\sqrt{x^2+4}-x\right)\frac{\sqrt{x^2+4}-x-2\phi_2}{4\phi_2 - 2\left(\sqrt{x^2+4}-x\right)}}\right]^{-1}$$

$$= \left[1 + \sqrt{1 - \frac{x}{2}\left(\sqrt{x^2+4}-x\right)}\right]^{-1}$$

$$= \left[1 + \frac{1}{2}\left(\sqrt{x^2+4}-x\right)\right]^{-1} = \frac{2}{\sqrt{x^2+4}-x+2}, \tag{4.72}$$

$$\eta_2 = 1 - \eta_1 = \frac{2}{\sqrt{x^2+4}+x+2}, \tag{4.73}$$

where in the fourth equality we multiply on both the numerator and the denominator of the fraction inside the square root by $\left(\sqrt{x^2+4}-x\right)$, and in the sixth equality we find that the expression inside the square root is actually a perfect square. Curiously enough, $\eta_{1,2}$ are monotonic functions in a simple one-parameter space. In other words, the gel composition at onset depends monotonically on one specific combination of the parameters in the form

$$x = \frac{f_1 - f_2}{\epsilon}. \tag{4.74}$$

The detailed relations are plotted in Fig. 4.4. They are nicely symmetric, and can also be neatly interpreted: a large f_α corresponds to strong intra-species aggregation for species α, and hence more species-i particles in the gel, whereas a large ϵ represents strong tendency of inter-species aggregation, and it follows that $x \to 0$ and therefore comparable number of both species in the gel.

4.2.3 Arbitrary population profiles

Gel onset time. Generally for arbitrary numbers of ι_1 and $\iota_2 = 1 - \iota_1$, eqn (4.40) yields the following second degree polynomial

$$\Delta \equiv 1 - (f_1\iota_1 + f_2\iota_2)\tau + \iota_1\iota_2\left(f_1f_2 - \epsilon^2\right)\tau^2 = 0, \tag{4.75}$$

whose solutions are

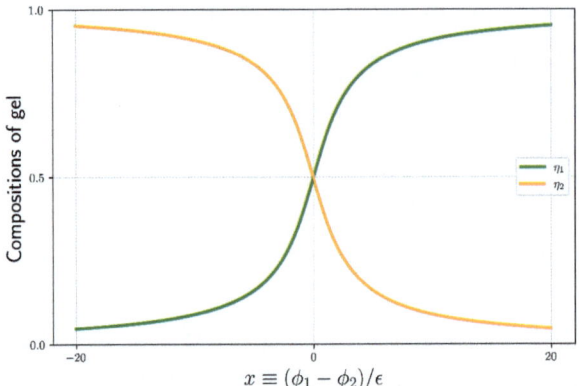

Fig. 4.4: The proportions of two species in the gel at onset time with respect to the composite parameter. The larger aggregation rate f_α for one species corresponds to its larger proportions in the gel.

$$\tau_g^{(1,2)} = \frac{f_1\iota_1 + f_2\iota_2 \mp \sqrt{(f_1\iota_1 - f_2\iota_2)^2 + 4\epsilon^2\iota_1\iota_2}}{2(f_1 f_2 - \epsilon^2)\iota_1\iota_2}. \qquad (4.76)$$

The reader is asked to verify this in a very simple exercise at the end of the chapter. We note that as in the identical population case, $\tau_g^{(1)} < \tau_g^{(2)}$, and there is a divergence on $\tau_g^{(2)}$ when the denominator of eqn (4.76) is zero. Hence, the root $\tau_g^{(1)}$ marks the gel onset of the system, while $\tau_g^{(2)}$ is physically relevant for $\epsilon = 0$, where it marks the gel of the species with lower same-species aggregation rate.

Gel composition at the onset. As explained before, the second moment provides information about the gel composition at the onset. Here we derive this onset composition for a 2-species system when N_1 is generally not equal to N_2. Following eqn (4.38), we can write down the second moment tensor for $D = 2$ as

$$\mathsf{m}_2(\tau) = [\det(\mathrm{diag}(\iota_1, \iota_2) - \mathsf{F}\tau)]^{-1} \begin{pmatrix} \phi_2(\tau)^2 & \epsilon\tau \\ \epsilon\tau & \phi_1(\tau)^2 \end{pmatrix}, \qquad (4.77)$$

where $\phi_\alpha(t)$ are defined as

$$\phi_\alpha(t) = \sqrt{\iota_\alpha^{-1} - f_\alpha \tau}, \quad \alpha = 1, 2. \qquad (4.78)$$

It follows that the gel composition at the onset is given by $\eta_1(\tau_g) = \phi_2(\tau_g)/\phi_0(\tau_g)$ and $\eta_2(\tau_g) = \phi_1(\tau_g)/\phi_0(\tau_g)$, with $\phi_0(\tau) = \phi_1(\tau) + \phi_2(\tau)$.

4.2.4 2-D gel curve solutions

Now let's turn our focus back to the explicit gelation solution of the 2-species system, given by eqn (4.48). We write it out explicitly for 2-species as

$$\varepsilon_1 = \iota_1 e^{-[f_1(\iota_1-\varepsilon_1)+\epsilon(\iota_2-\varepsilon_2)]\tau}, \tag{4.79}$$

$$\varepsilon_2 = \iota_2 e^{-[\epsilon(\iota_1-\varepsilon_1)+f_2(\iota_2-\varepsilon_2)]\tau}, \tag{4.80}$$

where again we take $f_1 = F_{11}$, $f_2 = F_{22}$, $\epsilon = F_{12} = F_{21}$. To look at gelation more explicitly we switch to a slightly different pair of variables:

$$G_1(\tau) = \iota_1 - \varepsilon_1(\mathbf{0}, \tau) = \iota_1 \left[1 - e^{-[f_1 G_1 + \epsilon G_2]\tau}\right] \tag{4.81}$$

$$G_2(\tau) = \iota_2 - \varepsilon_2(\mathbf{0}, \tau) = \iota_2 \left[1 - e^{-[\epsilon G_1 + f_2 G_2]\tau}\right]. \tag{4.82}$$

They represent the fraction of species 1 and 2 particles among the *total population* which are within the gel phase. The total size of gel (as a fraction of the entire population) is then

$$G(\tau) = 1 - \varepsilon_1 - \varepsilon_2$$
$$= 1 - \iota_1 e^{-(f_1 G_1 + \epsilon G_2)\tau} - \iota_2 e^{-(\epsilon G_1 + f_2 G_2)\tau}. \tag{4.83}$$

Having obtained the gel solution for all time, we can nonetheless return to look at the behaviours just after onset by expanding the solution at $G_\alpha \ll \iota_\alpha$. Taking logarithms of eqns (4.79) and (4.80) gives

$$\tau = -\frac{\ln(\varepsilon_1/\iota_1)}{f_1(\iota_1 - \varepsilon_1) + \epsilon(\iota_2 - \varepsilon_2)}, \tag{4.84}$$

$$\tau = -\frac{\ln(\varepsilon_2/\iota_2)}{\epsilon(\iota_1 - \varepsilon_1) + f_2(\iota_2 - \varepsilon_2)}. \tag{4.85}$$

We shall soon numerically plot the gel curves, but before that we first show another approach to analytically recover the gel-time composition starting from this pair of implicit solutions. We begin with equating the above two equations and obtain

$$[\ln(\varepsilon_1/\iota_1)][\epsilon(\iota_1 - \varepsilon_1) + f_2(\iota_2 - \varepsilon_2)] = [\ln(\varepsilon_2/\iota_2)][f_1(\iota_1 - \varepsilon_1) + \epsilon(\iota_2 - \varepsilon_2)], \tag{4.86}$$

i.e.

$$[\ln(1 - G_1/\iota_1)](\epsilon G_1 + f_2 G_2) = [\ln(1 - G_2/\iota_2)](f_1 G_1 + \epsilon G_2), \tag{4.87}$$

Immediately after gelation onset, $G_\alpha(\tau_g^+) \ll \iota_\alpha$, so we linearize the logarithms and get

$$-\frac{G_1(\tau_g^+)}{\iota_1}[\epsilon G_1(\tau_g^+) + f_2 G_2(\tau_g^+)] = -\frac{G_2(\tau_g^+)}{\iota_2}[f_1 G_1(\tau_g^+) + \epsilon G_2(\tau_g^+)]. \tag{4.88}$$

From our definition in Section 4.2.2, $\eta_\alpha = G_\alpha(\tau_g^+)$, $\alpha = 1, 2$. Hence, tidying up the equation we have

$$\epsilon\left[\eta_1^2/\iota_1 - \eta_2^2/\iota_2\right] = (f_1/\iota_2 - f_2/\iota_1)\eta_1\eta_2$$

$$\Rightarrow \left(\frac{\eta_1}{\eta_2}\right)^2 + \frac{f_1(\iota_1/\iota_2) - f_2}{\epsilon}\left(\frac{\eta_1}{\eta_2}\right) - \frac{\iota_1}{\iota_2} = 0. \qquad (4.89)$$

This gives a quadratic equation straightforward to solve for gel composition at onset. But before we carry out the calculations, we identify that, under the identical-population assumption $\iota_1 = \iota_2 = 1/2$, the linear coefficient is just

$$\frac{f_1 - f_2}{\epsilon} \equiv x. \qquad (4.90)$$

Hence, the equation can be simplified into

$$\left(\frac{\eta_1}{\eta_2}\right)^2 + x\left(\frac{\eta_1}{\eta_2}\right) - 1 = 0, \qquad (4.91)$$

and the solution is

$$\frac{\eta_1}{\eta_2} = \frac{\sqrt{x^2 + 4} - x}{2}, \qquad (4.92)$$

where we tacitly discarded the negative root. Now we check this expression against eqns (4.72, 4.73):

$$\eta_1 = \frac{2}{\sqrt{x^2 + 4} - x + 2}, \quad \eta_2 = \frac{2}{\sqrt{x^2 + 4} + x + 2},$$

which give

$$\begin{aligned}
\frac{\eta_1}{\eta_2} &= \frac{\sqrt{x^2 + 4} - x + 2}{\sqrt{x^2 + 4} + x + 2} \\
&= \frac{(\sqrt{x^2 + 4} - x + 2)(\sqrt{x^2 + 4} + x - 2)}{(\sqrt{x^2 + 4} + x + 2)(\sqrt{x^2 + 4} + x - 2)} \\
&= \frac{x^2 + 4 - (x - 2)^2}{(\sqrt{x^2 + 4} + x)^2 - 4} = \frac{4x}{2x^2 + 2x\sqrt{x^2 + 4}} = \frac{2}{x + \sqrt{x^2 + 4}} \\
&= \frac{\sqrt{x^2 + 4} - x}{2}, \qquad (4.93)
\end{aligned}$$

indeed identical to the result in eqn (4.92).

For the general result with an arbitrary population profile, one simply needs to insert the factor ι_1/ι_2 into the appropriate positions of eqn (4.93). The reader is asked to obtain the final form of it in the exercises.

Finally comes the plot of gel curves: below we plot the total gel size, i.e. $(1-\varepsilon_1-\varepsilon_2)$ against time for different ϵ values while fixing $f_{1,2}$. Figure 4.5 illustrates many of our previous discussions. First, we notice that at $\epsilon = 0$ the curve has two abrupt bends, corresponding to two separate gelations—note that these two gels never aggregate together. These two bends fall on $\tau_g^{(1,2)}$ exactly. As soon as we turn on ϵ, the second

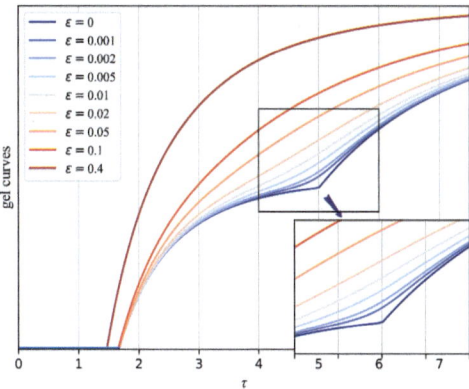

Fig. 4.5: Gel curves plotted at $f_{1,2} = 1.2,\ 0.4$.

bend disappears as $\tau_g^{(2)}$ becomes a so-to-speak virtual gelation point. Instead we have a smooth bump of gelation rate near virtual gelation. After the bump when the rate ticks up again, it indicates that species 2 now dominates the gelation growth. This bump becomes more apparent for smaller ϵ and greater difference between f_1 and f_2, i.e. for larger composite parameter x. As ϵ increases further, we can see the advancement of gelation point as well as the disappearance of that bump, and when ϵ is large enough the gel curve appears similar to a single-species one. At this point ϵ is comparable in value with f_α, so the numbers of both species grow at comparable time during gelation. Below we shall see in Fig. 4.6 that the same rule holds for 3-species systems.

4.3 Scenario 2: 3-species fusion and gelation

The 2-species model produced some simple and nice results. Next, we want to know if similar relations hold for 3 species. We follow the footsteps above and first write down m_2^{-1}, which itself is the solution of the homogeneous 3-D Riccati equation. Denoting $\sigma = \tau/3$, we have

$$\mathsf{m}_2^{-1} = 3 \begin{pmatrix} 1 - f_1\sigma & -\epsilon_{12}\sigma & -\epsilon_{13}\sigma \\ -\epsilon_{12}\sigma & 1 - f_2\sigma & -\epsilon_{23}\sigma \\ -\epsilon_{13}\sigma & -\epsilon_{23}\sigma & 1 - f_3\sigma \end{pmatrix}. \tag{4.94}$$

Now we write the determinant of $\mathsf{m}_2^{-1}/3 = (3\mathsf{m}_2)^{-1}$ as Δ again, and write down the determinants of its three 2-D submatrices

$$\begin{vmatrix} 1 - f_\alpha\sigma & -\epsilon_{\alpha\beta}\sigma \\ -\epsilon_{\alpha\beta}\sigma & 1 - f_\beta\sigma \end{vmatrix} \equiv \delta_{\alpha\beta}. \tag{4.95}$$

Then, we take the inverse of $(3\mathsf{m}_2)^{-1}$ explicitly and obtain

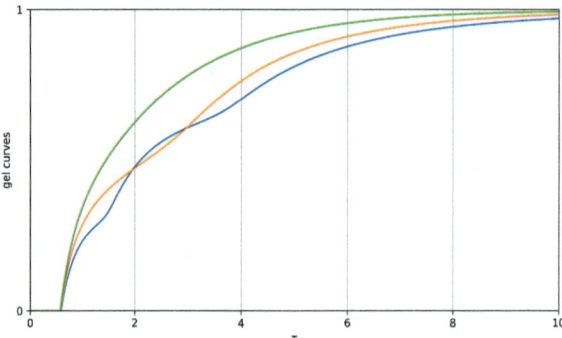

Fig. 4.6: 3-species gelation curves, each curve corresponding to a different set of parameters. Blue: $f_{1,2,3} = 5, 2, 0.8$, $\epsilon_{12,13,23} = 0.01, 0.005, 0.002$, respectively. Orange: $f_{1,2,3} = 5, 1, 0.8$, $\epsilon_{12,13,23} = 0.01, 0.5, 0.02$. Green: $f_{1,2,3} = 5, 1, 0.8$, $\epsilon_{12,13,23} = 0.5, 0.5, 0.02$. The curves have 2 to 0 kinks following the aforementioned qualitative rule on the number of positive roots of τ_g.

$$m_2 = \frac{\mu}{3\Delta}, \qquad (4.96)$$

where

$$\mu = \begin{pmatrix} \delta_{23} & (\epsilon_{23}\sigma)(\epsilon_{13}\sigma) + (\epsilon_{12}\sigma)(1-f_3\sigma) & (\epsilon_{12}\sigma)(\epsilon_{23}\sigma) + (\epsilon_{13}\sigma)(1-f_2\sigma) \\ (\epsilon_{23}\sigma)(\epsilon_{13}\sigma) + (\epsilon_{12}\sigma)(1-f_3\sigma) & \delta_{13} & (\epsilon_{13}\sigma)(\epsilon_{12}\sigma) + (\epsilon_{23}\sigma)(1-f_1\sigma) \\ (\epsilon_{12}\sigma)(\epsilon_{23}\sigma) + (\epsilon_{13}\sigma)(1-f_2\sigma) & (\epsilon_{13}\sigma)(\epsilon_{12}\sigma) + (\epsilon_{23}\tau)(1-f_1\sigma) & \delta_{12} \end{pmatrix}. \qquad (4.97)$$

Since τ_g is now a root of the cubic equation

$$\Delta(\tau_g) = 0 \qquad (4.98)$$

and hence harder to solve, we do not here pursue an exact evaluation. We can still, however, check that the diagonal entries, i.e. $\delta_{\alpha\beta}$ here, still measure the gel composition at gelation onset—this is left as an exercise. Writing this out more explicitly, the proportions of the three species in the gel are

$$\frac{\sqrt{\delta_{23}}}{\sqrt{\delta_{23}} + \sqrt{\delta_{13}} + \sqrt{\delta_{12}}}, \frac{\sqrt{\delta_{13}}}{\sqrt{\delta_{23}} + \sqrt{\delta_{13}} + \sqrt{\delta_{12}}}, \frac{\sqrt{\delta_{12}}}{\sqrt{\delta_{23}} + \sqrt{\delta_{13}} + \sqrt{\delta_{12}}}, \qquad (4.99)$$

respectively, evaluated at the gel time.

4.3.1 Evaluating gelation onset time

The cubic nature of eqn (4.98) makes its analytic solutions highly complicated, so a numerical evaluation of the gel time may be a more convenient choice. The physical root that corresponds to the real gelation onset time is the smallest positive root,[1] and implementing a numerical solver should be simple enough. Nevertheless, before we show the numerical results, there do exist some useful analytic results on a cubic equation that we can do without too much trouble, and it turns out that the number of positive roots, and hence the correct root for gelation, has a not-too-complicated dependency on the relative strengths of the cross-species aggregation rates $\epsilon_{\alpha\beta}$s.

First, we write down the explicit form of eqn (4.98):

$$(f_1\epsilon_{23}^2 + f_2\epsilon_{13}^2 + f_3\epsilon_{12}^2 - f_1f_2f_3 - 2\epsilon_{12}\epsilon_{13}\epsilon_{23})\sigma^3 + (f_1f_2 + f_1f_3 + f_2f_3 - \epsilon_{12}^2 \\ -\epsilon_{13}^2 - \epsilon_{23}^2)\sigma^2 - (f_1 + f_2 + f_3)\sigma + 1 = 0. \quad (4.100)$$

We start by defining three relative quantities

$$\xi_1 = \frac{\epsilon_{23}^2}{f_2f_3}, \quad \xi_2 = \frac{\epsilon_{13}^2}{f_1f_3}, \quad \xi_3 = \frac{\epsilon_{12}^2}{f_1f_2}. \quad (4.101)$$

Clearly they measure the relative strengths of the three possible pairs of cross-species aggregations compared with their respective intra-species aggregations. Then, we can express the cubic and quadratic coefficients A, B in terms of f_αs and ξ_αs:

$$A = f_1f_2f_3\left(\xi_1 + \xi_2 + \xi_3 - 2\sqrt{\xi_1\xi_2\xi_3} - 1\right) \quad (4.102)$$

$$B = f_1f_2(1 - \xi_3) + f_1f_3(1 - \xi_2) + f_2f_3(1 - \xi_1). \quad (4.103)$$

With the help of the Vieta theorem, we can discuss the distribution of all three roots, and hence the number of possible roots, as the coefficients take different signs. Since the linear coefficient $-(f_1 + f_2 + f_3)$ is always negative and the constant term 1 always positive, we deduce that the number of positive roots depends solely on the signs of A and B, as shown in the following table.

A	−	+	+	−
B	+	−	+	−
Nos. of positive roots	3	2	2	1

Table 4.1 The number of positive roots dependent upon A and B.

Since by definition $f_{1,2,3} \geqslant 0$, the signs of A, B depend purely on ξ_αs. With the use of some simple inequalities we can then find some exact criteria on ξ_α values that determine the number of positive roots. A detailed discussion and quantitative results will soon follow, but first we give a qualitative summary. The general trend is that, as

---

[1] Apart, of course, from the trivial case of decoupled systems, where multiple gels appear at different onset times.

ξ_αs are turned up, i.e. as we turn up cross-species aggregation rates, the number of positive roots gradually reduces from 3 to 1. This aligns with what we expect from the 2-D counterpart and shouldn't be too hard to make sense of.

Now comes the lengthy part. The critical value for $\xi_\alpha \equiv \epsilon_{\beta\gamma}^2/(f_\beta f_\gamma)$ is 1; it decides whether inter-species aggregation takes over ($\xi_\alpha > 1$) between species β, γ. Hence, we discuss the values of A, B for different values of ξ_α relative to 1. Since ξ_1, ξ_2, ξ_3 are completely symmetric, we assume that $\xi_1 \leqslant \xi_2 \leqslant \xi_3$ *wlog* (i.e. without loss of generality).

1. When $\xi_1, \xi_2 < 1$ and $\xi_3 > 1$, the value of B looks arbitrary, as its sign depends on ξ_αs as well as f_αs. However, the value of A is constrained:

$$\begin{aligned}A &= \xi_1 + \xi_2 + \xi_3 - 2\sqrt{\xi_1\xi_2\xi_3} - 1 \\ &\geqslant 2\sqrt{\xi_1\xi_2} - 2\sqrt{\xi_1\xi_2\xi_3} + \xi_3 - 1 \\ &= 2\sqrt{\xi_1\xi_2}(1 - \sqrt{\xi_3}) + (\sqrt{\xi_3} + 1)(\sqrt{\xi_3} - 1) \\ &= (\sqrt{\xi_3} + 1 - 2\sqrt{\xi_1\xi_2})(\sqrt{\xi_3} - 1).\end{aligned} \quad (4.104)$$

Since $\xi_3 > 1$, $\xi_1, \xi_2 < 1$, we have that $\sqrt{\xi_3} > 1$, and $2\sqrt{\xi_1\xi_2} < 2$, and therefore,

$$\sqrt{\xi_3} + 1 - 2\sqrt{\xi_1\xi_2} > 0, \quad \sqrt{\xi_3} - 1 > 0. \quad (4.105)$$

Thus,

$$A \geqslant (\sqrt{\xi_3} + 1 - 2\sqrt{\xi_1\xi_2})(\sqrt{\xi_3} - 1) > 0. \quad (4.106)$$

According to Table 4.1, this corresponds exactly to 2 positive roots. Hence, when there is only one ξ_α above 1 and two below, there are 2 positive roots for σ.

2. When $\xi_1 < 1$ and $\xi_2, \xi_3 > 1$, B still takes arbitrary values. Similar to above, it can be shown that $A > 0$ also (see exercises). Again, this corresponds exactly to 2 positive roots. Hence, when there are only two ξ_αs above 1 and one below, there are also 2 positive roots for σ.

The other two scenarios are a bit more involved.

3. When $\xi_1, \xi_2, \xi_3 < 1$, $B > 0$. The value of A is, however, undetermined, so there can be 3 or 2 positive roots. But can we say a bit more about exactly when we have 3 or 2 positive roots? Let's consider the function A. Since ξ_1, ξ_2, ξ_3 are completely symmetric in A, we for now regard ξ_2, ξ_3 as parameters and ξ_1 as the sole variable, in order to determine the condition between ξ_1 and $\xi_{2,3}$ such that A becomes positive/negative. Then A can be seen as a quadratic function with respect to $y_1 \equiv \sqrt{\xi_1}$. We rewrite A as

$$A(y_1) = y_1^2 - 2\sqrt{\xi_2\xi_3}y_1 - (1 - \xi_2 - \xi_3). \quad (4.107)$$

Clearly $A(y_1)$ has its minimum at $y_1 = \sqrt{\xi_2 \xi_3} < 1$, being

$$A(\sqrt{\xi_2 \xi_3}) = \xi_2 + \xi_3 - 1 - \xi_2 \xi_3 = -(1 - \xi_2)(1 - \xi_3) < 0. \tag{4.108}$$

This implies that there must exist a range for y_1, and thus ξ_1, in which $A < 0$ and there are in turn three positive roots. On the other hand, to decide whether A can ever become positive in this scenario, we need to first check the boundary values $A(0)$ and $A(\sqrt{\xi_2})$ (since $\xi_1 \leqslant \xi_2$ by assumption). We can see that

$$A(0) = \xi_2 + \xi_3 - 1 \tag{4.109}$$

depends on the exact values of $\xi_2 + \xi_3$, whereas

$$A(\sqrt{\xi_2}) = \left(2\xi_2 - 1 - \sqrt{\xi_3}\right)\left(1 - \sqrt{\xi_3}\right) < 0, \tag{4.110}$$

since by definition $\xi_{2,3} < 1$ and by assumption $\xi_2 \leqslant \xi_3 < \sqrt{\xi_3}$. This means that for $\xi_2 + \xi_3 < 1$, $A < 0$ for certain, while for $\xi_2 + \xi_3 \geqslant 1$, $A(y_1)$ has a zero (the smaller one) within the range $0 \leqslant y_1 \leqslant \sqrt{\xi_2}$. This zero is given by

$$y_1^{(0)} = \sqrt{\xi_2 \xi_3} - \sqrt{(1 - \xi_2)(1 - \xi_3)}. \tag{4.111}$$

For $0 \leqslant y_1 \leqslant y_1^{(0)}$, $A \geqslant 0$, and as $y_1^{(0)} \leqslant y_1 \leqslant \sqrt{\xi_2}$, $A < 0$. Thus, we conclude that,
(a) For $\xi_2 + \xi_3 < 1$, there are three positive roots.
(b) For $\xi_2 + \xi_3 \geqslant 1$, there are three positive roots when $\sqrt{\xi_1} > \sqrt{\xi_2 \xi_3} - \sqrt{(1 - \xi_2)(1 - \xi_3)}$, and two when $\sqrt{\xi_1} \leqslant \sqrt{\xi_2 \xi_3} - \sqrt{(1 - \xi_2)(1 - \xi_3)}$.
4. When $\xi_1, \xi_2, \xi_3 > 1$, $B < 0$. This is the opposite scenario to the one above, and there can be 2 or 1 positive roots. We adopt a similar approach as above and are able to conclude that there are two positive roots when $1 \leqslant \sqrt{\xi_1} \leqslant \sqrt{\xi_2 \xi_3} - \sqrt{(\xi_2 - 1)(\xi_3 - 1)}$, and one only when $\sqrt{\xi_1} > \sqrt{\xi_2 \xi_3} - \sqrt{(\xi_2 - 1)(\xi_3 - 1)}$. The reader is asked to prove this in the exercises.

To summarize, we can describe a qualitative trend of the number of positive roots of the gel-time equation in terms of the relative strength $\epsilon_{\alpha\beta}$ of inter-species aggregations. In general, the number decreases as $\epsilon_{\alpha\beta}$s are tuned up against $f_{\alpha,\beta}$, from 3 roots to 1. Meanwhile, the smallest positive root shifts towards $\sigma = 0$, corresponding to a speed-up of gelation following the increase in aggregation rate, while the other roots shift towards $\sigma \to \infty$. During this shift of roots, when one of the relative interaction strength $\xi_\alpha \equiv \epsilon_{\beta\gamma}^2/(f_\beta f_\gamma)$ is much smaller or larger than the other two, this scenario **always** corresponds to two positive roots.

In short, there is a close connection between the number of positive roots and the relative inter-intra-aggregation strengths. When all ξ_αs are small, 3 positive roots;

96 *Multi-body Physics 2.0: multi-species fusion and fission*

when one or two ξ_αs are significantly larger than the rest, 2 positive roots; when all ξ_αs are large, 1 positive root.

Finally, we attach a numerical plot of gelation onset time in Fig. 4.7, i.e. the smallest positive root, as a function of two of the six parameters, ξ_1, ξ_2. As mentioned above, it does indeed decreases as $\xi_{1,2}$ increase.

4.3.2 Onset gel composition

By substituting the gel time that we previously found numerically into the gel composition expressions in eqn (4.99), we are able to obtain the onset gel compositions and study how they depend on the parameters. Generally speaking, larger f_α and $\epsilon_{\alpha\beta}$ should correspond to a larger proportion of species α in the gel. Nonetheless, there exist many nuances due to the high complexity as a result of no less than 6 independent parameters. They are captured in the plots of gel composition against part of the parameter space in Fig. 4.8.

In the discussions below we mainly look at the behaviours of species 1 *wlog* due to symmetry, and how changing parameters shift the behaviours. The following results follow:

1. In both plots of Fig. 4.8, at $\epsilon_{13} = \epsilon_{23} = 0$ species 3 is decoupled from the rest of the system, whereas species 1 and 2 dominate the gel evenly. As we turn on ϵ_{13} while keeping ϵ_{23} small, a dip in the proportion of the leading component, i.e. species 1, can be observed, even though increasing ϵ_{13} also increases the aggregation rate of species 1. In other words, turning on cross-species aggregation between two species suppresses the leading species in the gel upon gelation onset, until the interaction becomes strong enough. As ϵ_{23} increases, the dip gradually disappears and species

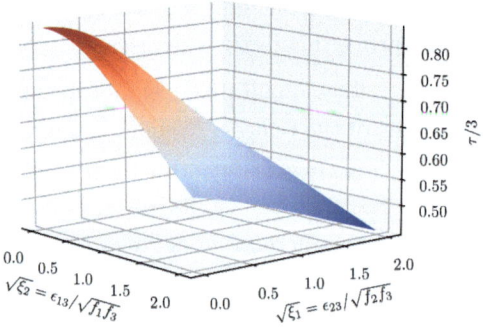

Fig. 4.7: Gelation onset time as a function of ξ_1, ξ_2. The rest of the parameters are set at $f_1 = f_2 = 1, f_3 = 0.25, \epsilon_{12} = 0.1$.

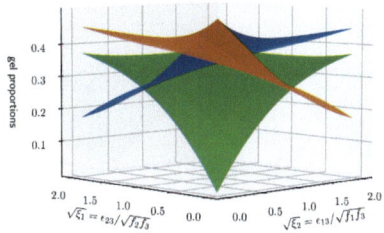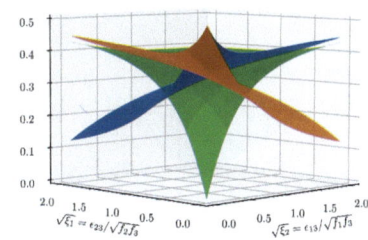

Fig. 4.8: The onset gel compositions against parameters ξ_1, ξ_2, whilst holding $f_1 = f_2 = 0.33$ and $\xi_3 = 0.33$. The left graph has $f_3 = 0.2$ and the right $f_3 = 0.33$. Here we set the aggregation parameters for both species 1 (blue) and 2 (orange) to be identical, and that the coupling between species 1 and 2 is constant, hence the symmetry. The increased proportionality of species 3 (green) on the right is the result of increased f_3 and manifests the general trend mentioned above.

1 in the gel becomes monotonically increasing with ϵ_{13}—this is shown in the 2-D cross-sectional diagrams in Fig. 4.9.

2. Larger values of f_1 and ϵ_{12} both push this dip to grow deeper and larger along ϵ_{13}. This qualitatively means that if we have a species 1 that dominates the gel composition, introducing cross-species interaction with a new species (in this case 3) weakens this domination, and the degree of suppression is positively related to the level of domination. See Figs 4.10 and 4.11.

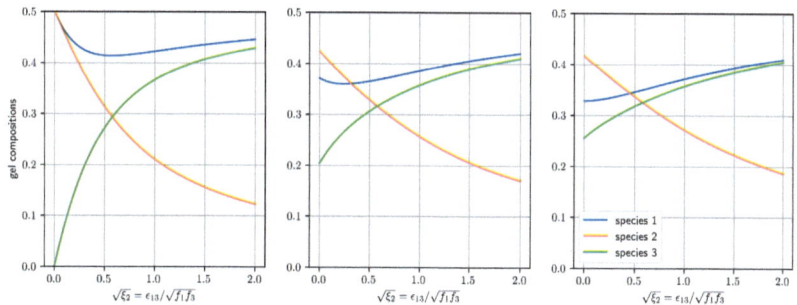

Fig. 4.9: Three 2-D cross-sections of gel compositions against $\sqrt{\xi_2} \propto \epsilon_{13}$, with $f_1 = f_2 = f_3 = 0.33$ and $\xi_3 = 0.33$. The ξ_1 values of the three plots are taken to be $0, 0.1, 0.2$ from left to right, respectively. This shows the disappearance of the dip as ϵ_{23} increases.

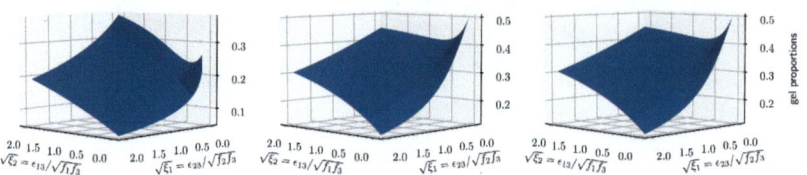

Fig. 4.10: Proportions of species 1 against $\sqrt{\xi_1}, \sqrt{\xi_2}$ with $f_{2,3} = \xi_3 = 0.33$, and 3 different f_1 values of 0.1, 0.33 and 0.75 from left to right. Note how the dip grows with larger f_1.

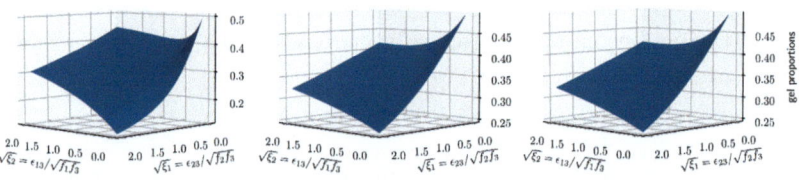

Fig. 4.11: Proportions of species 1 against $\sqrt{\xi_1}, \sqrt{\xi_2}$ with $f_{1,2,3} = 0.33$, and 3 different ξ_3 values of 0.33, 2.2 and 4.4 from left to right. Note how the dip grows with larger ξ_3, i.e. larger ϵ_{12}.

4.3.3 3-species fusion dynamics for arbitrary species sizes

For generic values of $\iota_{1,2,3} \neq 1/3$ the analyticity remains, albeit slightly more complicated. Equation (4.100) in the equal-population scenario becomes

$$1 - (f_1\iota_1 + f_2\iota_2 + f_3\iota_3)\tau + \left[(f_1f_2 - \epsilon_{12}^2)\iota_1\iota_2 + (f_1f_3 - \epsilon_{13}^2)\iota_1\iota_3 + (f_2f_3 - \epsilon_{23}^2)\iota_2\iota_3\right]\tau^2$$
$$+ \iota_1\iota_2\iota_3\left(f_1\epsilon_{23}^2 + f_2\epsilon_{13}^2 + f_3\epsilon_{12}^2 - f_1f_2f_3 - 2\epsilon_{12}\epsilon_{13}\epsilon_{23}\right)\tau^3 = 0. \qquad (4.112)$$

The solutions are given by

$$\tau^{(j)} = \frac{p_1 + a_j p_2 + a_j^* p_3}{p_0}, \quad j = 0, 1, 2 \qquad (4.113)$$

where all p_i are functions of the polynomial coefficients, which are in turn determined by the 3×3 F-matrix elements:

$$p_0 = \iota_1\iota_2\iota_3\left(f_1\epsilon_{23}^2 + f_2\epsilon_{13}^2 + f_3\epsilon_{12}^2 - f_1f_2f_3 - 2\epsilon_{12}\epsilon_{13}\epsilon_{23}\right)$$
$$p_1 = (f_1f_2 - \epsilon_{12}^2)\iota_1\iota_2 + (f_1f_3 - \epsilon_{13}^2)\iota_1\iota_3 + (f_2f_3 - \epsilon_{23}^2)\iota_2\iota_3$$
$$p_2 = \frac{p_1^2 + 3p_0p_6}{p_3}$$
$$p_3 = \left(p_4 + 3\sqrt{3p_5p_0^2}\right)^{1/3}$$
$$p_4 = 9p_0p_1p_6 + 2p_1^3 - 27p_0^2$$
$$p_5 = 27p_0^2 - 4p_1^3 - 4p_0p_6^3 - p_1^2p_6^2 - 18p_0p_1p_6$$
$$p_6 = f_1\iota_1 + f_2\iota_2 + f_3\iota_3. \tag{4.114}$$

The complex coefficients a_j point to the different roots and are given by:

$$a_0 = 1$$
$$a_1 = (-1)^{4/3}$$
$$a_2 = (-1)^{2/3}. \tag{4.115}$$

As before, the gel onset for the three-species system is the smallest positive root from the set derived in eqn (4.113).

The gel composition at the onset for $D=3$ follows the same treatment that we employed in the previous section for $D=2$. Solving for the diagonal elements of the second moment tensor, the resultant function ϕ_i that we use to calculate the composition as $\eta_\alpha = \phi_\alpha/\phi_0$ at the onset, is given by

$$\phi_\alpha = \left[1 - (f_\beta\iota_\beta + f_\gamma\iota_\gamma)\tau_g + \iota_\beta\iota_\gamma\left(f_\beta f_\gamma - \epsilon_{\beta\gamma}^2\right)\tau_g^2\right]^{1/2}, \tag{4.116}$$

where $\alpha, \beta, \gamma = 1, 2, 3$, $\alpha \neq \beta \neq \gamma$, and $\phi_0 = \phi_1 + \phi_2 + \phi_3$.

4.4 Gelation theory for general multi-species fusion systems

The multi-species theory we have developed so far haven't yet answered all the questions. Two of them can be particularly concerning: (1) for general dimensionality D, can we be more specific about gelation onset time τ_g than a determinant equation without a simple closed-form expression—or even just something computationally simpler? (2) We have yet to explain the sudden loss of physicality of all the other positive roots of τ_g than the smallest, as soon as any off-diagonal entry of F is turned on. In this section we aim to answer these concerns by developing the multi-species aggregation theory to its full potential.

4.4.1 A second encounter with the gelation onset time

To answer the first question, we begin with writing down the inverse of the second moment, m_2^{-1}, explicitly:

$$\mathsf{m}_2^{-1} = \begin{pmatrix} D - f_1\tau & -\epsilon_{12}\tau & \cdots & -\epsilon_{1D}\tau \\ -\epsilon_{12}\tau & D - f_2\tau & \cdots & -\epsilon_{2D}\tau \\ \vdots & \vdots & \ddots & \vdots \\ -\epsilon_{1D}\tau & -\epsilon_{2D}\tau & \cdots & D - f_D\tau \end{pmatrix}. \quad (4.117)$$

We extract a common factor $-\tau$ out of the matrix bulk:

$$\mathsf{m}_2^{-1} = -\tau \begin{pmatrix} f_1 - D/\tau & \epsilon_{12} & \cdots & \epsilon_{1D} \\ \epsilon_{12} & f_2 - D/\tau & \cdots & \epsilon_{1D} \\ \vdots & \vdots & \ddots & \vdots \\ \epsilon_{1D} & \epsilon_{2D} & \cdots & f_D - D/\tau \end{pmatrix} \quad (4.118)$$

and define the new matrix as

$$\omega = \begin{pmatrix} f_1 - D/\tau & \epsilon_{12} & \cdots & \epsilon_{1D} \\ \epsilon_{12} & f_2 - D/\tau & \cdots & \epsilon_{2D} \\ \vdots & \vdots & \ddots & \vdots \\ \epsilon_{1D} & \epsilon_{2D} & \cdots & f_D - D/\tau \end{pmatrix} = -\frac{1}{\tau}\mathsf{m}_2^{-1}. \quad (4.119)$$

Note that ω can be written as

$$\omega(\tau_g) = \mathsf{F} - \frac{D}{\tau_g}\mathbb{I}, \quad (4.120)$$

where F is exactly the rate matrix for multi-species aggregation—this is arguably the most important observation of this chapter. Since we know that at gelation onset, $\det(\mathsf{m}_2^{-1}(\tau_g)) = 0$, we can deduce that

$$\det(\omega(\tau_g)) = -\frac{1}{\tau_g^D}\det(\mathsf{m}_2^{-1}(\tau_g)) = 0. \quad (4.121)$$

By combining eqns (4.120, 4.121), we can see that eqn (4.121) is in fact the *characteristic equation* of the symmetric matrix F, written out explicitly as

$$\left|\mathsf{F} - \frac{D}{\tau_g}\mathbb{I}\right| = 0, \quad (4.122)$$

and $\lambda_\mu \equiv D/(\tau_g)_\mu$ are its eigenvalues. In other words, solutions for gelation onset time are the inverse of the eigenvalues of F, but note that this does not mean that all solutions are guaranteed to be physical.

This is a significant result. For one-species systems, we know that the gelation onset time is the inverse of scalar F. Here lies its natural extension. For 2- and 3-species systems, we already know that the physical gelation occurs at the earliest of the roots, so it is given by the largest eigenvalue $\lambda_{\max}$.

4.4.2 Diagonalization of F matrix and the eigendirections for gelation

The above section means that the multi-species gelation problem is essentially a diagonalization problem of the F matrix. Therefore, we define the diagonalization of matrix F as

$$\Lambda = \operatorname{diag}(\lambda_1, \ldots, \lambda_D), \tag{4.123}$$

and denote the corresponding eigenvector of λ_μ as $\boldsymbol{x}^{(\mu)}$. They satisfy

$$(\mathsf{F} - \lambda_\mu \mathbb{I})\boldsymbol{x}^{(\mu)} = \mathbf{0}. \tag{4.124}$$

Because F is symmetric, the eigenvectors are mutually orthogonal, i.e. $\boldsymbol{x}^{(\mu)} \cdot \boldsymbol{x}^{(\nu)} = \delta_{\mu\nu}$. Consequently, $\{\boldsymbol{x}^{(\mu)}\}$ form an orthonormal basis that spans the D-dimensional abstract space of $\boldsymbol{x}$. Moreover, the basis vectors of its dual space of size profile $\{\boldsymbol{k}^{(\mu)}\}$ must satisfy, by definition,

$$\boldsymbol{k}^{(\mu)} \cdot \boldsymbol{x}^{(\nu)} = \delta_{\mu\nu}. \tag{4.125}$$

These $\{\boldsymbol{k}^{(\mu)}\}$ basis vectors are physically significant—their components represent the relative **compositions** of each species at gelation onset, and thus we call them the **eigendirections for gelation**. For convenience we sometimes refer to $\boldsymbol{x}^{(\mu)}$s as eigendirections also, since they are related to $\boldsymbol{k}^{(\mu)}$ simply by an inverse. In order to show why they represent gelation composition, we return to eqn (4.27). Obviously the generative equation is frame-independent since all the terms in eqn (4.27) are scalars. Hence, we are free to transform it into the eigenbasis of F. In the rotated frame, eqn (4.28), i.e. the generalized Burgers' equation, now reads

$$\frac{\partial \varepsilon'_\mu}{\partial \tau} + \sum_{\nu=1}^{D} \lambda_\nu (\varepsilon'_\nu - \iota'_\nu) \partial_\nu \varepsilon'_\mu = 0, \tag{4.126}$$

where in this shorthand notation, $\partial_\mu = \partial/\partial x'_\mu$, $\boldsymbol{\iota} = \sum_{\mu=1}^{D} \iota'_\mu \boldsymbol{k}^{(\mu)}$, and $\boldsymbol{\varepsilon} = \sum_{\mu=1}^{D} \varepsilon'_\mu \boldsymbol{k}^{(\mu)}$. Now that the diagonal matrix Λ is the heterogeneity rate matrix under this frame. It means that these D linear combinations of species, i.e. the 'bundled' species $\boldsymbol{k}^{(\mu)}$s, are completely decoupled. This in turn means that as long as the initial condition contains only these 'bundled-up' clusters and their aggregates, i.e. $\propto \boldsymbol{k}^{(\mu)}$ (equivalent to 'pure' clusters under this rotated frame), then we always have $\partial_\beta \varepsilon'_i = 0$ if $i \neq j$, and hence,

$$\frac{\partial \varepsilon'_i}{\partial \tau} + \lambda_\mu (\varepsilon'_i - \iota'_i) \partial_\alpha \varepsilon'_i = 0. \tag{4.127}$$

Physically this means that gelation must retain these species bundles, i.e. occurring along these **eigendirections**. Therefore, the (inverses of) eigenvectors of F do indeed yield the gel composition at onset.

However, there remains the problem of physicality—all of $\{k^{(\mu)}\}$ are not necessarily physical eigendirections, since in principle there can be negative components in $k^{(\mu)}$, while we obviously do not allow for negative number of particles in a physical cluster. Fortunately, this is resolved by the *Perron–Frobenius theorem*, which states that for an irreducible non-negative matrix, its largest eigenvalue corresponds to an eigenvector whose components are all positive. By assumption, all entries of F represent aggregation rates ($F_{\alpha\beta} \geq 0$), so it must be non-negative. In order to apply the theorem to F we need to first investigate the conditions for its irreducibility.

To begin with, we recall the definition of matrix irreducibility. A matrix M is *reducible* if there exists a *permutation matrix* P such that matrix PMP^{-1} is block upper triangular, i.e. it can be written into the form

$$\mathsf{PMP}^{-1} = \begin{pmatrix} A & B \\ 0 & C \end{pmatrix} \tag{4.128}$$

for some A, B, C. However, a more straightforward definition to apply to our system is the following graphical one. Consider the matrix M as a graph representation. If the graph is connected, M is irreducible. For an asymmetric M, it represents a directed graph, and the connection requirement is such that starting from every vertex on the graph, it is possible to reach all the others. In other words, there cannot be sinks or sources. For symmetric matrices like F, on the contrary, the condition is more straightforward: since it represents an undirected graph, simple connection of all vertices suffices. Therefore, we are able to draw the two following conclusions from this graphical perspective:

1. There must be at least $2(D-1)$ non-zero $F_{\alpha\beta}$ values for the $D \times D$ matrix F to be irreducible.

2. F is reducible iff it is block diagonal (or can be permuted into a block diagonal form).

The second point above is very important—it follows that, after relabelling the species appropriately, we must be able to write any reducible F into the following form:

$$\mathsf{F} = \begin{pmatrix} F_1 & 0 & & & 0 \\ 0 & F_2 & & & \\ & & \ddots & & \\ & & & F_{d-1} & 0 \\ 0 & & & 0 & F_d \end{pmatrix}, \tag{4.129}$$

for some $2 \leq d \leq D$, where all $\mathsf{F}_s, \forall s = 1, 2, \ldots, d$ are irreducible. By doing this we have split the system into d decoupled subsystems, each of which irreducible, i.e. fully coupled. We can further extend the range of d to include the trivial case $d = 1$; it simply represents a fully coupled system in its own right.

Now we are ready to apply the Perron–Frobenius theorem to all F_ss. Each of these matrices must have a maximum eigenvalue, which we can freely relabel as $\lambda_1^{(s)}$, and its corresponding eigenvector $x_1^{(s)}$ must have all its components positive. Consequently, its dual vector $k_1^{(s)}$ is a physical eigendirection. Since $x_1^{(s)}$ is orthogonal to all the other eigenvectors due to symmetric matrix F_s, we can further deduce that all the other eigenvectors must contain at least one negative component (they cannot all be negative of course), and thus unphysical for gelation. We can conclude that for each irreducible submatrix F_s, there exists and only exists **1 physical eigendirection** for gelation, $k_1^{(s)}$. Its corresponding physical gelation onset time $\tau_g^{(s)}$, on the other hand, is given by the **largest eigenvalue** of F_s, as

$$\tau_g^{(s)} = \frac{D}{\lambda_1^{(s)}}. \tag{4.130}$$

For the whole D-species system, there are then d separate and independent gelation onsets, and their respective gelation times are $D/\lambda_1^{(s)}$. This exactly agrees with our lower-dimensional results, where decoupled systems aggregate separately. Moreover, this successfully explains why turning on inter-species aggregation instantaneously makes larger roots of τ_g unphysical—it is really that their corresponding eigendirections for gelation rotate into unphysical (negative) realms. Correspondingly, as inter-species couplings increase, the growth curve of the cohesive unit morphs from having $(D-1)$ kinks when inter-species couplings are faint, to gradually consolidating into one rapid cohesive growth, as shown in Fig. 4.12.

4.4.3 Effective gelation theory for multi-species fusion

We have established above that for a D-species aggregation system, physical gelation onsets occur at $\tau_g^{(\mu)} = D/\lambda_\mu$ if the corresponding eigenvector $x^{(\mu)}$ has no negative components. Recalling the scalar heterogeneous aggregation parameter F for 1-D systems, here we can define similar effective parameters

$$F_{\text{eff}}^{(\mu)} = \frac{\lambda_\mu}{D}. \tag{4.131}$$

These effective parameters may seem a bit redundant, but they become useful when dimensionality D is very high and aggregation parameters $\epsilon_{\alpha\beta}$s and f_αs are each of very similar values. For such a system there exists only one physical F_{eff}, i.e. there is only one physical gelation onset at $1/F_{\text{eff}}$. A good approximation of the gel curve can then be written as

$$\varepsilon = e^{-F_{\text{eff}}(1-\varepsilon)\tau}, \tag{4.132}$$

cf. eqn (4.50). To interpret this, we can take a coarse-grained perspective and regard it as an effective 1-D aggregation—remembering that the 1-D parameter F is in its

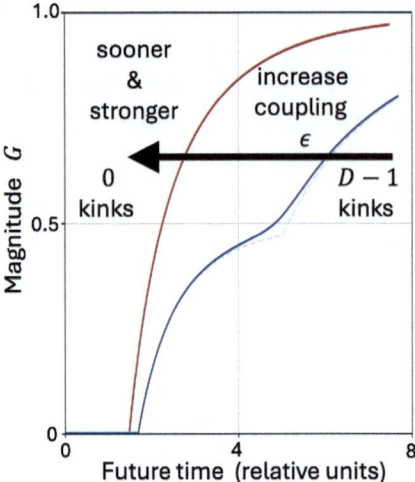

Fig. 4.12: Our theory's predicted cohesive unit magnitude G versus (scaled) time for $D = 2$ species system ($G = (1/2) \sum_{\alpha=1}^{2} G_\alpha(\tau)$). Cross-species aggregation terms (i.e. off-diagonal terms in 2×2 matrix **F**) all set to ϵ. As ϵ increases from 0 (grey dashed line) the physically observed onset moves to earlier times, and the number of kinks reduces successively from $D - 1 = 1$ to 0. This means the rise in G—and hence a good or bad 'surprise'—will occur sooner and more strongly. This explains why the real-world growth curves in Fig. 2.1 appear the way that they do.

nature a mean-field heterogeneity parameter itself. This construction shows that when D is very large, a coarse-grained model suffices and F_{eff} averages over all platforms (species) involved. This is shown in Figure 4.13, where the coarse-grained model agrees well with the exact gel curve.

In order to quantitatively explore the effective gelation theory, we work under the assumption $\epsilon_{\alpha\beta} = \epsilon$ and $f_\alpha = f$ for all α, βs. We then have

$$\omega \approx \begin{pmatrix} f & \epsilon & \cdots & \epsilon \\ \epsilon & f & \cdots & \epsilon \\ \vdots & \vdots & \ddots & \vdots \\ \epsilon & \epsilon & \cdots & f \end{pmatrix}. \tag{4.133}$$

By inspection, we can easily find that the normalized vector

$$v = \frac{1}{\sqrt{D}} \begin{pmatrix} 1 \\ 1 \\ \vdots \\ 1 \end{pmatrix} \tag{4.134}$$

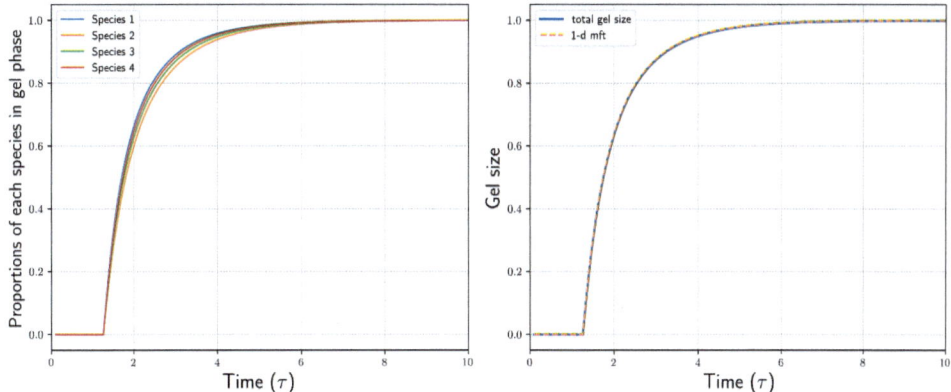

Fig. 4.13: Comparison of a 4-species aggregation with similar f_α and $\epsilon_{\alpha\beta}$ values with a 1-dimensional effective theory. The left panel shows proportions of each of the 4 species in gel, while the right panel compares directly the real gel curve with the mean-field approximation.

is an eigenvector of w with its corresponding eigenvalue being

$$\lambda = f + (D-1)\epsilon. \tag{4.135}$$

Perron–Frobenius theorem ensures that v is the only physical eigendirection for gelation, and thus there is only one gelation onset, at time $\tau_g = D/[f + (D-1)\epsilon]$, and correspondingly,

$$F_{\text{eff}} = \frac{f + (D-1)\epsilon}{D}. \tag{4.136}$$

When $D \to \infty$, $F_{\text{eff}} \approx \epsilon$, we can see that multi-species aggregation recovers to a cross-platform interaction process. Moreover, from eqn (4.48) we can further deduce that in this scenario the gelation curves for all D species satisfy the following equation:

$$\varepsilon_\alpha = e^{-[f(1/D-\varepsilon_\alpha) + \epsilon \sum_{\beta \neq \alpha}(1/D-\varepsilon_\beta)]\tau}, \quad \alpha, \beta = 1, \ldots, D. \tag{4.137}$$

These D equations are identical, so all of the D species must gel simultaneously (as we concluded earlier) and their gel curves must also be identical. Hence, the combined gel curves for the whole system must retain the simple shape of a 1-dimensional one.

To briefly summarize, this is the most important section of this chapter. We have now fully established a robust and powerful D-species aggregation theory, and all the key quantities of this multi-dimensional gelation process can be exactly calculated from the defining parameters of the system, and even better, they all come with clear physical interpretations. This is a proper multi-dimensional extension of the classic single-species model. Not only are the resulting growth curves able to explain the multi-kink empirical ones in Chap. 2, their derivative will yield multi-peak burst profiles akin to many real-world systems.

4.4.4 Species as abstract notion for coarse-graining heterogeneity

The 'species' in our model does not have to be physical—in other words, there do not have to be obvious microscopic distinctions between two species on top of the particle-level heterogeneity (i.e. different character vectors). Instead it is a statement about how different particles contribute to the collective system. In fact, since the notion of species represents a different (higher) level of heterogeneity, we can regard it as a way of coarse-graining, or renormalizing, our system. By labelling particles into different species, what we really achieve is grouping up similar particles as measured across the spectrum of heterogeneity, such that our mean-field treatment becomes less crude, averaging over a fraction of the population instead of the whole. This action of 'grouping-up' particles into D species then constructs a new scale of heterogeneity, and we consider them to encapsulate a higher level of heterogeneous characteristics. While the above mathematical framework holds for all $D \in \mathbb{Z}^+$, we have seen that for small $D = 2, 3$ it yields some explicit and analytic results, but for larger D analytics becomes complicated. When D becomes very large, close to the scale of N, the total particle number, we essentially retain most of the system's inherent heterogeneity without much renormalization, and we find that there is not much else we can do besides some descriptive formulae. It then becomes necessary to adopt the effective treatment described in Section 4.4.3, and the result for large D indeed recovers the mean-field theory from Chap. 3. The upshot is that this multi-species, or multi-level, aggregation theory really constructs a systematic treatment of aggregating particles from being identical ($D = 1$) to completely heterogeneous ($D = N$). In the next chapter, we will develop this point further into a highly general and very powerful theory of heterogeneity.

4.4.5 A different treatment to unequal species sizes

The abstract notion of species provides us with a neat trick to treat unequal species sizes. Since it is no longer required that each species has to be physically different, we can break up a physical species into several identical portions, i.e. the effective (sub-)species. As long as it retains the overall aggregation rates, this new classification of effective species is mathematically equivalent to the original mathematical representation. For example, let's assume a 2-species system with $\iota_1 = N_1/N = 1/3$ and $\iota_2 = N_2/N = 2/3$. This generates the same mathematics to an equally partitioned 3-species system (i.e. $\iota_{1',2',3'} = 1/3$), where, say, species $1'$ is identical to the former species 1, while species

$2', 3'$ **combined** are equivalent to the earlier species 2. We can write down the F matrices of the two equivalent systems for reference. For the 2-species system it is

$$F_{2d} = \begin{pmatrix} f_1 & \epsilon \\ \epsilon & f_2 \end{pmatrix}, \tag{4.138}$$

and for the effective 3-species one it is straightforward:

$$F_{3d} = \begin{pmatrix} f_1 & \epsilon & \epsilon \\ \epsilon & f_2 & f_2 \\ \epsilon & f_2 & f_2 \end{pmatrix}, \tag{4.139}$$

since species $2'$ and $3'$ should in reality aggregate with identical rate both within themselves and between each other.

Generally, to split a species with rate f into d identical effective species with rates f' and ϵ', we recall eqn (4.136) which provides a relation between them:

$$f = \frac{f' + (d-1)\epsilon'}{d}.$$

But we also require $f' = \epsilon'$ since in reality they are both 'intra-species' aggregation. Hence, we find that

$$f' = \epsilon' = f \tag{4.140}$$

is true universally. This can simplify the system considerably in both ways: we can either split some species to obtain identical populations, or merge species with approximately identical aggregation rates to lower the dimensions of the problem.

4.5 Introducing fission again: 2 species and beyond

Back in Section 3.7.2, we have seen an important model, where total fragmentation kicks in to counter the effect of aggregation and, furthermore, gelation. Here in the context of multiple species, can we once more extend our model to incorporate similar fragmentation phenomena? In this section, we try to start from 2 species and extend beyond. However, before we even start, we can already identify the following complexities compared to the vanilla EZ model:

1. For D species, an appropriate fragmentation model is not as straightforward as coalescence. For the latter, we normally just pick a binary model and that is representative enough, while there are numerous interesting fragmentation models by comparison. Even when we fix the fragmentation scheme to the 'complete fission' limit, the possibilities are still endless. There can be fission events that break up clusters across communities, or those that only target a particular species, or anything in between. It may be necessary to specify the particular models that we look into.

2. Moreover, in a D-species system there is mathematical difference between pure, mixed, and partially mixed clusters. The contribution of a certain species can go to 0, and there exist D different types of monomers, one for every species. These lead to complexities in the calculations.

Among the numerous possible scenarios, two exhibit nice analyticity as well as physical interpretations. Here we describe these two scenarios under 2 species, before generalizing the theory into arbitrary D dimensions.

4.5.1 Complete fragmentation: 'global censorship'

We start with a simple fragmentation model, where a cluster fragments completely and all constituent particles are disconnected into monomers. In order to formulate the equations, we assume that just like the F matrix for aggregation, fragmentation rates are also taken to be species-dependent (φ_1, φ_2): in reality, this could mean that a fragmentation starts from or centres around one specific particle, whose species determines its rate. We can thus write down the following Smoluchowski-like master equations for this model:

$$\frac{d\nu_s}{d\tau} = \frac{F_{\alpha\beta}}{2} \sum_{\substack{u_1+v_1=s_1 \\ u_2+v_2=s_2}} \iota_\alpha \nu_\beta \nu_u \nu_v - F_{\alpha\beta}\nu_s s_\alpha \iota_\beta - \frac{\varphi_\alpha}{2}\nu_s s_\alpha, \tag{4.141}$$

where $\alpha, \beta = 1, 2$ and we follow the summation convention. The vectorial size appears in the suffix $s = (s_1, s_2)$, and the fragmentation rate in this model φ_α is also vectorial. As before, ι_α is the fraction of species-α particles within the whole population. We can then read off the fragmentation kernel for this model

$$\kappa_f(s) = \varphi_\alpha s_\alpha. \tag{4.142}$$

Note that eqn (4.141) is only valid for non-monomers ($s_1 + s_2 \geqslant 2$), while the monomers follow the following equation:

$$\frac{d\nu_\alpha^{(1)}}{d\tau} = -F_{\alpha\beta}\nu_\alpha^{(1)}\iota_\beta + \frac{\varphi_\beta}{2}\sum_{\substack{s_1=0,s_2=0 \\ s_1+s_2\geqslant 2}}^{\infty} s_\beta s_\alpha \nu_s, \tag{4.143}$$

where we note that the summation convention does *not* sum over $\alpha = 1, 2$, which represent the species of the monomers in question, but only sums over $\beta = 1, 2$. $\nu_\alpha^{(1)}$ denotes the number of species-α monomers. Then, the steady-state equations can be derived:

$$F_{\alpha\beta}\sum_{\substack{u_1+v_1=s_1 \\ u_2+v_2=s_2}} \iota_\alpha \nu_\beta \nu_u \nu_v = (2F_{\alpha\beta}\iota_\beta + \varphi_\alpha)\nu_s s_\alpha, \tag{4.144}$$

$$\varphi_\beta \sum_{\substack{s_1=0, s_2=0 \\ s_1+s_2 \geqslant 2}}^{\infty} s_\beta s_\alpha \nu_s = 2F_{\alpha\beta} l_\beta \nu_\alpha^{(1)} \text{ (no sum in } \alpha\text{).} \tag{4.145}$$

For eqn (4.144), we sum up on both sides all s except the two types of monomers:

$$F_{\alpha\beta} \sum_{u_1=0, u_2=0}^{\infty} u_\alpha \nu_u \sum_{v_1=0, v_2=0}^{\infty} v_\beta \nu_v = (2F_{\alpha\beta} l_\beta + \varphi_\alpha)\left(\iota_\alpha - \nu_\alpha^{(1)}\right)$$

$$\Rightarrow F_{\alpha\beta} \iota_\alpha l_\beta \equiv \bar{F} = (2F_{\alpha\beta} l_\beta + \varphi_\alpha)\left(\iota_\alpha - \nu_\alpha^{(1)}\right)$$

$$= (2\bar{F} + \bar{\varphi}) - (2F_{\alpha\beta} l_\beta + \varphi_\alpha)\nu_\alpha^{(1)}$$

$$\Rightarrow \bar{F} + \bar{\varphi} = (2F_{\alpha\beta} l_\beta + \varphi_\alpha)\nu_\alpha^{(1)}, \tag{4.146}$$

where we define the *average rates* $\bar{F}, \bar{\varphi}$ as

$$\bar{F} = F_{\alpha\beta} \iota_\alpha l_\beta \tag{4.147}$$

and

$$\bar{\varphi} = \varphi_\alpha \iota_\alpha, \tag{4.148}$$

respectively. For eqn (4.145), using the definition of the second moment tensor $m_{\alpha\beta}^{(2)}$, we have

$$\varphi_\beta m_{\alpha\beta}^{(2)} - \varphi_\alpha \nu_\alpha^{(1)} = 2F_{\alpha\beta} l_\beta \nu_\alpha^{(1)} \text{ (no sum in } \alpha\text{)}$$

$$\Rightarrow \varphi_\beta m_{\alpha\beta}^{(2)} = (2F_{\alpha\beta} l_\beta + \varphi_\alpha)\nu_\alpha^{(1)} \text{ (no sum in } \alpha\text{)}$$

Now we do sum up both $\alpha = 1, 2$ and obtain

$$\sum_{\alpha=1}^{2} \varphi_\beta m_{\alpha\beta}^{(2)} = (2F_{\alpha\beta} l_\beta + \varphi_\alpha)\nu_\alpha^{(1)}. \tag{4.149}$$

Comparing eqns (4.146) and (4.149), we have

$$(\bar{F} + \bar{\varphi}) = \sum_{\alpha=1}^{2} \varphi_\beta m_{\alpha\beta}^{(2)}. \tag{4.150}$$

Recall that in Chap. 3 we defined a mean-field connection probability in the 1-species EZ model. Similarly, for this 2-species model, the mean-field connection probability P can be defined as

$$P(\tau) = \frac{\sum_{s_1=0, s_2=0}^{\infty} n_s \left[\binom{s_1}{2} + s_1 s_2 + \binom{s_2}{2}\right]}{\binom{N}{2}} = \frac{\sum_{\alpha=1}^{2}\sum_{\beta=1}^{2} M_{\alpha\beta}^{(2)} - N}{N(N-1)} \approx \frac{M_2 - N}{N^2}, \tag{4.151}$$

where we define
$$M_2 = \sum_{\alpha=1}^{2}\sum_{\beta=1}^{2} M_{\alpha\beta}^{(2)} = N \sum_{\alpha=1}^{2}\sum_{\beta=1}^{2} m_{\alpha\beta}^{(2)}. \tag{4.152}$$

We can see that this definition counts the general connection regardless of species. Together with eqn (4.150), we can see that the 1-D result of steady state

$$P = \frac{\bar{F}}{N\bar{\varphi}} \tag{4.153}$$

is perfectly recovered iff $\varphi_1 = \varphi_2 = \bar{\varphi}$, in which case $\sum_{\alpha=1}^{2}\varphi_\beta m_{\alpha\beta}^{(2)} = \bar{\varphi}M^{(2)}/N$. There does otherwise exist discrepancy, unfortunately. The reason lies in the fact that there are 3 independent entries in $m_{\alpha\beta}^{(2)}$ but only 2 equations. Though limited, it does prove that the 1-species theory can be strictly extended to 2 species in the constraint case of $\varphi_1 = \varphi_2$, and furthermore, if φ_1 and φ_2 are not too different, work as a good approximation.

Under this further assumption of identical components of fragmentation kernel $\varphi_1 = \varphi_2 = \varphi$, the kernel of this 'complete fragmentation' reduces to

$$\kappa_s = \varphi(s_1 + s_2) \tag{4.154}$$

which is dependent on the total cluster size only. Contextually, this implies that there is a sort of cross-platform grand 'coalescence police' that monitors solely the total cluster size and cannot distinguish, say, a $(6,1)$ cluster and a $(5,2)$ cluster, and break clusters based on this size: so it is a 'global censorship' kind of fragmentation model. The upshot is therefore that, for a very simple fragmentation model where particles on all platforms get disconnected once a cluster fragments with the same rate regardless of types, the mean-field connectivity probability can be generalized from the 1-species result.

4.5.2 Partial fragmentation: 'censorship by platform'

We now have a generalized EZ model for $D = 2$ that gives a nice and simple result. However, the above model could be too crude—or too powerful—in reality. For example, social media platforms may have their own different community-wide censorship standards, and this means we may need species-specific fragmentation rates that only impact that exact species, while a cross-platform 'super censorship' is difficult to achieve. In this section, we devise a more sophisticated model for such behaviours.

The model. In a system comprised of two types (1 and 2) of particles, they coalesce following the same binary rule as always, and fragment according to the following expressions:

$$(x,y) \xrightarrow{\varphi_1 x} x(1,0) + (0,y), \quad x \geq 1, x+y \geq 2$$

or
$$\xrightarrow{\varphi_2 y} (x,0) + y(0,1), \quad y \geqslant 1, x+y \geqslant 2. \tag{4.155}$$

To compare, for 'complete fragmentation' these two possible paths happen altogether (i.e. 'and'), with $\varphi_1 = \varphi_2 = \bar{\varphi}$.

A key difference in this model is that we need to take care of a whole new type of cluster in addition to monomers—the *pure* clusters $(x,0)$ and $(0,y)$ for $x,y \geqslant 2$, since now they can be created from fragmentations. This results in five different Smoluchowski-like ODEs for these different types, and we write them down below.

- Mixed clusters s, $s_\alpha \geqslant 1$, $\alpha = 1,2$.

$$\frac{d\nu_s}{d\tau} = \frac{F_{\alpha\beta}}{2} \sum_{\substack{u_1+v_1=s_1 \\ u_2+v_2=s_2}} u_\alpha v_\beta \nu_u \nu_v - \frac{2F_{\alpha\beta}\iota_\beta + \varphi_\alpha}{2} \nu_s s_\alpha. \tag{4.156}$$

Unsurprisingly, this looks identical to eqn (4.141) with only the applicable s being different.

- Pure clusters $(s,0)$, $s \geqslant 2$.

$$\frac{d\nu_{(s,0)}}{d\tau} = \frac{F_{11}}{2} \sum_{u+v=s} uv\nu_{(u,0)}\nu_{(v,0)} - \frac{2F_{1\beta}\iota_\beta + \varphi_1}{2}\nu_{(s,0)}s + \frac{\varphi_2}{2}\sum_{\beta=1}^{\infty} \beta\nu_{(s,\beta)}, \tag{4.157}$$

where the last term is the gain of the cluster from the mechanism in eqn (4.156).

- Pure clusters $(0,s)$, $s \geqslant 2$.

$$\frac{d\nu_{(0,s)}}{d\tau} = \frac{F_{22}}{2} \sum_{u+v=s} uv\nu_{(0,u)}\nu_{(0,v)} - \frac{2F_{2\beta}\iota_\beta + \varphi_2}{2}\nu_{(0,s)}s + \frac{\varphi_1}{2}\sum_{\beta=1}^{\infty} \beta\nu_{(\beta,s)}, \tag{4.158}$$

similarly as above.

- Monomers $(1,0)$.

$$\frac{d\nu_{(1,0)}}{d\tau} = -F_{1\beta}\iota_\beta \nu_{(1,0)} + \frac{\varphi_2}{2}\sum_{\beta=1}^{\infty}\beta\nu_{(1,\beta)} + \frac{\varphi_1}{2}\sum_{\substack{\beta_1=0,\beta_2=0 \\ \beta_1+\beta_2\geqslant 2}}\beta_1^2\nu_{(\beta_1,\beta_2)}, \tag{4.159}$$

where the last term is the gain of the cluster from the mechanism in eqn (4.156).

- And finally, monomers $(0,1)$.

$$\frac{d\nu_{(0,1)}}{d\tau} = -F_{2\beta}\iota_\beta \nu_{(0,1)} + \frac{\varphi_1}{2}\sum_{\beta=1}^{\infty}\beta\nu_{(\beta,1)} + \frac{\varphi_2}{2}\sum_{\substack{\beta_1=0,\beta_2=0 \\ \beta_1+\beta_2\geqslant 2}}\beta_2^2\nu_{(\beta_1,\beta_2)}. \tag{4.160}$$

Since here we have 5 equations, and so calculating their steady states respectively likely results in chaos, we utilize the Kronecker delta to combine them into one:

$$\frac{d\nu_s}{d\tau} = \frac{F_{\alpha\beta}}{2} \sum_{\substack{u_1+v_1=s_1 \\ u_2+v_2=s_2}} u_\alpha v_\beta \nu_u \nu_v - \frac{2F_{\alpha\beta}\iota_\beta + \varphi_\alpha}{2} \nu_s s_\alpha + \delta_{s_1,0}\frac{\varphi_1}{2} \sum_{\beta=1}^{\infty} \beta \nu_{(\beta,s_2)}$$

$$+ \delta_{0,s_2}\frac{\varphi_2}{2} \sum_{\beta=1}^{\infty} \beta \nu_{(s_1,\beta)} + \delta_{s_1,1}\delta_{s_2,0}\frac{\varphi_1}{2} \sum_{\substack{\beta_1=0,\beta_2=0 \\ \beta_1+\beta_2\geqslant 2}}^{\infty} \beta_1^2 \nu_{(\beta_1,\beta_2)}$$

$$+ \delta_{s_1,0}\delta_{s_2,1}\frac{\varphi_2}{2} \sum_{\substack{\beta_1=0,\beta_2=0 \\ \beta_1+\beta_2\geqslant 2}}^{\infty} \beta_2^2 \nu_{(\beta_1,\beta_2)} + \delta_{s_1,1}\delta_{s_2,0}\frac{\varphi_1}{2} s_1 \nu_s + \delta_{s_1,0}\delta_{s_2,1}\frac{\varphi_2}{2} s_2 \nu_s.$$

(4.161)

Note that the final two terms are added since monomers cannot fragment, but their 'fragmentations' have been counted in the second term under the summation convention. These two terms compensate for the excessive subtraction. It is exactly these two terms that make the calculation interesting. We only look at the first one, $\delta_{s_1,1}\delta_{s_2,0}(\varphi_1/2)s_1\nu_s$, since the other follows by symmetry. Here $s_1 = 1$, so it follows that $s_1^2 = s_1 = 1$. Thus, it can be included in the fifth term, i.e. the summation term for monomer gain due to fragmentation:

$$\delta_{s_1,1}\delta_{s_2,0}\frac{\varphi_1}{2} \sum_{\substack{\beta_1=0,\beta_2=0 \\ \beta_1+\beta_2\geqslant 2}}^{\infty} \beta_1^2 \nu_{(\beta_1,\beta_2)} + \delta_{s_1,1}\delta_{s_2,0}\frac{\varphi_1}{2} s_1 \nu_s = \delta_{s_1,1}\delta_{s_2,0}\frac{\varphi_1}{2} \sum_{\beta_1=0,\beta_2=0}^{\infty} \beta_1^2 \nu_{(\beta_1,\beta_2)}$$

$$= \delta_{s_1,1}\delta_{s_2,0}\frac{\varphi_1}{2} m_{11}^{(2)}.$$

(4.162)

It follows that we can simplify the equation into

$$\frac{d\nu_s}{dt} = \frac{F_{\alpha\beta}}{2} \sum_{\substack{u_1+v_1=s_1 \\ u_2+v_2=s_2}} u_\alpha v_\beta \nu_u \nu_v - \frac{2F_{\alpha\beta}\iota_\beta + \varphi_\alpha}{2} \nu_s s_\alpha + \delta_{s_1,0}\frac{\varphi_1}{2} \sum_{\beta=1}^{\infty} \beta \nu_{(\beta,s_2)}$$

$$+ \delta_{0,s_2}\frac{\varphi_2}{2} \sum_{\beta=1}^{\infty} \beta \nu_{(s_1,\beta)} + \delta_{s_1,1}\delta_{s_2,0}\frac{\varphi_1}{2} m_{11}^{(2)} + \delta_{s_1,0}\delta_{s_2,1}\frac{\varphi_2}{2} m_{22}^{(2)}.$$

(4.163)

Now after tidying up the equation, we as usual take the steady-state condition $\dot{\nu}_s = 0$ and sum it up directly for all s:

$$\frac{F_{\alpha\beta}}{2}\iota_i\iota_j - \frac{2F_{\alpha\beta}\iota_\beta + \varphi_\alpha}{2}\iota_\alpha + \frac{\varphi_1}{2} \sum_{\beta=1}^{\infty}\sum_{s_2=1}^{\infty} \beta\nu_{(\beta,s_2)} + \frac{\varphi_2}{2} \sum_{s_1=1}^{\infty}\sum_{\beta=1}^{\infty} \beta\nu_{(s_1,\beta)}$$

$$+ \frac{\varphi_1}{2} m_{11}^{(2)} + \frac{\varphi_2}{2} m_{22}^{(2)} = 0$$

$$\Rightarrow \bar{F} - (2\bar{F} + \bar{\varphi}) + \varphi_1 \frac{N_1^{\text{mixed}}}{N} + \varphi_2 \frac{N_2^{\text{mixed}}}{N} + \varphi_1 m_{11}^{(2)} + \varphi_2 m_{22}^{(2)} = 0$$

$$\Rightarrow \bar{F} + \bar{\varphi} = \frac{1}{N}\left(\varphi_1 M_{11}^{(2)} + \varphi_1 N_1^{\text{mixed}} + \varphi_2 N_2^{\text{mixed}} + \varphi_2 M_{22}^{(2)}\right), \quad (4.164)$$

where

$$N_\alpha^{\text{mixed}} = \sum_{s_1=1}^{\infty} \sum_{s_2=1}^{\infty} s_\alpha n_{(s_1,s_2)} \quad (4.165)$$

is defined to be the number of all type-α particles in *mixed* clusters.

Equation (4.164) seemingly does not provide us with any more information, but there is more to it. It actually looks very similar to eqn (4.150) above, only with N_α^{mixed} replacing $M_{12}^{(2)}$. However, by definition,

$$N_\alpha^{\text{mixed}} = \sum_{s_1=1}^{\infty} \sum_{s_2=1}^{\infty} s_\alpha n_{(s_1,s_2)} < \sum_{s_1=1}^{\infty} \sum_{s_2=1}^{\infty} s_\alpha s_{(3-\alpha)} n_{(s_1,s_2)} = M_{12}^{(2)}. \ (\alpha = 1, 2) \quad (4.166)$$

This important relation results in an inequality for this model, although we cannot obtain an equality with respect to the connection probability P. From eqns (4.164) and (4.166),

$$(\bar{F} + \bar{\varphi}) < \sum_{\beta=1}^{2} \varphi_\alpha m_{\alpha\beta}^{(2)}. \quad (4.167)$$

Equation (4.150) in the previous model here becomes an inequality. In order to compare this with the 'global censorship' scenario, we again consider $\varphi_1 = \varphi_2 = \varphi$ to get an analytic result. For real-world applications, this can be interpreted as each platform having similar detection abilities towards clusters. Then we obtain

$$\frac{M^{(2)}}{N} > \frac{\bar{F} + \varphi}{\varphi}, \quad (4.168)$$

followed by

$$P = \frac{M^{(2)} - N}{N^2} > \frac{\bar{F}}{N\varphi}. \quad (4.169)$$

To summarize the result, although P cannot be analytically calculated in the 'censorship-by-platform' model, we have derived that the probability of connection must be strictly higher than that of the 'global censorship' model, provided similar fragmentation probabilities for each platform. This makes sense theoretically, since here a cluster is not broken completely once detected: there can be remainders of the broken cluster on the other platform, only to be detected and eliminated again. For the more general case where φ_α are not equal, though we cannot find an exact relationship for P, just as discussed in Section 4.5.1, the comparison of eqns (4.150) and (4.167) still qualitatively indicates that $M_{\alpha\beta}^{(2)}$ of the 'censorship by platform' model are larger than those of the 'global censorship' model, and hence the probability P. This means that a decentralized source of fission is by definition less effective than a global one, hence less effective in preventing gelation.

4.5.3 Higher dimensional fission

Of course we don't want to stop at 2 species—just like before—and our goal is to extend the model into even higher dimensions to be able to describe heterogeneous behaviours akin to Fig. 1.2. As shown in the previous section, the partial fragmentation model does not produce an analytic solution after all, and note that the number of equations at D dimensions is $2^D + D - 1$, accounting for all the mixed, pure, and partially pure (e.g. $(x_1, 0, \ldots, x_D)$) clusters together with monomers. Thus, we will remain focused on the 'global censorship' model. Now if we look back at Section 4.5.1, we will notice that none of the derivations there assumed that $D = 2$ except for the summation limits, which can be trivially generalized into higher dimensions. Therefore, the 'global censorship' model is indeed generalizable, and we write down below the defining equations of this model in D dimensions:

$$\frac{d\nu_s}{d\tau} = \frac{F_{\alpha\beta}}{2} \sum_{\substack{u_k+v_k=s_k \\ \forall k=1,\ldots,D}} u_\alpha v_\beta \nu_u \nu_v - F_{\alpha\beta} \nu_s s_{\alpha l \beta} - \frac{\varphi}{2} \nu_s \sum_{\beta=1}^{D} s_\beta, \tag{4.170}$$

$$\frac{d\nu_\alpha^{(1)}}{d\tau} = -F_{\alpha\beta}\nu_\alpha^{(1)} l_\beta + \frac{\varphi}{2} \sum_{\substack{s_\beta=0 \ \forall \beta=1,\ldots,D \\ \sum_{\beta=1}^{D} s_\beta \geq 2}}^{\infty} s_\alpha \left(\sum_{\beta=1}^{D} s_\beta \right) \nu_s. \text{ (no sum in } \alpha\text{)} \tag{4.171}$$

Note that here $F_{\alpha\beta}$ are $D \times D$ matrix elements while φ, being constant, is the fragmentation probability due to the 'global censorship'. The same basic result then readily follows:

$$P(\tau) = \frac{\bar{F}}{N\varphi}. \tag{4.172}$$

4.6 Interacting shocks: a simplified approach

So far, we have developed a full mathematical theory of D species of interacting entities undergoing fusion and also fission—which was our aim as stated in Chaps 1 and 2. In the language of fluid dynamics that we adopted in Chap. 3, they can be represented by D interacting shock waves, where each shock wave $i = 1, 2, \ldots, D$ follows a specific generative PDE (or a component of a vectorial equation). We have seen in eqn (4.48) that only the toy model of pure aggregation with a product kernel can be solved analytically, and all the rest would be too hard to solve exactly. Previous discussions try to walk around this limitation, and we have found a number of nice properties of the models using various tricks, without actually solving these equations. But when it comes to these defining equations of the system itself, what shall we do to quickly grasp its core properties?

It turns out that we can greatly simplify the task here, to *approximately* describe the interactive system, as well as to reproduce and explain the data. To do this, we need to develop an approximate form for the mathematical expression for the size of each single shock wave in time $S_i(t) = N_i(t) - \mathcal{E}_i(t)$. Let's recall eqn (4.48) once more—from there we can see that the interactions between species in the master equations have been transformed into the interactions between functions ε_αs, i.e. the shockwaves, by their presence in all the equations. It then follows that, in order to put forward a simpler theory on this matter, we essentially want to write down a set of new equations for interacting shock waves, and this should preferably be linear in order to keep things as simple as possible. In short: D coupled equations for D components S_is. In the context of online behaviour, for example, we see from the data that the in-built communities (species) are each rich in membership and content and can overlap in topics, and so it is no surprise therefore that they are interested in each other's content, either to support it, or criticize it. So they interact.

The starting point for our approximate theory, is to linearize the one-species solution. Recall that eqn (3.81) specifies the equation for the gel curve

$$s(\tau) = 1 - e^{-s(\tau)\tau} \tag{4.173}$$

under normalized units, where $s = S/N = (N - \mathcal{E})/N$. We can develop an even simpler closed-form approximation that does the job for exploring coupling: consider the simple linear differential equation $\dot{s} = r_s(1 - s)$ and modify it to have time start from the shock wave onset time $t_{sw} = t_c$ (or the respective dimensionless variables $\tau_{sw} = \tau_c$ as we had previously). The equation becomes

$$\dot{s}_i = H(\tau - 1) r_s (1 - s_i), \tag{4.174}$$

and the solution can be written as

$$s_i(\tau) = H(\tau - 1) \left[1 - e^{-r_s(\tau - \tau_c)} \right]. \tag{4.175}$$

where $H(\ldots)$ is the well-known Heaviside function, which is not strictly a differentiable function at τ_c but can nonetheless be treated so in a piece-wise way. This turns out to be a very good approximation to the actual shock wave solution obtained numerically, as shown explicitly in Fig. 4.14.

We can also justify this analytically. We can expand eqn (4.173) using the knowledge of the shock wave transition time $\tau_c = 1$ (in these units) as follows, using a simple Taylor series:

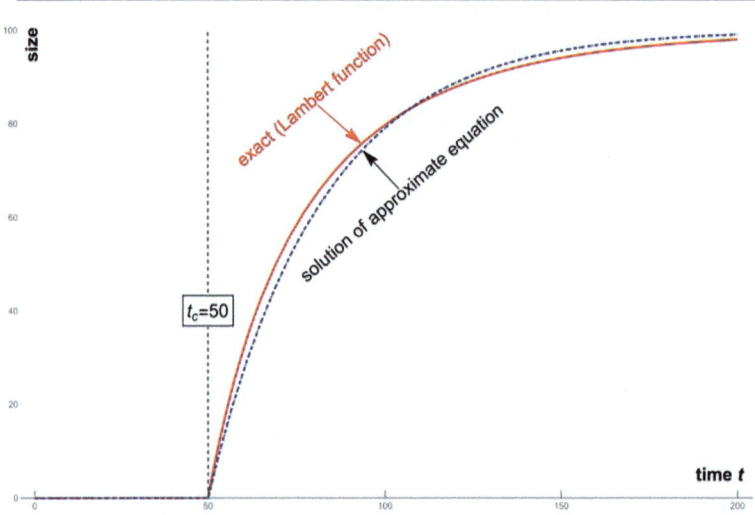

Fig. 4.14: Demonstration of how the shock wave exact solution (at least, for the simple case where $N(t)$ is constant) can be well approximated by the far simpler exponential form in the text.

$$s(\tau) = 0 \quad \tau < 1 \tag{4.176}$$

$$s(\tau) = 2(\tau - 1) - 8(\tau - 1)^2/3 + \ldots \quad \tau \geqslant \tau_c \tag{4.177}$$

$$s(\tau) = 1 - e^{-\tau} - \tau e^{-2\tau} + \ldots \quad t \to \infty, \tag{4.178}$$

and so for large τ with $\tau \gg \tau_c$,

$$s(\tau) \approx 1 - e^{-(\tau - \tau_c)} \tag{4.179}$$

so the expression for any τ is approximately

$$S(\tau) = H(\tau - \tau_c)\left[1 - e^{-(\tau - \tau_c)}\right], \tag{4.180}$$

which is in turn the exact solution of the following simple linear PDE as discussed above:

$$\dot{s}(\tau) = H(\tau - \tau_c)[1 - s(\tau)] \ . \tag{4.181}$$

Extending this to multiple species is simple enough: if we then put multiples of these approximate shock wave equations together, with their own respective onsets $\tau_{c,i}$, and we let them interact through some linear coupling terms, we get simpler coupled PDEs for the various shock wave sizes $\{S_i(t)\}$, and the coupling generates various types of deviations from the single shock wave shapes, as shown in Fig. 4.15. We could even add a simple decay term with rate constant d_i to mimic the fragmentation process,

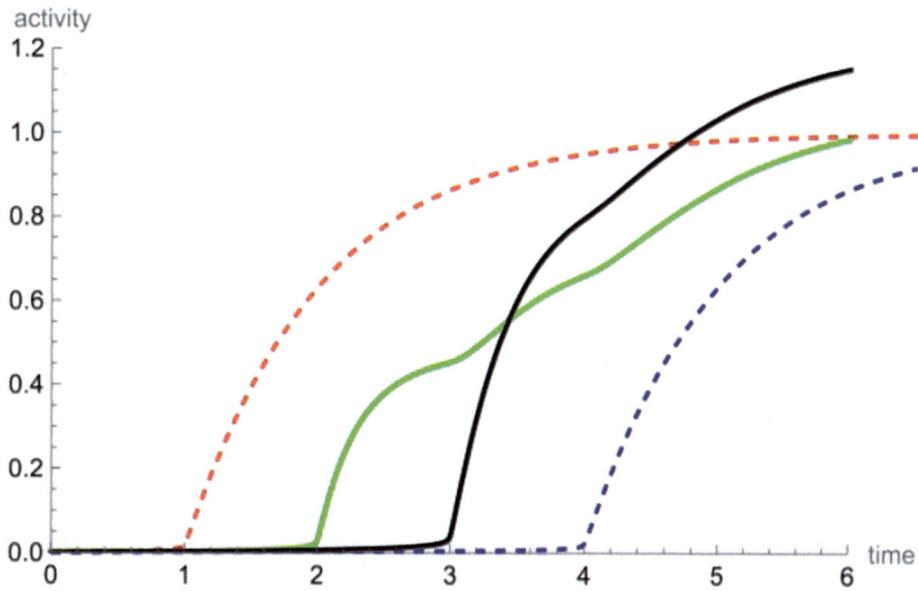

Fig. 4.15: Example showing how the coupled linear PDEs can give the type of rich shapes observed both empirically and in the exact model, where multiple shock waves evolve around the same time.

and hence reduction in shock wave size over time.[2] The resulting equations look like the following:

$$\dot{S}_i = H(t - t_{c,i}) \left[a_i (N_i - S_i(t)) - d_i S_i(t) + \sum_{j=1,\ j\neq i}^{D} F_{ij} (S_j(t) - S_i(t)) \right], \quad (4.182)$$

for $i = 1, \ldots, D$, where N_i, as before, denotes the total number of particles of species i, i.e. the *final* gel size of species i; a_i represents the strength of shock wave creation from its native species i; d_i is the aforementioned decay strength; F_{ij} represents the coupling strength between shock wave species i and j. Equation (4.182) is no longer written in dimensionless units for straightforwardness. This reproduces remarkably well, as shown in Fig. 4.16, the interacting shock wave curves observed from real data. This approximation is advantageous because (1) it is minimal and interpretable; (2) it can be applied at various levels of aggregation; (3) it produces output curves that are very similar to empirically observed curves; and (4) it can be used to explore intervention points.

---

[2] Strictly speaking, this oversimplification models exactly the aforementioned chipping model, but for a minimal model it serves our purpose well.

118 *Multi-body Physics 2.0: multi-species fusion and fission*

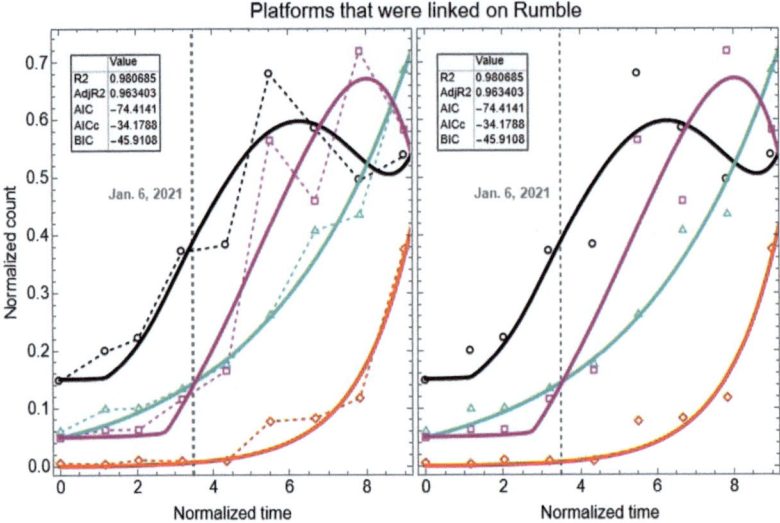

Fig. 4.16: Empirical data (red diamonds, Telegram; cyan triangles, Rumble; pink squares, YouTube; black circles, Other) showing the number of Telegram, Rumble, YouTube, and Other communities linked in posts in Rumble communities. Solid lines represent the corresponding shock wave model fit, i.e. solid red line represents the approximation for the YouTube shock wave, etc. The grey dashed line marks the 6 January riot, and goodness-of-fit metrics are displayed in the upper left-hand corner. Left plot has dashed lines connecting data points to aid in tracking empirical data. Source: data from Manrique, Huo, El Oud, Zheng, Illari, and Johnson (2023).

However, there is a big question to discuss: how is eqn (4.182) connected to the exact formulation as in eqn (4.48)? After all we obtained this from direct generalization of the 1-species linear approximations, so does this result agree with our first-principle D-species theory at all?

The good news is that direct comparison between eqns (4.48) and (4.182) demonstrates similarities between the structures of the two equations. Nevertheless, a significant difference is in how we include the other species S_j. In the exact yet aggregation-only model, the other species are included in completely symmetric positions as S_i, while this does not hold in eqn (4.182). However since eqn (4.182) is highly generic in that the parameters can take arbitrary values, F_{ij} does not even have to be positive, let alone symmetric, for example. This makes it possible to construct a similar structure to eqn (4.48) by taking F_{ij} to be negative and symmetric in i, j, and then balance the constants accordingly. Moreover, eqn (4.182) leads to further generalization possibilities that we are not able to include analytically in eqn (4.48), such as fragmentation and even competition between these entities, representable by the direct subtraction

$(S_j - S_i)$. These behaviours cause the irregular shapes in the gel curves as shown in Fig. 4.15, not observed in eqn (4.48). Speaking of species-wise competition, we will make a thorough investigation in the next chapter. Here we have it then: an empirical or phenomenological equation that describes our complex systems rather well. Before we close this chapter, we rewrite eqn (4.182) in a more general form:

$$\dot{S}_i = H(t - t_{c,i}) \left[A_i + \sum_{j=1}^{D} F_{ij} S_j(t) \right]. \qquad (4.183)$$

By writing it in this form, we sacrifice the motivation which, as it turns out, is not too indicative anyway, but can slim the number of parameters down to the fewest possible.

It is also worth mentioning that the resulting linear coupled PDEs for D shock waves $S_i(t)$s represent an intriguing and novel generalization of the coupled equations used to model well-known arms races in conflicts of all sorts including economic ones—but now involving humans, technology, and AI. This comes with the benefit that we have provided a first-principles rigorous derivation behind them and hence we can delve into how they can be systematically generalized and when they will break down. This is in contrast to the practice in the traditional arms-race literature which simply claims verbally that such equations are reasonable without a fully detailed generative explanation.

4.7 Predictions of super-shocks

Finally we mention briefly future super-shock attacks which are another major implication of this new shock-formation mathematics. We do this in the context of the conflict scenario surrounding Israel as discussed in Chap. 2 and shown in Fig. 2.7, and we use simpler symbols for the couplings. If inter-species (inter-allegiance, Fig. 4.17(A)) couplings increase beyond 7 October values, i.e. if the $\epsilon_{\Omega\Omega'}$'s are no longer much smaller than f_Ωs, a 'super-shock' is predicted to emerge in which the D shocks rise together. We mentioned this piling effect back in Section 5.1.5. Hence, a super-shock attack will arrive earlier and will be more lethal than 7 October (Fig. 4.17(B)). It will also be less attributable to a single adversary, potentially causing new geopolitical problems (Fig. 4.17(C)). Its future formation time can be obtained approximately using eqn (4.136)

$$t_c^{\text{super-shock}} = \left(\frac{f}{f + (D-1)\epsilon} \right) t_c^{7 \text{ Oct}} \qquad (4.184)$$

for the scenario of large and similar inter-allegiance couplings ($\epsilon_{\Omega\Omega'} \approx \epsilon$) compared to the small 7 October values. $t_c^{7 \text{ Oct}}$ is the formation time for a future 7 October-like attack, and $f_\Omega \approx f$. This equation means $t_c^{\text{super-shock}} < t_c^{7 \text{ Oct}}$, hence the super-shock attack will arrive earlier than any future 7 October-like attack; and it will arrive

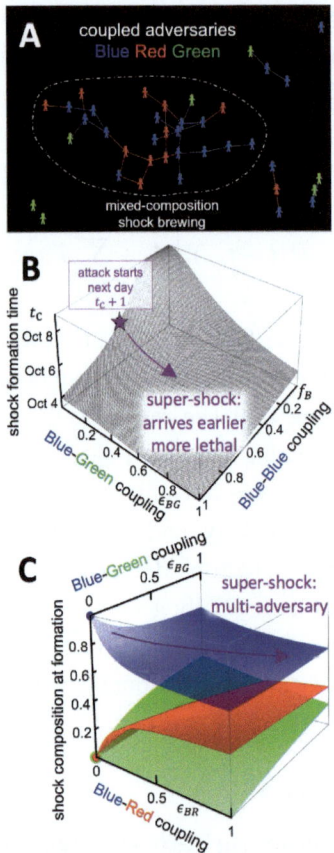

Fig. 4.17: Predictions from the shock-formation mathematics for $D = 3$ allegiance types (species). A: Snapshot from the $D = 3$ simulation provided in Fig. 2.8. Allegiance types shown as Blue (e.g. Hamas), Red (e.g. Hezbollah), Green (e.g. Palestinian Islamic Jihad, PIJ). B and C: Effect of increasing couplings within and between allegiance types. For strong inter-allegiance coupling (e.g. Hamas with PIJ) the shock and hence attack arrives earlier, is more lethal, and has a multi-adversary composition. It becomes a super-shock. See online article https://arxiv.org/abs/2409.02816 for full details.

increasingly early as the number of species (e.g. terrorist organizations) D increases and/or the coupling between them ϵ increases. Conversely, its risk of occurrence will be decreased by interventions that decrease the inter-allegiance coupling and hence their co-fusion.

As we discussed mathematically in Section 4.5, globalized (i.e. multi-adversary) surveillance and intervention is the best strategy going forward. Of course such a real-world analysis has limitations. This book purposely takes a 'crude look at the whole' as championed by physics Nobel Prize laureate Murray Gell-Mann, so that we can calculate consequences in a rigorous way. The good empirical agreement that this provides suggests that the net effect of the many other missing details cancels out to some degree, e.g. our results are valid irrespective of whether 7 October involved Hezbollah officially, or just some fighters with allegiance to it and/or others. More generally, our analysis establishes a concrete connection to fusion–fission studies across the animal kingdom and human tribes. Those fields could provide fresh insight into why fusion–fission might be so beneficial to fighters in term of resource optimization and adaptability, which may then inform more effective disruption schemes that can operate earlier on the microscale.

Exercises

4.1 Write down the master equation for the modified sum kernel as shown in eqn (4.10), and transform it into its corresponding generative equation.

4.2 Show that the master equation eqn (4.20) does transform into eqn (4.27). (**Hint:** use summation convention.)

4.3 Prove that the dynamics of 2 types of monomers for a 2-species fusion system is indeed given by eqns (4.21) and (4.22).

4.4 This question is on 2-species aggregation with an arbitrary population profile (ι_1 of the total population being species 1 and $\iota_2 \equiv 1 - \iota_1$ being species 2).

(a) Verify that the onset time is indeed given by eqn (4.76).
(b) From eqn (4.89) obtain the gel composition $\eta_{1,2}$ at onset.
(c) Verify that your result is correct by calculating the composition once more, this time using the second-moment method, as outlined in Section 4.2.3.

4.5 Recall how we showed that for 2-species systems, the square root of diagonal entries of the (normalized) second moment $\sqrt{\mu_g^{(11)}}$, $\sqrt{\mu_g^{(22)}}$ represents the proportions of each species in the gel at onset. Verify that the same applies for three-species systems. (**Hint:** take species 1 and 2 for example wlog—you simply need to verify that

$$\delta_{23}\delta_{13} = [(\epsilon_{23}\tau)(\epsilon_{13}\tau) + (\epsilon_{12}\tau)(n^{-1} - f_3\tau)]^2 \text{.)} \quad (4.185)$$

4.6 This question considers the root analysis of a 3-species aggregation system.

(a) Show that for $\xi_1 < 1$, and $\xi_{2,3} > 1$, $A \equiv \xi_1 + \xi_2 + \xi_3 - 2\sqrt{\xi_1\xi_2\xi_3} - 1 > 0$, and hence in this scenario, there are two positive roots for gelation onset time.
(b) Prove the result for $\xi_{1,2,3} > 1$ in Section 4.3.1.

4.7 For a 2-species aggregation system with heterogeneous matrix

$$F = \begin{pmatrix} 0.5 & 0.2 \\ 0.2 & 0.2 \end{pmatrix},$$

(a) calculate its normalized gelation time;
(b) calculate its gelation composition at onset time by diagonalizing F matrix, and compare it against the result from eqn (4.73).

4.8 For the following heterogeneous matrix

$$F = \begin{pmatrix} 0.4 & 0.2 & 0 \\ 0.2 & 0.3 & 0.2 \\ 0 & 0.3 & 0.2 \end{pmatrix},$$

is it reducible to lower dimensions? Find the gelation onset time(s) and gel composition(s) at onset(s) for its corresponding aggregation system.

4.9 For the following heterogeneous matrix

$$F = \begin{pmatrix} 2.6 & 0 & 0.8 & 0 & 0 \\ 0 & 2.45 & 0 & 0.95 & 1.1 \\ 0.8 & 0 & 1.4 & 0 & 0 \\ 0 & 0.95 & 0 & 2.85 & 0.7 \\ 0 & 1.1 & 0 & 0.7 & 2.7 \end{pmatrix},$$

corresponding to a 5-species aggregation system,

(a) is it reducible to lower dimensions?
(b) If so, write down the reduced subsystems with their respective F_ss.
(c) Find the gelation onset times and gel compositions at onsets for its corresponding aggregation system.

4.10 Transform a multi-species gelation system from its canonical coordinate system ξ_α to a new coordinate system ξ'_α, where the heterogeneity matrix F becomes diagonal Λ. In particular, find the transformation rules for the generating functions and the heterogeneity matrix. (**Hint:** this would make a very simple differential geometry problem in general relativity.)

4.11 In Section 4.4.3 we have demonstrated the possibility of finding an effective gel curve for a system with identical f_αs and ϵ_{ij}s, respectively. From eqn (4.137), derive the explicit form of the gel curve expression.

4.12 This question is on the multi-species model with fragmentation.

(a) First let's consider the 'global-censorship' model. Obtain its generative equation from eqn (4.141). Can you derive the steady-state condition from there?
(b) For the 'censorship-by-platform' model, also obtain the generative equation from the master equations from eqn (4.163). By definition you should find that when setting $\xi_\alpha = 0$ in the generative PDE, eqn (4.164) should be recovered.

4.13 Derive the expression for the onset time of a potential 'super-shock', as shown in eqn (4.184).

5
Multi-body Physics 3.0: general fusion–fission theory for time-dependent heterogeneous systems

In our search for physics that can describe the collective behaviour and the empirical patterns presented in Chap. 2, we have so far formulated a simple theory for the fusion and fission of heterogeneous entities, and how it leads to a dynamical phase change known as gelation. We have seen that partitioning the heterogeneous population into D coarse-grained 'species' preserves some of its heterogeneity, manifested by the 'bumps' (kinks) that appear in the gel growth curves as seen empirically in Chap. 2—and consequently the multi-peak features in its derivative which can be used to model bursts (recall Fig. 1.6 for $D = 2$). Nevertheless, there are still some important limitations of our theory 2.0 from Chap. 4 that we will now explain and fix.

This chapter takes the mathematics in Chap. 4, shows its limitations, and then reveals how it can be generalized. By the end of this chapter, we will finally have a powerful framework and theory for all kinds of real-world fusion–fission interactions—including most importantly the human–machinery–AI systems of interest in this book (Fig. 1.2).

But first we need to make an important point. The level of detail with which we analyse the growth profile of the cohesive unit (G) and its time-derivative ($\dot{G}$) in this chapter may appear excessive at first sight. However these details—when observed empirically as for example in Fig. 2.7 or in Fig. 2.1—can provide key insight into how much heterogeneity is in the system, the role that it plays in the system's macroscopic behaviour, and hence future surprises and risk.

Specifically, this chapter analyses in depth the key question: When we see a bump (kink) appear in the empirical data for the growth of a cohesive unit (supposing we are measuring the macroscopic variable G as in Fig. 1.6, left panel and Fig. 2.7, upper panel)—or equivalently, when we see a peak appear in the empirical data for a burst (supposing we are measuring the macroscopic variable $\dot{G}$ as in Fig. 1.6, right panel and

Fig. 2.7, lower panel)—what is this telling us about how the system actually works? So for example, even if a given human–machinery–AI system has very heterogeneous entities that cannot a priori be classified into a small number of species, does it actually behave as a small $D = 2$ species system? As we show in this chapter, these details provide us with a form of spectroscopy of what might otherwise be a black-box system—including AI itself in the example in Fig. 2.1(c). In short, measuring and modelling the precise shape of the G and $\dot{G}$ can play a crucial role in unlocking how future systems of humans-machinery-AI actually 'work' (Fig. 2.1(a)(b)(d)) including AI itself (Fig. 2.1(c)). Therefore we will pay a lot of attention to the shape of G (what we call its bumps or kinks) and hence $\dot{G}$ (its corresponding peaks, since each bump or kink in G can generate a peak-like shape in $\dot{G}$) and we will seek analytic results wherever we can. The theoretical formulae that we derive can then be easily compared against future human–machine–AI systems' empirical data to find—for example—how many species of entity are effectively driving the system's macroscopic behaviours.

As in Chaps 3 and 4, we may refer to a cohesive unit as a **gel**, or a **giant connected component (GCC)**—or even as a **shock** in an inviscid-like fluid because of the form of its generative equation. All these terms mean the same thing at the macroscopic level: a **cohesive unit**. Also because of a similar diversity in terminology across the literature, we may use words such as **coalescence, aggregation, coagulation**, etc. interchangeably with **fusion**, and **fragmentation, disaggregation, degradation**, etc. interchangeably with **fission**.

5.1 Motivation: later onsets and bumps grow more slowly

To start this chapter, we take a closer look at the multi-species gelation curves produced in Chap. 4, i.e. Figs 4.5 and 4.6. A careful check reveals that later bumps (kinks) all have slower growth rates than the original onset. To see this mathematically, we will now examine the derivative $\dot{G}(\tau)$, recalling that τ is just scaled time.

5.1.1 Multi-species gelation growth rate

Recall from Chap. 4 the multi-species generalized Burgers' equation, eqn (4.28)

$$\frac{\partial \varepsilon_\alpha}{\partial \tau} + F_{\beta\gamma}(\varepsilon_\beta - \iota_\beta)\partial_\alpha \varepsilon_\gamma = 0,$$

where $\iota_\beta = 1/D$, and its gelation solution at $x = 0$,

$$G_\alpha(\tau) = \left(1 - e^{-\sum_\beta F_{\alpha\beta} G_\beta(\tau)\tau}\right)/D \tag{5.1}$$

$$G(\tau) = 1 - \frac{1}{D}\sum_{\alpha=1}^{D} e^{-\sum_\beta F_{\alpha\beta} G_\beta(\tau)\tau}, \tag{5.2}$$

which follow by direct derivation from eqn (4.137).

Instead of taking the derivatives of eqn (5.1), we note that the generative eqn (4.28) itself already contains the time derivatives. We simply need to substitute $G_\alpha(\tau) = \iota_\alpha - \varepsilon_\alpha(\bm{x}=0,\tau)$ into that equation:

$$\frac{dG_\alpha}{d\tau} = F_{\beta\gamma}G_\beta[\partial_\gamma\varepsilon_\alpha]_{\bm{x}=0}. \tag{5.3}$$

The partial derivative looks familiar, since it is how we defined the second moment tensor

$$(m_2)_{\alpha\gamma} = [\partial_\gamma\varepsilon_\alpha]_{\bm{x}=0}$$

in the previous chapter. The problem is that the tensor is only well-defined before gelation, and then it becomes singular. We did previously look at all the roots of $\det\{m_2\}$, including those after the physical gelation, without considering its domain, but those are 'virtual points' after all, and only help us gauge roughly how multi-species gel curves develop. Is there a way to formally expand the definition of m_2? There certainly should be, since all the other factors in eqn (5.3) are well-defined, except exactly at the instant of gelation onset.

We have to go all the way back to the defining equations of m_2. Recall that it is directly obtained by taking one more partial derivative of ε_α from eqn (4.28) and setting $x_\alpha = 0$. Having found the exact solution of ε_α, we do not need to solve the matrix Riccati equation (eqn (4.32)) all over again. From eqn (4.48), we take logarithms on ε_α to get

$$\ln\frac{\varepsilon_\alpha}{\iota_\alpha} = -\sum_\beta F_{\alpha\beta}(\iota_\beta - \varepsilon_\beta)\tau. \tag{5.4}$$

Now we take the partial derivative with respect to x_γ and set $\bm{x} = 0$:

$$\frac{1}{\varepsilon_\alpha(\bm{0},\tau)}m_{2\gamma\alpha} = \sum_\beta F_{\alpha\beta}m_{2\gamma\beta}\tau. \tag{5.5}$$

Note that we do *not* adopt the summation convention here—instead, the sum is denoted with the $\sum$ sign. Applying the symmetry of m_2 and substituting in $G_\alpha(\tau)$, we obtain

$$(1 - G_\alpha)^{-1}m_{2\alpha\gamma} = \sum_\beta F_{\alpha\beta}m_{2\beta\gamma}\tau. \tag{5.6}$$

Written in matrix form, this is equivalent to

$$\text{diag}\left[(\iota_\alpha - G_\alpha)^{-1}\right]\mathsf{m}_2 = \mathsf{F}\mathsf{m}_2\tau. \tag{5.7}$$

Finally, the general solution of m_2 for all time is

$$\mathsf{m}_2(\tau) = \left(\text{diag}\left[\iota_\alpha - G_\alpha(\tau)\right]^{-1} - \mathsf{F}\tau\right)^{-1}. \tag{5.8}$$

Second moment tensor, revisited. Now that we have unlocked the complete form of m_2, it is time to review its properties. To begin, it is not surprising that it recovers the old expression as in eqn (4.38) when $\tau < \tau_g$, i.e. before gelation, since $G_\alpha(\tau < \tau_g) = 0$. An interpretation, if not too far-fetched, is to regard this additional $-G_\alpha(\tau)$ term as excluding the particles in the gel phase, when calculating the second moment of the system, which is essentially a 'solution'. Furthermore, since m_2 is now a valid physical solution for the whole duration of gelation, by its definition of summing up strictly positive terms $\boldsymbol{k k} n_{\boldsymbol{k}}$, all of its entries, and naturally its determinant $\det(m_2)$, must be strictly positive. This in turn proves once more that the virtual gelation points we discussed throughout Chap. 4 are truly virtual and unphysical.

We now write down the implicit expression for $d\boldsymbol{G}/d\tau$:

$$\frac{d\boldsymbol{G}}{d\tau} = \mathsf{m}_2 \mathsf{F} \boldsymbol{G} = \left(\mathrm{diag}\,[\iota_\alpha - G_\alpha(\tau)]^{-1} - \mathsf{F}\tau \right)^{-1} \mathsf{F}\boldsymbol{G}. \tag{5.9}$$

Just like our previous encounters with m_2, the complexity of the expressions rapidly grows with the number of species, and for large D we have to numerically evaluate eqn (5.9). But as always, we can at least try to obtain an analytical result for the simplest case of $D = 2$. For each species we have

$$\frac{dG_1}{d\tau} = f_1 G_1 m_{11} + \epsilon G_1 m_{21} + \epsilon G_2 m_{11} + f_2 G_2 m_{21}$$
$$= (f_1 G_1 + \epsilon G_2) m_{11} + (\epsilon G_1 + f_2 G_2) m_{21} \tag{5.10}$$

$$\frac{dG_2}{d\tau} = f_1 G_1 m_{12} + \epsilon G_1 m_{22} + \epsilon G_2 m_{12} + f_2 G_2 m_{22}$$
$$= (f_1 G_1 + \epsilon G_2) m_{12} + (\epsilon G_1 + f_2 G_2) m_{22}. \tag{5.11}$$

Our ultimate goal here is the growth rate of the entire gel

$$\frac{d G}{d\tau} \equiv \dot{G}_1 + \dot{G}_2. \tag{5.12}$$

We first define the vector

$$\boldsymbol{X} \equiv \begin{pmatrix} X \\ Y \end{pmatrix} = \mathsf{F}\boldsymbol{G}. \tag{5.13}$$

Equivalently, in components,

$$X = f_1 G_1 + \epsilon G_2$$
$$Y = \epsilon G_1 + f_2 G_2. \tag{5.14}$$

This is mathematically simply a conversion of basis. Recall that the expressions for $G_{1,2}$ are

$$G_1 = \left(1 - e^{-f_1 G_1 \tau} e^{-\epsilon G_2 \tau}\right)/2 = \left(1 - e^{-X\tau}\right)/2 \tag{5.15}$$

$$G_2 = \left(1 - e^{-\epsilon G_1 \tau} e^{-f_2 G_2 \tau}\right)/2 = \left(1 - e^{-Y\tau}\right)/2, \qquad (5.16)$$

and m_2^{-1} can hence be further simplified into

$$m_2^{-1} = 2 \begin{pmatrix} e^{X\tau} - f_1\tau/2 & -\epsilon\tau/2 \\ -\epsilon\tau/2 & e^{Y\tau} - f_2\tau/2 \end{pmatrix}. \qquad (5.17)$$

Take the inverse and we have

$$\begin{aligned} \frac{dG_1}{d\tau} &= \frac{1}{2\Delta}\left[X(e^{Y\tau} - f_2\tau/2) + Y(\epsilon\tau/2)\right] \\ \frac{dG_2}{d\tau} &= \frac{1}{2\Delta}\left[X(\epsilon\tau/2) + Y(e^{X\tau} - f_1\tau/2)\right], \end{aligned} \qquad (5.18)$$

where Δ is the determinant of $(2m_2)^{-1}$:

$$\Delta = (f_1 f_2 - \epsilon^2)(\tau/2)^2 - \left(f_2 e^{X\tau} + f_1 e^{Y\tau}\right)(\tau/2) + e^{X\tau}e^{Y\tau}. \qquad (5.19)$$

Finally, we sum up $\dot{G}_{1,2}$ to obtain

$$\begin{aligned} \frac{dG}{d\tau} &= \frac{Xe^{Y\tau} + Ye^{X\tau} + [X(\epsilon - f_2) + Y(\epsilon - f_1)]\tau/2}{2\Delta} \\ &= \frac{Xe^{Y\tau} + Ye^{X\tau} - (f_1 f_2 - \epsilon^2)G\tau/2}{2\left[(f_1 f_2 - \epsilon^2)(\tau/2)^2 - (f_2 e^{X\tau} + f_1 e^{Y\tau})(\tau/2) + e^{X\tau}e^{Y\tau}\right]}, \end{aligned} \qquad (5.20)$$

where in the last equality we used eqn (5.14) to solve for

$$G = G_1 + G_2 = \frac{X(\epsilon - f_2) + Y(\epsilon - f_1)}{\epsilon^2 - f_1 f_2}. \qquad (5.21)$$

Equation (5.20) is our exact expression for $dG/d\tau$ when $D = 2$. It requires only the gel sizes $G_{1,2}$ and time τ, all of which we know exactly.

5.1.2 Special case: $\epsilon = 0$

The mathematics above in Section 5.1.1 is too complicated to evaluate analytically everywhere to prove our observation, so we try to find evidence heuristically. We start with the simplest model: a decoupled $D = 2$-species system. Trivially the system is a sum of 2 single-species subsystems:

$$G_1 = \frac{1}{2}\left(1 - e^{-G_1 f_1 \tau}\right), \quad G_2 = \frac{1}{2}\left(1 - e^{-G_2 f_2 \tau}\right), \qquad (5.22)$$

$$G(\tau) \equiv G_1(\tau) + G_2(\tau) = 1 - \frac{1}{2}\left(e^{-G_1 f_1 \tau} + e^{-G_2 f_2 \tau}\right). \qquad (5.23)$$

From Chap. 4 we know that this system has two onsets, at time instants $\tau_1 = 2/f_1$ and $\tau_2 = 2/f_2$ respectively. We assume *wlog* that $f_1 > f_2$. For such a decoupled system we can conveniently consider each species independently as a $D=1$ system. For a single-species growth curve

$$G(\tau) = 1 - e^{Gf\tau}, \tag{5.24}$$

its growth rate $dG/d\tau$ right after onset $\tau_g = 1/f$ is

$$\left.\frac{dG}{d\tau}\right|_{\tau_{g+}} = -\left.\frac{d(e^{-Gf\tau})}{d\tau}\right|_{\tau_{g+}} = \left.\frac{d(Gf\tau)}{d\tau}\right|_{\tau_{g+}} e^{-G(\tau_{g+})}$$

$$\approx \left(G(\tau_{g+})f + \left.\frac{dG}{d\tau}\right|_{\tau_{g+}}\right)(1 - G(\tau_{g+})), \tag{5.25}$$

where in the last step of eqn (5.25) we Taylor expand the exponential e^{-G_1} since $G_1(\tau_{g+}) \ll 1$ just after onset. This gives

$$G(\tau_{g+})\dot{G}(\tau_{g+}) \approx fG(\tau_{g+})(1 - G(\tau_{g+})) \approx fG(\tau_{g+}), \tag{5.26}$$

again keeping linear order of $G(\tau_{g+})$. Finally

$$\dot{G}(\tau_{g+}) \approx f \equiv 1/\tau_g. \tag{5.27}$$

Note that the τ_{g+} notation is necessary since $G(\tau)$ is not smooth and bends at τ_g. We see that indeed, for a single-species system at least, the growth rate right after gelation onset is simply inversely proportional to the onset time.

This nice $D=1$ result can be directly translated to the decoupled 2-species system, for the first onset at least. Soon after $\tau_1 = 2/f_1$,

$$\dot{G}(\tau_{1+}) \approx \frac{1}{2\tau_1}, \tag{5.28}$$

where the prefactor $1/2$ is simply due to normalization. For the second onset, though, it is a little more complicated:

$$\dot{G}(\tau_{2+}) \approx \frac{1}{2\tau_2} + \dot{G}_1(\tau_{2+}). \tag{5.29}$$

Here we have a small caveat: although τ_1^{-1} is strictly larger than τ_2^{-1}, the extra contribution from species 1 in $\dot{G}(\tau_{2+})$ can mess up our assumption. Fortunately, we are familiar with the gel curve and know that $\dot{G}_1$ quickly decays asymptotically to 0, so as long as the 2 onsets are well separated then $\dot{G}(\tau_{2+}) < \dot{G}(\tau_{1+})$ will be true. Now if we turn on the inter-species parameter ϵ, the second onset turns into a smooth bump when ϵ is small. It would be very hard to calculate $\dot{G}$ analytically somewhere inside the bump as we can no longer apply small-quantity approximations, but with the decoupled limit now known, we at least have a heuristic idea that the bump does grow more slowly than the initial onset—excluding of course the trivial case when f_1 and f_2 are close in values (trivial because the faster growth is due to two surges of gelation growth coinciding together, a sort of 'piling' effect). We will have a closer look into this piling effect in a moment and see how it affects the gelation growth in the system.

5.1.3 General case: arbitrary parameter values

Studying the onset for a generic 2-species system is feasible. Starting from eqn (5.20), we now try to analytically tackle $\dot{G}(\tau_{g+})$ and see if we can generalize eqn (5.28). As before, the first step is to linearize the small quantities. At τ_{g+}, X, Y, G are all just above 0, so we linearize them all and obtain

$$\dot{G}(\tau_{g+}) \approx \frac{1}{2} \frac{X + Y - (f_1 f_2 - \epsilon^2) G \frac{\tau_g}{2}}{\left[(f_1 f_2 - \epsilon^2) \frac{\tau_g^2}{4} - (f_2 + f_1) \frac{\tau_g}{2} + 1 \right] + (X + Y)\tau_g - (f_2 X + f_1 Y) \frac{\tau_g^2}{2}}, \quad (5.30)$$

where for notational convenience we omit all the arguments τ_{g+} after the functions X, Y, G. We immediately identify that the first term in the denominator within the square parenthesis is just $\det \mathsf{m}_2^{-1}(\tau_g)$ and is just 0 according to eqn (4.53). It is then simplified into

$$\dot{G}(\tau_{g+}) \approx \frac{1}{2\tau_g} \frac{X + Y - (f_1 f_2 - \epsilon^2) G(\tau_g/2)}{(X + Y) - (f_2 X + f_1 Y)(\tau_g/2)}, \quad (5.31)$$

where we extracted a common factor $\tau_g/2$ from the denominator. Now eqn (5.31) has a coefficient that is exactly the RHS in eqn (5.28), i.e. inversely proportional to onset time. If we can simplify the fraction into some well-behaved quantity, we can successfully prove our assumption, that later onsets are always slower for a generic 2-species system. But how to approach it without any explicit expressions of X, Y, or G?

First of all, we reduce the number of variables in the fraction by substituting in eqn (5.21):

$$\begin{aligned}
\dot{G}(\tau_{g+}) &\approx \frac{1}{2\tau_g} \frac{X + Y + [(\epsilon - f_2)X + (\epsilon - f_1)Y](\tau_g/2)}{(X + Y) - (f_2 X + f_1 Y)(\tau_g/2)} \\
&= \frac{1}{2\tau_g} \left[1 + \frac{\epsilon(X + Y)(\tau_g/2)}{(X + Y) - (f_2 X + f_1 Y)(\tau_g/2)} \right] \\
&= \frac{1}{2\tau_g} \left[1 + \epsilon \left(\frac{2}{\tau_g} - \frac{f_2 X + f_1 Y}{X + Y} \right)^{-1} \right],
\end{aligned} \quad (5.32)$$

where we recursively identify the equivalent elements on the numerator and the denominator and cancel them out, in order to obtain the simplest dependence on X

and Y and see if we can do anything about it. It turns out that this dependence is a simple fraction $(f_2 X + f_1 Y)/(X+Y)$ homogeneous in terms of X and Y, and thus we only need to find out the ratio X/Y.

X and Y here are both evaluated just after gelation onset, and we have indeed looked into some related quantities—in Section 4.2.2 we found exactly the ratio between $G_1(\tau_{g+})$ and $G_2(\tau_{g+})$ (denoted there as $\eta_{1,2}$), and to translate that into X/Y should be easy: they just differ by a linear transformation after all. From eqns (4.72) and (4.73) we have

$$\frac{G_1(\tau_{g+})}{G_2(\tau_{g+})} = \frac{\sqrt{x^2+4}+2+x}{\sqrt{x^2+4}+2-x}, \tag{5.33}$$

where $x = (f_1 - f_2)/\epsilon$. Now we are confident that $\dot{G}(\tau_{g+})$ can be expressed by a simple combination of the parameters F_{ij}. Before we proceed with calculations, one more note about eqn (5.33): the square root in the notation is not unfamiliar. Expressing eqn (5.33) with the 3 original parameters, we have

$$\frac{G_1(\tau_{g+})}{G_2(\tau_{g+})} = \frac{\sqrt{(f_1-f_2)^2+4\epsilon^2}+2\epsilon+f_1-f_2}{\sqrt{(f_1-f_2)^2+4\epsilon^2}+2\epsilon-f_1+f_2}. \tag{5.34}$$

In addition, recall from eqn (4.54) that the expression of τ_g is

$$\tau_g = \frac{(f_1+f_2) - \sqrt{(f_1-f_2)^2+4\epsilon^2}}{(f_1 f_2 - \epsilon^2)}.$$

The square root is the same—not too surprising as we derived one from the other, but just a reminder to relate them during the following derivations. Denote $\delta = \sqrt{(f_1-f_2)^2+4\epsilon^2}$ and we can set off:

$$\frac{f_1 Y + f_2 X}{Y + X} = \frac{f_1(\epsilon+f_2)G_1 + f_2(\epsilon+f_1)G_2}{(\epsilon+f_1)G_1 + (\epsilon+f_2)G_2}$$
$$= \frac{f_1(\epsilon+f_2)(\delta+2\epsilon+f_1-f_2) + f_2(\epsilon+f_1)(\delta+2\epsilon-f_1+f_2)}{(\epsilon+f_1)(\delta+2\epsilon+f_1-f_2) + (\epsilon+f_2)(\delta+2\epsilon-f_1+f_2)}. \tag{5.35}$$

Regrouping the terms in both the numerator and the denominator, we have

$$\frac{f_1 Y + f_2 X}{Y + X} = \frac{(\delta+2\epsilon)[f_1(\epsilon+f_2)+f_2(\epsilon+f_1)] + (f_1-f_2)(\epsilon f_1 - \epsilon f_2)}{(\delta+2\epsilon)(2\epsilon+f_1+f_2) + (f_1-f_2)(f_2-f_1)}$$
$$= \frac{(\delta+2\epsilon)[f_1(\epsilon+f_2)+f_2(\epsilon+f_1)] + \epsilon(f_1-f_2)^2}{(\delta+2\epsilon)(2\epsilon+f_1+f_2) + (f_1-f_2)^2}. \tag{5.36}$$

Again, similar terms have appeared in both the numerator and the denominator, and we hope to use the same method as before to simplify. We try to construct a term that is ϵ times the denominator:

$$\frac{f_1 Y + f_2 X}{Y + X} = \epsilon + \frac{(\delta+2\epsilon)[f_1(\epsilon+f_2)+f_2(\epsilon+f_1) - \epsilon(2\epsilon+f_1+f_2)]}{(\delta+2\epsilon)(2\epsilon+f_1+f_2) + (f_1-f_2)^2}$$

$$= \epsilon + \frac{(\delta + 2\epsilon)(2f_1f_2 - 2\epsilon^2)}{(\delta + 2\epsilon)(2\epsilon + f_1 + f_2) + (f_1 - f_2)^2}. \tag{5.37}$$

In the remaining fraction, the numerator has become simple enough, but we still need to find a way to factorize the denominator. We observe that there is a $4\epsilon^2$ term and a $(f_1 - f_2)^2$ term, two components of δ^2—maybe that can lead us to factorization:

$$\frac{f_1 Y + f_2 X}{Y + X} = \epsilon + \frac{2(\delta + 2\epsilon)(f_1 f_2 - \epsilon^2)}{\delta(2\epsilon + f_1 + f_2) + 2\epsilon(f_1 + f_2) + [4\epsilon^2 + (f_1 - f_2)^2]}$$

$$= \epsilon + \frac{2(\delta + 2\epsilon)(f_1 f_2 - \epsilon^2)}{2\epsilon(f_1 + f_2) + \delta(2\epsilon + f_1 + f_2) + \delta^2}$$

$$= \epsilon + \frac{2(\delta + 2\epsilon)(f_1 f_2 - \epsilon^2)}{(2\epsilon + \delta)(f_1 + f_2 + \delta)}$$

$$= \epsilon + \frac{2(f_1 f_2 - \epsilon^2)}{f_1 + f_2 + \delta}. \tag{5.38}$$

This is much simpler than before. Substituting this back into eqn (5.32), we obtain

$$\dot{G}(\tau_{g+}) \approx \frac{1}{2\tau_g} \left[1 + \epsilon \left(\frac{2}{\tau_g} - \epsilon - \frac{2(f_1 f_2 - \epsilon^2)}{f_1 + f_2 + \delta} \right)^{-1} \right]$$

$$= \frac{1}{2\tau_g} \left[1 + \epsilon \left(\frac{2(f_1 f_2 - \epsilon^2)}{f_1 + f_2 - \delta} - \frac{2(f_1 f_2 - \epsilon^2)}{f_1 + f_2 + \delta} - \epsilon \right)^{-1} \right]$$

$$= \frac{1}{2\tau_g} \left[1 + \epsilon \left(2(f_1 f_2 - \epsilon^2) \frac{2\delta}{(f_1 + f_2 - \delta)(f_1 + f_2 + \delta)} - \epsilon \right)^{-1} \right]$$

$$= \frac{1}{2\tau_g} \left[1 + \epsilon \left(4(f_1 f_2 - \epsilon^2) \frac{\delta}{(f_1 + f_2)^2 - \delta^2} - \epsilon \right)^{-1} \right]. \tag{5.39}$$

Finally, substituting in $\delta^2 = (f_1 - f_2)^2 + 4\epsilon^2$, we have

$$\dot{G}(\tau_{g+}) \approx \frac{1}{2\tau_g} \left[1 + \epsilon \left(4(f_1 f_2 - \epsilon^2) \frac{\delta}{4f_1 f_2 - 4\epsilon^2} - \epsilon \right)^{-1} \right]$$

$$= \frac{1}{2\tau_g} \left(1 + \frac{\epsilon}{\delta - \epsilon} \right)$$

$$= \frac{1}{2\tau_g} \left(1 + \frac{1}{\sqrt{x^2 + 4} - 1} \right). \tag{5.40}$$

We have successfully found $\dot{G}(\tau_{g+})$ for a generic 2-species system, and if we take a closer look at it, we will find that it is an intriguing result: since $\sqrt{x^2 + 4} \in [2, \infty)$, the parenthesized factor must have a value within the range $[1, 2]$. In other words, the factor $\left[1 + (\sqrt{x^2 + 4} - 1)^{-1} \right]$ doesn't determine the onset gradient but only moderates it as a multiplier between 1 and 2. At the end of the day, it is mainly determined by our old friend, the inverse of onset time τ_g, once again proving that later onsets have slower growth rates.

Let's take a closer look at the multiplier $\left[1+(\sqrt{x^2+4}-1)^{-1}\right]$. It depends on a measurement of the relative strength of the 3 parameters, $x = (f_1 - f_2)/\epsilon$, and is dimension-free in terms of the parameters. By contrast, τ_g has a dimension $[F^{-1}]$, so it must be determined not only by their relative strengths but also the absolute magnitudes of the parameters. This is the deeper logic of why τ_g^{-1} is the determining factor while $\left[1+(\sqrt{x^2+4}-1)^{-1}\right]$ is merely a moderator. To see how exactly it moderates the gradient, we look at the extremes:

1. When $\epsilon \ll f_{1,2}$ and $f_1 \neq f_2$, this is asymptotically the decoupled case we discussed in the previous section. In this case $x \to \infty$, so $\left[1+(\sqrt{x^2+4}-1)^{-1}\right] \to 1$, and the gradient tends to $1/(2\tau_g)$, which agrees with the previous discussions.

2. When $\epsilon \gg f_{1,2}$, $x \to 0$ and $\left[1+(\sqrt{x^2+4}-1)^{-1}\right] \to 2$. The gradient turns out to be $1/\tau_g$, identical to the $D = 1$ case also discussed previously. This makes perfect sense as well, since when inter-species aggregation prevails, the system is very well mixed and hence can simply be seen as an approximate single-species system.

3. Finally, we consider the trivial case $f_1 = f_2$, and its approximate condition $|f_1 - f_2| \ll \epsilon$. In this scenario we also have $x = 0$ and $\dot{G}(\tau_{g+}) = 1/\tau_g$, regardless of ϵ value. This is due to the aforementioned piling effect and hence two separate surges merge into one.

5.1.4 Causality: an interpretation

The following two pairs of plots in Fig. 5.1 illustrate our discussions above.

Here then is an interpretation of this 'later onsets/bumps grow slower' phenomenon that complements the mathematical explanations earlier in this section. Equation (4.28), which defines multi-species aggregation, implies that the aggregation process for all species commences simultaneously. Hence, later onsets are associated with a

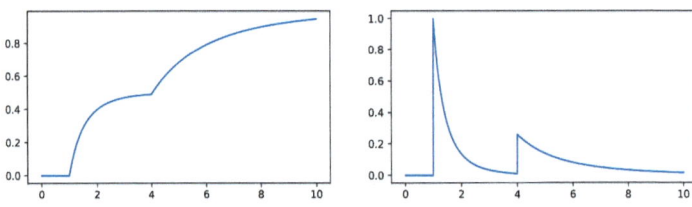

Fig. 5.1: A possible shape of gel curve and its time derivative in our 2-species theory. Notice the much smaller gradient at the second onset.

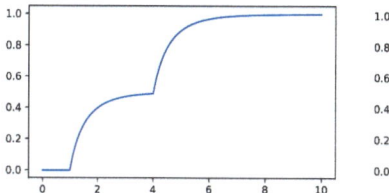 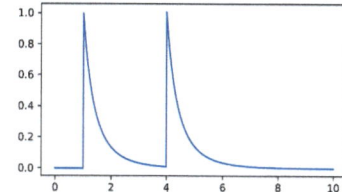

Fig. 5.2: This gel curve is impossible to obtain from the theory in the case of non-zero inter-species interactions. It is plotted by stacking together two 1-species gel curves with a shift of time.

smaller growth rate, which can be expressed as a combination of growth rates $F_{\alpha\beta}$ as we showed previously. This is asymptotically the case for the smooth bumps as well. Consequently, as shown in Fig. 5.2, the piling effect aside, the only way to have a late yet rapid onset is to delay the start of the corresponding aggregation altogether.

Simultaneity is therefore the cause of the 'later onsets/bumps grow slower' phenomenon. We can perhaps see this as a sign that we are studying some truly related species in one system, rather than a bunch of unrelated subsets of a population that just happen to be in the proximity of the measurement. Within this system, heterogeneous particles are batched up into species by their assembly rates, reflected by their onset time and growth rates. This is to say that increasingly slower growths as time progresses show a causal relationship between the species—slower growing species must appear later than the faster growing ones.

Nevertheless, the causal argument cannot preclude plausible alternatives that could potentially create late and rapid onsets or re-entrants while also respecting causality. What if, for example, we have new species joining the system later in time, so that their assembly is delayed? Can our theory accommodate this possibility at all?

This is the motivation for all the generalizations later in this chapter. But before we look into that, there is one more thing to discuss: re-entrant bumps and when they appear.

5.1.5 Phase diagram of bumps

When should we see a bump in the empirical data for the growth of a cohesive unit (i.e. if measuring G)—or equivalently, when should we see a peak in the empirical data for a burst (i.e. if measuring $d/dG \,/t$)? Qualitatively this is easy to answer: when inter-species rates are small there are bumps, because entities tend to aggregate more with their own species, and vice versa. But when exactly are the transitions from the situation of $(D-1)$ bumps all the way down to 1 bump?

134 *Multi-body Physics 3.0: general fusion–fission theory for time-dependent heterogeneous systems*

We may be tempted to recall our number of roots discussion in Chap. 4 and naively identify the number of positive bumps with the number of positive roots. Unfortunately this is not the case since all but 1 of the roots is 'virtual', and as we have shown earlier the equation has to be modified after the onset and the virtual roots do not in any case show themselves. From the various gel curves we have plotted so far, we empirically know that this transition occurs when the ϵ_{ij}s are very small; much smaller than for the transitions of the number of positive roots. However, just like all of our other treatments with these implicit solutions of the gel curves, no analytic solutions are available past the instances of onset. We hence turn to numerical methods, noting that counting the number of bumps in G is equivalent to counting the number of peaks in the function $\mathrm{d}G/\mathrm{d}\tau$.

This is still too complicated for $D \geqslant 3$ due to the sheer number of degrees of freedom, so we return to the good old 2-species model. Since the absolute magnitudes of the parameters only dictate how soon gelation onsets, the problem is essentially a 2-parameter one. At $D = 2$ examining $\mathrm{d}G/\mathrm{d}\tau$ is simple: we look at the second derivative $\mathrm{d}^2G/\mathrm{d}\tau^2$ after onset and see if it ever becomes positive again. This is because the peak at onset is abrupt and singularly positive, and immediately after

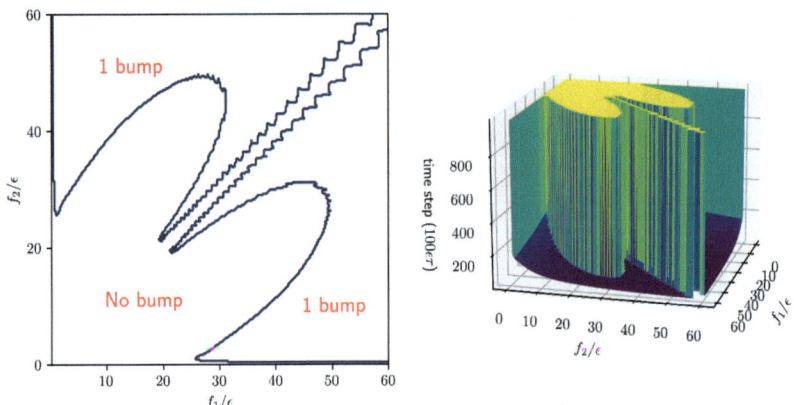

Fig. 5.3: Plots showing whether a bump is observed during a 2-species gelation process. The contour plot on the left shows the boundary of 1-bump and 2-bump zones in the 2-D parameter space. The surface plot on the right shows the time step when the bump emerges. The yellow surface means that the bump emerges at infinitely late time, and the discontinuity is exactly at the zone boundary.

Motivation: later onsets and bumps grow more slowly

it stays negative until there is a bump. We know the expression of $dG/d\tau$ exactly but that of $d^2G/d\tau^2$ is too complicated to employ, and we may as well numerically differentiate $dG/d\tau$. The result is plotted as a phase diagram-like graph in Fig. 5.3.

How can we make sense of this rather non-trivially shaped diagram qualitatively? To start with, let's list the main factors of bump emergence, and then try to associate different zones of this 'bump plot'. There are 3 main factors:

1. Relative strength of ϵ. If ϵ is large enough, as we have mentioned many times, inter-species aggregation becomes significant. So the whole system has a cohesive growth without any bumps.

2. The difference between f_1 and f_2. This factor is exactly what we were discussing in Sections 5.1.2 and 5.1.3. When ϵ is very small the system can almost be seen as 2 decoupled species. In this case the strength of f_1 and f_2 also determines the abruptness of onset growth: the larger f_α is, the earlier species α has an abrupt growth and the faster it grows. And vice versa. Depending on the exact parameter values, the slower-growing species may have its growth peak (bump) hidden in the overall growth curve—see below for a discussion.

3. The 'piling up' effect. If f_1 and f_2 are similar, then even if ϵ is very small, there is no bump due to superposition of the growth curves of both species. We have seen this effect at Section 5.1.2 when it induces a later yet faster second onset. By contrast if f_1 is much larger (or smaller) than f_2, then even for some rather large ϵ we can still see the bump.

In Fig. 5.4 we divide the bump plot in Fig. 5.3 into 6 zones, and attribute its different zones to these 3 factors. The most straightforward one is the 'piling up' effect: zone D around the diagonal is due to that, and also zone C ('anti-piled-up', growth of the 2 species too separated so a bump, however small, emerges). Factor 1 manifests itself in the main portion of the plot: it is responsible for zones A (ϵ too big) and B (ϵ too small). Zones E and F are the 'transition zones' in between, and can be explained by combining factors 2 and 3. Now we consider the $dG_{1,2}/d\tau$ peaks and how adding them up contributes to the total gel growth. For zone E, f_1 and f_2 are different but not enough to reach zone C, and neither of them is big enough to reach zone B. In other words, we are adding up a moderately abrupt peak with a smoother one slightly later in time, and so for the ranges of parameters in zone E there appears to be no bump. By contrast for zone F, f_1 and f_2 are very close in value, but not so close as to pile up entirely as in zone D. From the perspective of the $dG/d\tau$ peaks, we are adding up two sharp peaks with a small time interval between them: as long as the time interval is not too small (to reach zone D), we would see a bump. Additionally there is one

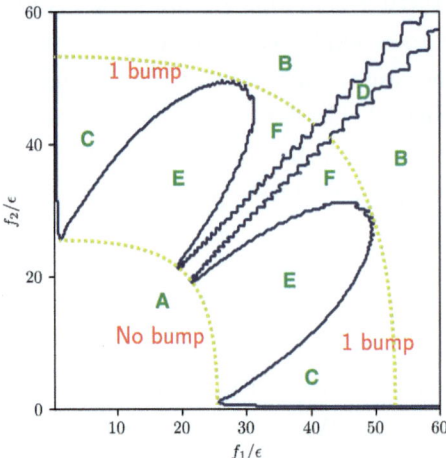

Fig. 5.4: The bump plot divided into zones in the 2-dimensional parameter space. Each zone can be qualitatively explained in terms of the factors behind bump emergence.

final trivial zone which is around the two axes: they trivially have no bump since we can almost see them as single species, with the other one effectively turned off.

5.2 General theory of bursts

These findings highlight the limitations of the mean-field theory developed so far: the gel growth is monotonic, and onsets and bumps all become less abrupt the later they occur. This behaviour of G has the consequence that $\dot{G}$ looks like a burst with successively smaller peaks, i.e. up to $D-1$ successively smoother kinks in G means that $\dot{G}$ has up to $D-1$ successively smaller peaks.

However, real-world bursts—and even those in the physical world such as Gamma Ray Bursts—show more general behaviour with re-entrant peaks (like revivals) that can reach peak heights even larger than the initial peak. Such bursts are a common feature of real-world systems and represent something substantially more than any single event. Such a burst ($\dot{G}$), like the cohesive unit itself (G), can be a surprise in terms of its arrival, and its subsequent evolution, though short-lived, shows structure that can give insight into the internal fusion–fission dynamics. This is shown explicitly in the empirical example of Fig. 2.7. Although most of our data in Chap. 2 happened to measure G directly, not its derivative $\dot{G}$, it is quite possible that future human–technology–AI systems will directly produce measurements of $\dot{G}$. Of course the two are essentially the same, since one is the differential of the other, but a difference in the number of peaks and their positions is a much easier visual test of a theory than a bump (kink) depth. So it is the bursts ($\dot{G}$) that are more likely to be used as a test of the theory going forward.

So how can our theory for $\dot{G}$ be generalized so that it can reproduce general bursts with a multi-peak structure such that later peaks can be even higher and sharper than earlier ones. Or equivalently, how can our theory for G be generalized so that it can reproduce general cohesive unit growths with bumps (kinks) of any abruptness at any stage after initial onset?

This is what we develop and demonstrate in the rest of the chapter, since this will provide us with our sought-after general theory for interacting heterogeneous human–technology–AI systems. But to achieve this, let's first take a step back from the complex multi-species kinetics and briefly look at a single species system in equilibrium, i.e. the well-known Erdős–Rényi random graph. For a complete discussion, textbooks like Newman (2018) provide a detailed introduction. Here we simply provide a short summary of the basics just to set the stage for the discussions that follow. As its name suggests, a random graph is a type of graph which is composed of vertices and edges. We already stated that a cohesive unit is equivalent to a giant connected cluster of nodes, noting that we do not keep track of the detailed links at the mesoscale (recall Figs 1.4–1.5). But we will now establish this deeper connection between graph theory and our mean-field kinetic theory. From Newman, a random graph is defined to be a graph with some of its properties fixed while others are random. The Erdős–Rényi random graph is one of the simplest models: it assumes a fixed set of n vertices (nodes), and its edges are formed randomly between pairs of nodes with probability p. The random graph is denoted as $G(n, p)$. With n fixed, the other fundamental quantity, the number of edges m, is given by the expression

$$\langle m \rangle = p \binom{n}{2}, \tag{5.41}$$

i.e. edge probability times all possible pairs gives the average number of edges in an Erdős–Rényi random graph. With this we can find the next essential network property, the **degree**, as well as the **mean degree**, of a node. The degree of a node is defined as the number of edges on a node, and the mean degree is naturally its average across all nodes within the network. For a network with n nodes and m edges, since an edge connects 2 nodes, the mean degree is simply $2m/n$. This with eqn (5.41) gives the mean degree c of an Erdős–Rényi random graph:

$$c = \left\langle \frac{2m}{n} \right\rangle = \frac{2p}{n} \binom{n}{2} = p(n-1). \tag{5.42}$$

The result makes perfect sense: in an Erdős–Rényi random graph, a node may be connected to all $(n-1)$ other nodes, each with a probability p. c is a key quantity in the random graph, and we shall see that a critical value of c denotes a phase transition.

Our kinetic model is far from being in equilibrium, but gelation can be seen as a dynamical phase transition process as we now explain. Think of a random graph in equilibrium. There are no new links joining nodes, nor are there old links getting deleted. We do not consider any dynamic process on a random graph, but simply control the network by tuning the parameter c, which determines connectivity within the graph. The level of connectivity defines 2 different phases: when c is small, links are sparse and nodes are largely disconnected, whereas when c becomes large enough, a non-negligible fraction of the nodes becomes connected and form what we call the *Giant Connected Component* (GCC) as shown explicitly in the simulation in Fig. 1.5. With this qualitative picture, we now study the mathematics of GCC formation. We assume that there is indeed a GCC in the random graph under mean degree c. We also assume that the probability that a node is in the GCC is G, and that its opposite is $u = 1 - G$. For a random graph, it is straightforward to see that G is simply the size of the GCC (relative to the total number of nodes). Our goal is to formulate an expression for G in terms of c, but it is easier to make sense of it by going bottom-up. To begin, what does it mean if a node is within the GCC? Naturally it means that the node is connected to at least one other node in the GCC. This does not help in terms of the expression. However, looking at this in reverse, for a node *not* inside of the GCC, it means that, for every other node in the graph ($(n-1)$ of them),

1. either they are not connected,
2. or they are connected but neither part of the GCC.

We now express this mathematically. For point 1, it is simple $(1-p)$, where p is the defining parameter of edge probability, and for point 2 it can be represented by pu. Altogether, it means that the probability for a node not in the GCC is $(1-p+pu)^{n-1}$, where the power of $(n-1)$ is counting all the other nodes. But by definition, this is exactly u, i.e.

$$u = [1 - p(1-u)]^{n-1}. \tag{5.43}$$

Substituting the mean degree c as the parameter with $c = p(n-1)$, and substituting in $G = 1 - u$, we have

$$1 - G = \left[1 - \frac{Gc}{n-1}\right]^{n-1}. \tag{5.44}$$

In the limit of $n \to \infty$, we can see that the RHS is by definition the exponent e^{-Gc}, and hence

$$G = 1 - e^{-Gc}. \tag{5.45}$$

When n is large but not infinite, this relation is approximate.

Now in eqn (5.45), when we take $c = F\tau$, it is exactly equivalent to eqn (5.2) at $D = 1$. This makes physical sense, since the action of randomly picking nodes to connect means that clusters (i.e. connected components) get connected to each other at a rate proportional to their sizes, which was the logic behind the product kernel of aggregation. Equation (5.45) has a second-order phase transition at $c = 1$. For $c < 1$ $G = 0$ (i.e. there is no GCC). For $c > 1$ GCC emerges. This is the physical basis of the terminology 'dynamical phase transition'. We can therefore see our system as the mathematically equivalent model of a phase transition, with the parameter c now realized dynamically by $F\tau$. Now we can really appreciate why our aggregation-based theory works: it is another way of formulating the network problem, which is the essence behind most of the problems that we study. But it has the additional power, beyond the network approach above, of easily incorporating dynamics and far-from-equilibrium scenarios.

5.2.1 1D aggregation as a uniform traversal of parameter space of random graph

We have identified $F\tau = c$ as providing a connection between our kinetic fusion theory and the Erdős–Rényi random graph. Since F is just a rate parameter that scales inversely with time τ, we can regard the gelation process simply as traversing through the whole parameter space of the mean degree c in a random graph, at a constant speed F, starting from $c = 0$ at $\tau = 0$. At $\tau = 1/F$, we reach the critical point $c = 1$ and hence our dynamical phase transition begins. As a timescale, it controls the shape of our shock curve: larger F contracts the curve so that its growth is earlier and more rapid, while smaller values extend it out and flatten it (see Fig. 5.5). Previously we showed this by analysing the $G(\tau)$ function, but here we see another way of looking at it: we can see the parameter space, and thus the fusion curve with respect to the mean-degree parameter is fixed. It is the rate at which the system traverses the space that finalizes the shape of gel dynamics. For this reason, the gelation onset time and growth rate are closely related. The sooner a system gels, the faster it grows.

5.2.2 Non-uniform traversing speed

We have shown that for this uniform traversal case, the underlying dynamics is actually the same for all the different F values—just scaled. But nothing stops us from traversing the parameter space c at any speed we like, even non-uniform ones, because ultimately we are just trying to establish a mapping between time τ and mean degree c, and we have made no assumptions about the nature of this mapping. Imagine walking through a standard curve $G = 1 - e^{Gc}$, i.e. the left plot in Fig. 5.5, but now you can walk at arbitrary speed, or even go backwards, thus deforming the curve arbitrarily. The only

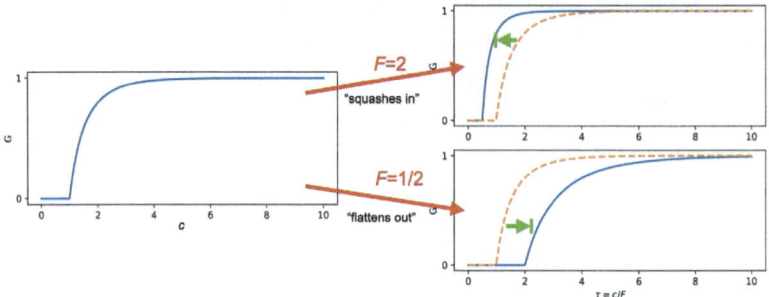

Fig. 5.5: Schematics showing how the constant F scales the phase diagram, producing gelation curves.

requirement is that whatever comes out cannot violate the kinetics, i.e. it must obey the master equations. Recall in eqn (3.35) that the master equation actually admits time-dependent mean-field rates $F(\tau)$. If we can solve that, we should be able to map the result again to the parameter space.

There is no need to go all the way back to the master equation since its transformation into the generative equation does not tamper with τ space. So we start with the generative PDE, eqn (3.62), with variable $F(\tau)$:

$$\frac{\partial \varepsilon}{\partial \tau} + F(\tau)(\varepsilon - 1)\frac{\partial \varepsilon}{\partial x} = 0.$$

We stress again that this is equivalent to the master equations of $\{n_k\}$s, and thus a faithful dynamic description of the kinetic process. This time we try to solve it for general time-dependent rates $F(\tau)$. We recall that the geometric interpretation of the method of characteristics is to utilize the vector $\left(1, F(\tau)(\varepsilon(x,\tau) - 1)\right)$ being orthogonal to the gradient $\nabla \varepsilon$ at any point (x, τ). This is true regardless of the τ dependence of $F(\tau)$, since the vector is fix-valued at any point (x, τ) whatsoever. Hence, the characteristics method is perfectly valid here: we find a curve on which $\varepsilon(x, \tau) = c$ is constant, and the vector $\left(1, F(\tau)(\varepsilon(x,\tau) - 1)\right)$ must be tangential to the curve:

$$\frac{dx}{d\tau} = F(\tau)\Big(\varepsilon(x,\tau) - 1\Big), \qquad (5.46)$$

at any point (x, τ) on the curve. It then follows that we can integrate it along the curve as usual, and since $\varepsilon(x, \tau) = A$ is constant along the curve, we have

$$x = (A - 1)\int_A F(\tau)d\tau. \qquad (5.47)$$

Then, define the antiderivative of $F(\tau)$ as

$$\frac{d\Phi(\tau)}{d\tau} = F(\tau), \tag{5.48}$$

and the solution can be simplified into

$$x = (\varepsilon - 1)\Phi(\tau) + C(\varepsilon), \tag{5.49}$$

where $C(\varepsilon)$ is a function depending only on ε, to be determined from the initial condition of the system. Taking the usual all-monomer initial condition this then gives

$$(\varepsilon - 1)\Phi(\tau) - \ln \varepsilon = x. \tag{5.50}$$

Finally, the gel solution that this yields takes the following form,

$$G = 1 - e^{-G\Phi(\tau)}. \tag{5.51}$$

This looks exactly like eqn (5.45) with $c = \Phi(\tau) \equiv \int_0^\tau F(\tau')\,d\tau'$, and we have proven that there is no need to keep F at a constant rate. Moreover, we can see that the dynamic function $\Phi(\tau)$ is just the path on the parameter space traversed by the system as it advances to the phase transition, i.e. it is the set of all the c values, being mapped to by different time instances. It contains all the history of the 'dynamic profile' of the system, i.e. at what time the particles are more prone to assemble, and at what time they are not.

The practical possibilities for $\Phi(\tau) \geqslant 0$ are endless. When it starts from the origin at $\tau = 0$, it represents the system building up from scratch with all monomers. When it starts from $\Phi(0) > 0$, it simply means a more clustered initial condition, indicating that we are closer to the phase change. It can stay at 0 until some time τ_s later, when aggregation is turned on; or it can grow with time until a point when it starts decaying, meaning that *binary disassembly* takes place instead of assembly. It probably cannot be discontinuous, since that would leave a gap in the system's history of evolution—but not necessarily so, as it may be able to represent some interactions that happen in a larger timescale. For example, a series of sparse, rare but explosive complete fission events, as we have discussed in Section 3.7.2, can potentially be approximated by discontinuous jumps in $\Phi(\tau)$ from above 1 to almost 0. All in all, this leads to a very powerful description of all kinds of plausible assembly pathways. Some examples are plotted out in Fig. 5.6. The key problem is then not *whether* there exists a descriptive theory for a real system, but *which* is an accurate one. A toy example is shown in Fig. 5.12(a), where we demonstrate that a step-function-shaped $F(\tau)$ is a faithful representation of a homogeneous fusion–fission system, as we rightfully assume.

5.3 *D*-species fusion–fission and multi-layer random graphs

So far this discussion has focused on a single species. But 1-species fusion–fission is too simple from the perspective of parameter space: it does not get more complex than

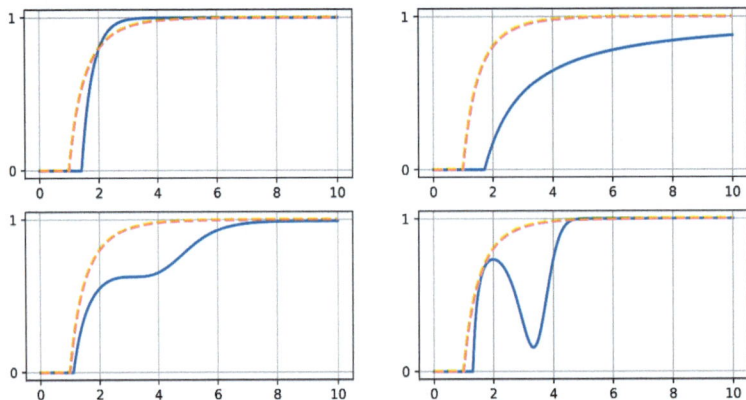

Fig. 5.6: Examples of gel curves corresponding to some general $\Phi(\tau)$ functions, with the orange dashed line having $\Phi(\tau) = \tau$, i.e. traversing at constant speed 1. (a) Upper left: $\Phi(\tau) = \tau^2/2$, $F(\tau) = \tau$; (b) upper right: $\Phi(\tau) = \ln \tau$, $F(\tau) = \tau^{-1}$; (c) lower left: $\Phi(\tau) = (\tau + \sin \tau)/2$, $F(\tau) = (1 + \cos \tau)/2$; (d) lower right: $\Phi(\tau) = (3/5)H(\tau - 1)[(\tau - 1)^3 - 5(\tau - 1)^2 + 7(\tau - 1)]$, $F(\tau) = (9\tau^2 - 48\tau)/5 + 12$ for $\tau > 1$, where $H(\tau - 1)$ is the Heaviside step function.

going back and forth on a 1-D straight line (a ray, technically) that is the range of $\Phi(\tau)$. It does, however, give us some insights about what to do in the general case of D species. Since there are $D(D+1)/2$ free parameters for D species, a good guess is that it wouldn't be too far from traversing a $D(D+1)/2$-dimensional parameter space. But in order to build an intuition, we once again start from the simplest higher-dimensional system: the redundant 2-dimensional system of 2 decoupled species with constant rate tensor $\mathsf{F} = \mathrm{diag}(f_1, f_2)$.

The advantage of a decoupled 2-species model is that we can regard them both effectively as 2 simple 1-D fusion–fission models, and conceptually as one holistic 2-dimensional system. We have already seen the equations of gel dynamics in eqn (5.22). As in the 1-D case, we first consider the 2 species separately and define $c_1 = f_1 \tau$ and $c_2 = f_2 \tau$. Then this pair of independent 1-D systems should admit all kinds of traversing processes $\Phi_\alpha(\tau) = c_\alpha$, $\alpha = 1, 2$, as we discussed previously. For its network counterpart, it correspondingly represents 2 disjoint simple (i.e. single-layer) random graphs each with mean degree c_α. The two networks are otherwise disconnected, and related only via a common time τ. Now we return to the 2-D perspective and plot $G(c_1, c_2)$ against a 2-D plane of orthogonal $c_{1,2}$. What we obtain should be the phase diagram for the entire family of decoupled 2-species systems (see Fig. 5.7).

Now this is much more interesting. As we did with $D = 1$ species, we can now traverse arbitrarily in a 2-dimensional parameter space and each run generates a specific

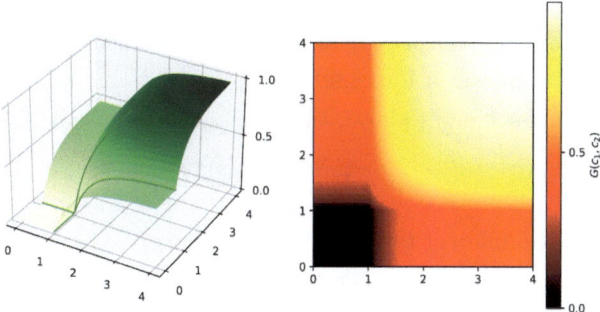

Fig. 5.7: Plots of $G(c_1, c_2)$ for a decoupled 2-species system. We can see from both plots that lines $c_1 = 1$ and $c_2 = 1$ are two critical values of phase transition.

assembly path. In Fig. 5.8 we show a simple straight-line path and compare it with its corresponding gel curve, with which we are more familiar.

At this point, the extension to $\epsilon \neq 0$ seems natural—it simply adds in one more dimension in the parameter space to traverse. Bear in mind that now the defining equations are

$$G_1 = \frac{1}{2}\left(1 - e^{-G_1 c_{11}} e^{-G_2 c_{12}}\right), \quad G_2 = \frac{1}{2}\left(1 - e^{-G_1 c_{12}} e^{-G_2 c_{22}}\right), \quad (5.52)$$

$$G \equiv G_1 + G_2 = 1 - \frac{1}{2}\left(e^{-G_1 c_{11}} e^{-G_2 c_{12}} + e^{-G_1 c_{12}} e^{-G_2 c_{22}}\right). \quad (5.53)$$

The calculation is considerably more complicated, but the idea is still the same: we now traverse a 3-dimensional parameter space corresponding to the mean degrees of a 2-layered *multilayer network*. This simply means that instead of 2 independent vanilla

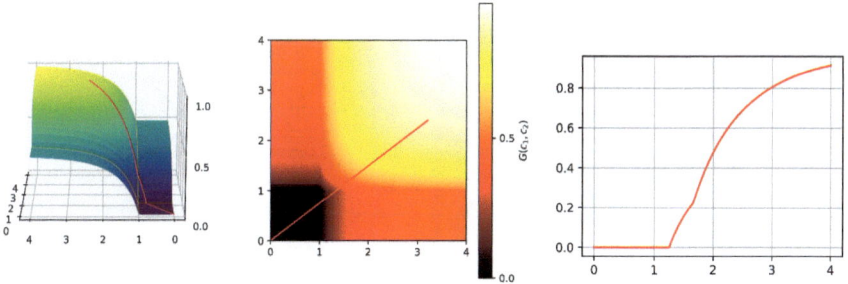

Fig. 5.8: A time-dependent path across the 2-dimensional parameter space, generates a gelation process. Here $f_1 = 0.8$, $f_2 = 0.6$. Generally for any decoupled 2-species fusion–fission, we are free to traverse an arbitrary path across the 2-dimensional parameter space shown in the centre panel, and each traversal corresponds to a distinct pattern of fusion–fission.

Erdős–Rényi random graphs as in the decoupled case, the 2 graphs now become 2 layers of nodes within a larger network, where links can be formed inside both layers as well as across layers, with each type of link being governed by a separate mean degree $c_{\alpha\beta}$. An example is shown in Fig. 5.9.

As for the simpler models above, the heterogeneous fusion (fission) rates are now

$$F_{\alpha\beta} = \frac{\mathrm{d}\Phi_{\alpha\beta}(\tau)}{\mathrm{d}\tau} \tag{5.54}$$

for each of the three different types of fusion (fission). We have been writing the terms 'fusion' and 'fission' side by side, because this construction allows for the possibilities that some of the matrix elements $F_{\alpha\beta} < 0$. From this parameter space construction, we see that since $\Phi_{\alpha\beta}(\tau) = c_{\alpha\beta} \geq 0$, the binary fusion (or fission) rate $F_{\alpha\beta}\, \mathrm{d}\Phi_{\alpha\beta}/\mathrm{d}\tau$ cannot always be negative. However, when it does become negative, it simply means that different species tend to break away from one another at that moment, and we hence go through a period of binary fission for the specific interaction between species α and β.

Hence at the end of this section, we can now write down the even more general form of the generalized Burgers' equation

$$\frac{\partial \varepsilon_\alpha}{\partial \tau} + F_{\beta\gamma}(\tau)(\varepsilon_\beta - \iota_\beta)\partial_\gamma \varepsilon_\alpha = 0, \tag{5.55}$$

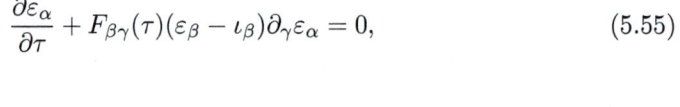

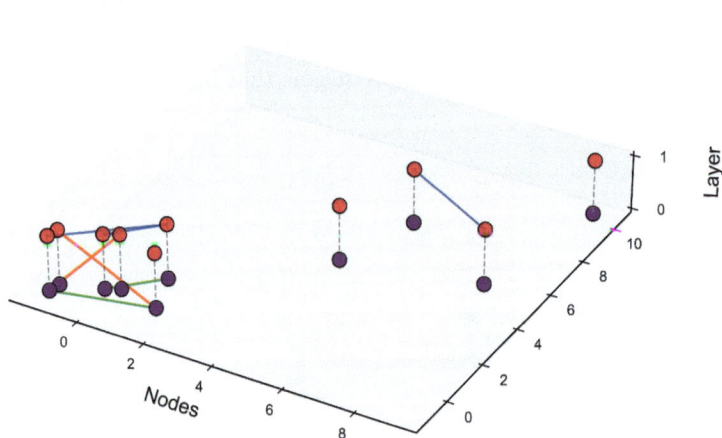

Fig. 5.9: A 2-layered multilayer network which serves as the network topology of the 2-species kinetics.

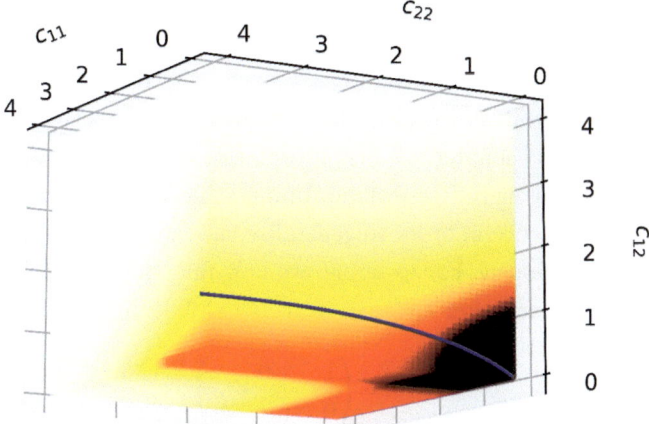

Fig. 5.10: The 2-species parameter space, with the heatmap representing the G value. Note that the degree of freedom here is 3 not 2, due to the existent coupling between species. Blue curve indicates a possible traversing curve, with $\Phi(\tau) \equiv (\Phi_{11}(\tau), \Phi_{22}(\tau), \Phi_{12}(\tau)) = (4\tau/5, 3\tau/5, \ln(\tau+1))$.

where $F_{\alpha\beta}(\tau) \equiv \dot{\Phi}_{\beta\gamma}(\tau)$ are the aggregation rates between species β and γ, which can be negative (i.e. fission). For D species $\alpha = 1, \ldots, D$, and there exist $D(D+1)/2$ independent $F_{\beta\gamma}(\tau)$ functions which traverse a $D(D+1)/2$-dimensional parameter space (a 2-D example is shown in Fig. 5.10), and the corresponding graph topology is a D-layered multilayer network. Their exact solutions are a direct generalization of eqns (5.1) and (5.2):

$$G_\alpha(\tau) = \left(1 - e^{-\sum_\beta \Phi_{\alpha\beta}(\tau) G_\beta(\tau)}\right)/D \tag{5.56}$$

$$G(\tau) = 1 - \frac{1}{D} \sum_{\alpha=1}^{D} e^{-\sum_\beta \Phi_{\alpha\beta}(\tau) G_\beta(\tau)}. \tag{5.57}$$

5.3.1 Asynchronous start of aggregation

As we have seen, the only constraint on the value of $F_{\beta\gamma}(\tau)$ is that

$$\Phi_{\beta\gamma}(\tau) \equiv \int_0^\tau F_{\beta\gamma}(\tau')\,d\tau' \geq 0 \tag{5.58}$$

hence the theoretically plausible $F_{\beta\gamma}(\tau)$ functions are endless. The simplest generalization that comes to mind concerns the simultaneity of fusion for all species. If we review our model in Chap. 4 using the newly defined parameter-space set of quantities $\Phi_{\alpha\beta}\tau$, we quickly notice that it simply looks like

$$\Phi(\tau) = \mathsf{F}\tau. \tag{5.59}$$

In other words, the fusion pathways on each of the $D(D+1)/2$ types of interactions begin simultaneously, and are strictly linear throughout the fusion process. Reflected on the $D(D+1)/2$-dimensional parameter space, the fusion pathway is strictly a straight line starting from the origin. This intuitively explains the question we raised at the beginning of this chapter: all fusion processes begin simultaneously and retain their fusion rates. So generally speaking, bumps that emerge later necessarily have slower growth rates. Hence in order to generalize this, we can simply turn on different interactions asynchronously. For example, we now find that Fig. 5.2 can be obtained from the extended theory after all by choosing the following $\mathsf{F}(\tau)$ tensor:

$$\mathsf{F}(\tau) = \begin{pmatrix} 1 & 0 \\ 0 & H(\tau - 2) \end{pmatrix}, \qquad (5.60)$$

where $H(x)$ is the Heaviside step function which is discontinuous at $x = 0$. Correspondingly the dynamics on the parameter space become

$$\Phi(\tau) = \begin{pmatrix} \tau & 0 \\ 0 & (\tau - 2)H(\tau - 2) \end{pmatrix}. \qquad (5.61)$$

Its corresponding path in the parameter space is shown in Fig. 5.11. This greatly expands the permitted routes leading to gelation. We can have new (coupled) species

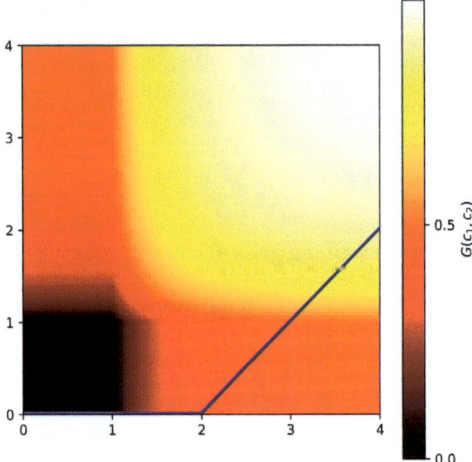

Fig. 5.11: The gel curve shown on a contour plot of the parameter space with the time-dependent F matrix given by eqn (5.60).

joining the system, and also old species leaving. We previously had an argument based on causality. Now we have expanded the causal possibility beyond a naive simultaneity. This even allows for a dynamical population effect through dynamic $F_{\beta\gamma}(\tau)$s. We also recall that we discussed how to translate an arbitrary species-size problem into a multi-species problem in Section 4.4.5. There we explained that a species can be used to simply represent some pocket of a population that appears, and that heterogeneity is actually an abstract concept. In the same spirit, we can introduce or exclude particular pockets of a population later in time by switching on or off their respective species. In theory, we could even divide up a population of size N into N species so as to simulate a continuous change in a population (instead of changing by pockets), but computationally it would be too hard and so we always opt for coarse-graining the system. We say more on this in the next section.

5.3.2 Species as a dynamic notion

We have already seen in Section 4.4.4 the possibility of defining arbitrary species as nothing more than a series of abstract collections. Then in Section 4.4.5 we mentioned a special mathematical treatment of unequal species sizes. There we saw that one species can be recast into an arbitrary number of equivalent ones, among which all $F_{\beta\gamma}$ values are identical. Together with dynamic $F_{\beta\gamma}(\tau)$s, this means that species are no more than a dynamic, or temporary, notion. For a heterogeneous system where each of the N particles is different, a more precise perspective of the system is to regard it as having N fundamental or 'atomic' species. Conceptually we can imagine that, at some time instant τ_0, d particles may decide to act together, thereby forming a 'union species' μ. Within these d particles, all the aggregation rates are identically $F_{\mu\mu}(\tau_0)/d$ (no sum). In terms of the parameter space, this synergy period would correspond to the system traversing in the *diagonal direction* of the d-dimensional subspace. Then when some particles decide to leave this species, the diagonal traversal is contracted into the diagonal of its lower dimensional subspace.

We note that this is by no means contradicting our earlier coarse-grain picture. In fact, it is broader in that it admits the coarse-grained perspective (species formation) but puts it in a dynamic setting, allowing the previously fixed space of species to morph and change. Moreover, in real calculations taking N fundamental species would be unrealistic anyway, and there do indeed exist highly similar (or more importantly, similarly *acting*) particles for us to coarse-grain in the first place in order to simplify computations. Consequently, we are more likely to deal with the dynamic merging and splitting of just a handful of a priori coarse-grained species in real human–technology–AI systems and across the real world.

This direction of generalization opens up to some very interesting and relevant interpretations. Imagine a system of heterogeneous humans, technology, and AI entities (Fig. 1.2) with a fundamental underlying motif of becoming coherent (fusion) or incoherent (fission). These are the basic, microscopic interactions that all entities have, regardless of any higher-level architectures. It is like a fundamental driving force. Over time, what happens at a larger scale is that the individuals are *collectively* deciding on the rates and intensities of these interactions in a group (cluster) setting. These collective behaviours will depend on a wide range of factors: individual heterogeneity and its evolution over time, as well as external factors. All of these factors shape the dynamic $\mathsf{F}(\tau)$ and in turn determine how the same interactions can give rise to very different emergent phenomena. In our theory, we never need to touch the fusion–fission construction in the master equations: instead, different forms of $\mathsf{F}(\tau)$ can produce different dynamics of the clusters and hence different behaviours for the cohesive unit(s) that may emerge.

This generalized mesoscopic theory will therefore be crucial for understanding the observed behaviour of any human–technology–AI system, whether that is through the empirical observation of G, $\dot{G}$, or some other observable function of these. Visually, it will produce up to $D-1$ kinks in G that are not necessarily successively smoother, and hence $\dot{G}$ has up to $D-1$ peaks which can be of general height. Hence this can form the basis of a theory of the profile of bursts which are a ubiquitous feature of activity involving humans, and hence will be ubiquitous in current and future human–technology–AI systems.

5.4 The x-space

Finally we note one last direction for possible generalization. We have seen the variable x, and its higher-dimensional generalization $\boldsymbol{x}$, throughout Chaps 3–5. Mathematically it defines the reciprocal space of cluster sizes, but operationally it appears so far as redundant, apart from its $x \to 0$ value. However, the parameter-space construction that we just studied provides another possibility. If there exists a phase-diagram-like parameter space for $x = 0$, there must be one for each $x \neq 0$. But how should we interpret them?

We go back to the definition of the generating functions, and for simplicity here we take $D = 1$ *wlog*:

$$\varepsilon(x, \tau) = \sum_{k=1}^{\infty} k n_k e^{-kx}. \tag{5.62}$$

Just like the time variable τ which scales inversely with rate F, the variable x scales inversely with k. This makes large values of x meaningless, because all terms with large

x converge to 0. However, when $x \sim 1/k$ it gets interesting. Let's take $x = 1/k_0$ for a large positive integer k_0, and then look into all the terms of eqn (5.62)

$$a_k(x) = kn_k e^{-kx}. \tag{5.63}$$

For $k \ll k_0$, $a_k(k_0^{-1}) \approx kn_k$, whereas for $k \gg k_0$, $a_k(k_0^{-1}) \approx 0$. Therefore we have roughly

$$\varepsilon(k_0^{-1}, \tau) \sim \sum_{k=1}^{k_0} kn_k \tag{5.64}$$

where we use the term 'roughly' because we have neglected terms close to k_0. Nevertheless, this still behaves like a finite cutoff of our otherwise infinite system, and the mathematics is baked deeply in the theory.

In other words, our intuition is that by picking a small but positive x value, we will be able to fit the gel curves for a finite population—which of course is the case of *all* real-world systems including all human–technology–AI systems—more closely

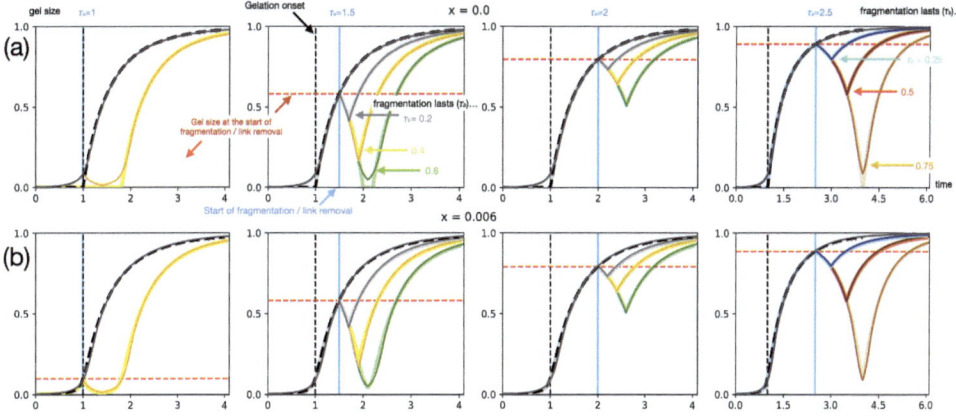

Fig. 5.12: Gelation curves with link removal periods, for different starting points τ_a (shown by blue vertical lines) and durations of removal τ_b (i.e. the decreasing period). For the theory we take $F(\tau) = 1$ for $0 \leqslant \tau < \tau_a$ and $\tau \geqslant (\tau_a + \tau_b)$, and $F(\tau) = -1$ for $\tau_a \leqslant \tau < (\tau_a + \tau_b)$. Simulation results are in dark, thin dotted-curves while analytic results from our theory are shown in light, thick smooth curves. As the gelation onset time is fixed at $\tau = 1$ (black vertical line), the plots show the effects of introducing a fragmentation/link removal stage during gelation, in contrast with the no-fragmentation (black dashed curves) scenario with which we are familiar. From left to right, fragmentation begins later as τ_a increases as shown in the blue vertical lines. Curves in different colours indicate different durations of fragmentation periods, as noted in the graph. Plots (a) and (b) are fitted with different x values: in (a) $x = 0$ while in (b) $x = 0.006$. The former ($x = 0$) gives the formally correct sizes of the gel and accurately fits the simulations at late time, whereas the latter ($x = 0.006$) fits the smooth onset in finite systems better.

than with our theoretical, infinite formal theory. Since all our simulations are executed with finite populations, this straightforwardly means that we should be able to fit the simulations better by picking a small x value.

To see if this actually happens, Fig. 5.12(b) compares the analytical results using a non-zero x value against the simulations. The simulations were performed with $N = 1000$ particles while $k_0 = x^{-1}$ was taken to be $1000/6 \approx 167$, which is a fraction of the total population. We see that the initial agreement between the theory and simulation is much better than for the ideal, infinite scenario shown in Fig. 5.12(a), though at late times the theoretical growths lag behind for all chosen parameters.

This leads to the following interpretation: x, or rather x^{-1}, behaves like an *activation* parameter which seems to govern the proportion of entities in the system that actively take part in forming the cohesive unit (i.e. gelation). Since it gauges the size of the system, a good fit for an empirical human–technology–AI system may involve a time-dependent x, most probably decreasing over time all the way to 0. We stress that this interpretation is not a proven fact, but more observational in nature. The immediate question is why the 'finite cutoff' is in exponential form, rather than a more precise step function or a continuous alternative like the logistic function. While it does seem that e^{-kx} is a poorer cutoff function than the ones above, it at least gives rise to nice fluid-dynamical mathematics, which is neatly analytic, while none of the others yields that. Additionally, as shown in Fig. 5.12, it does work phenomenologically, hence at the very least, this would be a good working approximation for finite systems. Intriguingly, it also takes a similar shape to the probability factors in statistical ensembles where the statistical parameter β, i.e. the inverse temperature, plays the same role as our 'inverse size' x. This leads us to wonder whether the exponential factors can be proven from a probability point of view. A strict proof appears difficult since in statistical ensembles, the exponentials can be obtained from not-too-complicated combinatorics. Here, however, if we look at the corresponding probabilistic network structures for the mean-field cluster, we see that the difficulty lies in the sheer number of network configurations which heavily exceed a simple exponential function. Hence the microscopic justification of this 'magical' factor of e^{-kx} is still an open question.

5.5 Summary

This chapter provides the key generalized theory that we sought, so it is worth providing a special summary to put it in context of what came before. In Chap. 3 we first introduced our simple mean-field theory that describes fusion and fission behaviours in heterogeneous systems. Two chapters later, we have generalized the dimensionality of the heterogeneity and extended our theory in both the temporal space of τ and the inverse-size space of $\boldsymbol{x}$. The former (τ space) is now identified as the parameter (phase) space, whereas the latter can be roughly seen as the inverse of 'active population' space. Both of these generalizations bring in new non-trivial responses in the emergence of the cohesive unit(s) and hence macroscopic surprise such as bumps of general size in G and equivalently peaks of general height in $\dot{G}$, as seen in the empirical systems of Figs 2.1 and 2.7. With these advances, we are better equipped to deal with specific real-world human–technology–AI system behaviours. For example, when we are able to clearly identify a very few (say 2–4) species, we can study the system with the full D-species theory, perhaps using the trick mentioned in Section 5.3.2. The rates $F_{\alpha\beta}(\tau) \equiv d\Phi_{\alpha\beta}/d\tau$ can be taken to be constant or slowly varying to simplify the problem, and its time dependence can also be used to describe late-joining species, etc. When there are too many species to consider exactly, we can always renormalize the system into fewer *effective* species for convenience. We also found that when studying early-time behaviours for finite systems, it is at least empirically favourable to take slightly greater-than-zero values of x.

Picking the right theory now becomes the crucial task. A suitable theory for a particular human–technology–AI system should reflect a balance between computational efficiency, accurate phenomenology, and physical interpretability. For example, the fits in Fig. 2.1 are for the simplest form $D = 2$ with no fission, and yet work surprisingly well. Extending this to $D = 3$ etc. would improve the fits even further.

Exercises

5.1 Convince yourself that eqn (5.7) is equivalent to eqn (5.6) in component form.

5.2 For a single-species aggregation system with rate f,

(a) write down its second-moment function $m_2(\tau)$ as a function of time and gel size $G(\tau)$.

(b) Numerically plot $m_2^{-1}(\tau)$ against time.

(c) In Section 4.6 we described an exponential approximation to the gel curve $\hat{G}(\tau) = 1 - e^{r_s(1/f - \tau)}$ for $\tau \geqslant 1/f$. Find a suitable r_s value that accurately approximates the onset behaviours.

(d) Now use the approximate $\hat{G}(\tau)$ from part (c) to plot by hand an approximate $m_2^{-1}(\tau)$. Compare it with the exact shape.

5.3 For a 2-species aggregation-only system with identical population sizes,

(a) write down explicitly the determinant of the second-moment tensor as a function of time, in terms of $F_{\alpha\beta}$ and G_α.

(b) Numerically plot it against time for $f_1 = 1, f_2 = 2/3, \epsilon = 0.1$ and observe its trend. Does it ever drop below 0, and how many zeros does it admit?

152 Exercises

5.4 In Fig. 5.2 we showed a gel curve that is impossible from our original theory. As we have seen later in the chapter however, we can obtain it through the extensive theory after all. Describe its corresponding fusion system.

5.5 This question contains some further properties of the Erdős–Rényi random graph and aims for a better understanding of this classic model.

  (a) Prove that for an Erdős–Rényi random graph $G(n,p)$, its degree distribution follows a binomial distribution, i.e. the probability of finding a node connecting to k others is given by

  $$p_k = \binom{n-1}{k} p^k (1-p)^{n-1-k}. \qquad (5.65)$$

  (b) For very large $n \to \infty$, use Stirling's approximation to express p_k in terms of the mean degree c, as defined in 5.42. As you may recall, this is exactly the property of the Poisson distribution which is the large-number limit of the binomial distribution.

5.6 In this question we look into time-dependent fusion rates $F(\tau)$. In a single-species system, its population starts off aggregating at a constant rate $F(\tau) = 2/3$ for $0 \leqslant \tau < 3$. However, from $\tau = 3$ onwards, a periodic oscillation $F(\tau) = 2/3 \cos(2\pi\tau)$ emerges reflecting a collective fluctuation in the buildup of fusion.

  (a) Write down the implicit expression of the gel size.
  (b) Find the path $c = \Phi(\tau)$ that the system traverses in the parameter space.
  (c) With the help of the standard gelation plot $G(c)$, plot by hand the gelation curve of this particular system.

5.7 In this question we look at a fusion system whose heterogeneity is dynamical at the species level. We begin with a homophily system. At $\tau = 3$, however, half of the population suddenly obeys heterophily. From then on, there is little interaction left between these two halves of the population.

  (a) Write down the fusion rates of this system throughout the fusion process.
  (b) Treating the system as effectively 2-species, write down the time-dependent $\mathsf{F}(\tau)$ matrix.
  (c) Draw the path in a 2-species parameter space that the system traverses. Beware of the dimensionality.
  (d) With the help of the standard gelation plot $G(c)$, plot by hand the gelation curve. Explain qualitatively the growth in each period of time.

6
Online wars: bad-actor AI and beyond

This chapter marks a transition, in that it focuses on applications of the fusion–fission ideas to the real world of nascent human–technology–AI systems. Specifically, we consider the looming threat of bad actors using AI to generate harms across social media globally. Given that this topic is of broad general interest, including to policymakers, we have written this chapter so that it can be read alone without requiring technical details from the preceding chapters. For this reason, there are necessarily some brief verbal summaries of earlier discussions.

Guided by our lab's detailed empirical mapping of the online multi-platform battlefield [1] this chapter offer answers to the key questions of what bad-actor–AI activity will likely dominate, where, when—and what might be done to control it at scale. To test the usefulness of our results in this chapter for predicting bad-actor AI use, in mid-2023 we posted a prediction made by our results in a publicly available preprint —which then appeared in print in early 2024 (Johnson *et al.*, 2024), well ahead of the U.S. and other elections. In it, we applied a dynamical Red Queen analysis that we explain in this chapter, based on our prior studies of cyber and automated algorithm attacks. It yielded a mathematical prediction of an escalation to daily bad-actor–AI activity by mid 2024, just ahead of U.S. and other global elections. We note that this prediction ended up happening, according to widespread security and media reports. Though obviously just a single example, this gives us some confidence in the future real-world applications of our results.

We end the chapter with an exactly solvable model of the observed bad-actor (and future bad-actor AI) community clustering dynamics, using the ideas from Chap. 3. This model and its results, which originated in an earlier collaboration with Alex Dixon to whom we are very grateful, yields a Policy Matrix which quantifies the outcomes and trade-offs between two potentially desirable outcomes: containment of future bad-actor–AI activity versus its complete removal. We also give explicit plug-and-play formulae for associated risk measures.

[1] See https://donlab.columbian.gwu.edu.

Hence in contrast to many current discussions of AI threats that are based on verbal arguments, our conclusions in this chapter are built from our lab's uniquely detailed mapping of the current online bad-actor battlefield, combined with a rigorous first-principles mathematical description of its empirical behaviours. Much of the material in this chapter is adapted from Chap. 3 (Johnson et al., 2024) to which we refer for fuller details about the figures and prior references.

6.1 Background

Even prior to the introduction of the newest Generative Pre-trained Transformer tools (e.g. GPT-4, ChatGPT), projections suggested that by 2026, 90 per cent of online material would be produced by Artificial Intelligence (AI) (2022a). As elections approach in numerous countries over the coming years (Milmo and Hern, 2023; Hsu and Myers, 2023), and as evidence mounts that real-world violence is increasingly connected to toxic digital content (2023a,), the threat of malicious misuse (in whatever form that takes (Hoffmann and Frase, 2023)) looms larger. Furthermore, ongoing wars such as those between Israel and Hamas and Russia and Ukraine have heightened the visibility and activity of malicious actors online, including the spread of mis/disinformation and orchestrated campaigns.

The European Union currently leads efforts on regulation through measures like the 'Digital Services Act' and the 'AI Act' (2023d, 2021), obligating so-called 'Very Large Online Platforms' (e.g. Facebook) to conduct risk assessments regarding harmful behaviours on their services (2023b). At first glance, this focus on large platforms seems logical: these platforms hold massive user bases, and the assumption is that harmful outliers exist on some imagined 'fringe' (Benninger, 2022; Rodrigo and Klar, 2021; Hsu, 2022; Dewey, 2022)(2019, 2022b). However, our mapping of the current bad-actor landscape online shows this assumption is wrong. To go beyond such guesswork, effective policies to combat bad-actor–AI activities demand a nuanced, data-driven understanding of the digital battlefield—which is what our analysis supplies.

Research conducted by collaborative teams from Meta and academic institutions have showed that even without AI, the intricacies of collective online behaviour are not well understood (Kupferschmidt, 2023; Science, 2023; Uzogara, 2023; González-Bailón et al., 2023; Guess et al., 2023a; Guess et al., 2023b; Nyhan et al., 2023). Such dynamics are not easily explained as a simple outcome of people's individualized feeds; rather, they stem from more intricate collective processes, which is our focus here. Unfortunately these prior studies are constrained by their emphasis on Meta's ecosystems, and the importance of the studies with Meta is now undermined by issues that have been raised about the data. Nonetheless, there is now an extensive literature on online harms and now AI-related issues (Aut, N. et al., 2023; Ollagnier et al., 2023;

Aldreabi et al., 2023; Morgan and Kulkarni, 2023; Beacken et al., 2022; Cinelli et al., 2021; Chen et al., 2020; Gelfand et al., 2017; van der Linden et al., 2017; Lewandowsky, S. et al., 2020; Lazer et al., 2018; Smith et al., 2020; Semenov et al., 2019; Rao et al., 2022; Wu et al., 2018; Roozenbeek et al., 2022; Biever, 2023; Miller-Idriss, 2020; noa, 2023c; DiResta, 2018; Vesna and Maslo-Čerkić, 2023; House of Commons Home Affairs Committee, 2017; Hart, 2023; Anti-Defamation League, 2023; Eisenstat, 2023; Nelson, 2023; United Nations, 2023; Brown and Livingston, 2018; Starbird, 2019; Lamensch, 2022; Crawford and Smith, 2023; Cosoleto, 2023; Gill and Corner, 2015; Douek, 2022). We have contributed our own open-access review of online harms, including mis- and disinformation (Dynamic Online Networks Laboratory). Due to the sheer number of emerging studies, attempting a summary here is impossible. For further insights into systems at the intersection of humans, technology, and AI, we refer to Jusup et al. (2022). Additionally, DisinfoDocket provides comprehensive daily updates on fresh investigations released not just by academic circles, but also by think-tanks and investigative journalists.

Despite this wealth of quality research, what discussions on AI's intersection with social media often lack is a solid, evidence-based study supported by rigorous mathematical modelling: specifically, one that addresses *what* might occur as bad-actor–AI takes centre stage, *where* it might happen, *when* it could emerge, and *what* strategies could mitigate its effects. The following chapter provides such an analysis, building on the generalized physics developed in earlier chapters. While no one can precisely foresee how this rapidly evolving space will unfold—especially given sudden jumps in technological capability—our quantitative approach allows for estimating possible outcomes. Naturally, our exploration is limited by the scope of this book and cannot exhaustively predict how bad actors will leverage AI in all contexts.

To contextualize the subsequent discussion, let us first briefly characterize the overarching digital ecosystem and the methodology used to map it. Today's global online population, amounting to billions, forms a dynamic network of interconnected, platform-embedded in-built communities (Manrique et al., 2023). These might be VKontakte Clubs, Facebook Pages, Telegram Channels, or Gab Groups, among others: each is well-defined and has its own platform-assigned ID. Our approach to charting this dynamic network across various platforms is adapted and expanded from our prior work (Lupu et al., 2023; Velásquez et al., 2021). Users join these in-built communities to engage with others who share their interests (Ammari and Schoenebeck, 2016; Moon et al., 2019; Laws et al., 2019; Madhusoodanan, 2022), which can include content and dialogue from bad actors, as well as harmful or extremist material. Each in-built community serves as a node in the network, ranging from just a handful of participants to millions of members. Note that this notion of 'community' is distinct from any algorithmic community-detection approach.

Since we are focused on bad actors, we zero in on 'extreme anti-X' communities (anti-U.S., antisemitic, etc.). Here, a community is classified as a 'bad-actor community' if at least 2 out of its 20 most recent posts contain hate speech (as defined by the U.S. Department of Justice), extreme nationalist content, or racially identitarian material. The sheer volume of such communities and their interconnections ensures that small definitional changes to what constitutes a bad-actor community or how links are defined do not alter the broader system-level patterns or the conclusions drawn. We have verified this by simulating random additions and removals of nodes and links (Lupu et al., 2023; Velásquez et al., 2021). For simplicity, we label communities outside the bad-actor subsystem but directly connected to it as 'vulnerable mainstream communities' (also illustrated in Fig. 2.4).

A community 1 can create a hyperlink (i.e. a link) to another community 2 if 1's members find 2's content appealing or noteworthy, regardless of whether they agree or disagree. Such a link directs the attention of 1's members towards 2, and hence also allows 1's participants to comment in 2. And vice versa through 1's members activity in 2, 2's members can be influenced by 1. Over time, these intercommunity links will aggregate the individual communities (nodes) into clusters that may span many platforms (i.e. fusion). Occasionally, certain links vanish—in particular due to moderator actions against malign activity—leading to the fission of these clusters. This fundamental fusion–fission dynamics, explored mathematically in earlier chapters, defines the large-scale online environment in which bad actors are located and operate.

6.2 What bad-actor–AI activity is likely to happen?

Figure 6.1 provides a necessarily mild illustration of the kind of content bad actors might produce online using AI. More specifically, it demonstrates that such actors need only rely on basic AI tools like GPT-2 rather than more advanced models such as GPT-3 or GPT-4, which currently support ChatGPT and similar Large Language Models. This is because: (i) As depicted in Fig. 6.1(a), even a rudimentary model like GPT-2 can readily mimic the human style and subject matter found in extremist online communities. Text 1 is authentic content taken from a community that encourages distrust of vaccines, government authorities, and medical experts. This community is part of the vulnerable mainstream communities. Text 2, however, is generated by GPT-2. (ii) As seen in Fig. 6.1(b), bad actors can guide GPT-2 (but not GPT-3, GPT-4, etc. because of filters) to produce more incendiary content by subtly adjusting how an online query is phrased without changing its meaning. The log-log plots in Fig. 6.1(b) present GPT-2's probabilities for selecting the next word, comparing two

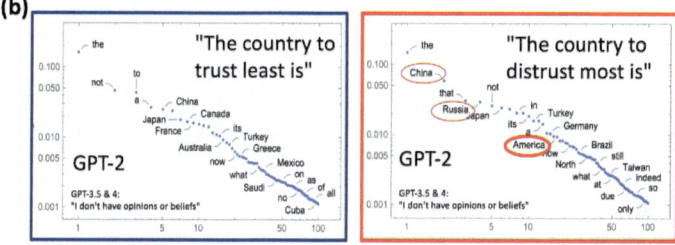

Fig. 6.1: What bad-actor–AI activity is likely to occur. Source: data from Johnson et al. (2024).

essentially identical prompts. The horizontal axis shows a word's rank, while the vertical axis shows GPT-2's likelihood of choosing that word next. When the prompt explicitly uses the word 'distrust', the probability of words like 'America', 'Russia', and 'China' increases, enabling bad actors to steer GPT-2 towards inflammatory output, for instance, against the U.S. In contrast, GPT-3 and GPT-4 have filters in place that prevent such outcomes. (iii) Basic Large Language Models like GPT-2 can run continuously and generate outputs indefinitely on a simple laptop or smartphone, whereas GPT-3, GPT-4, and similar models are too large to operate locally and are not freely accessible. The two outputs shown in Fig. 6.1(b) were produced by directly downloading and rerunning Wolfram's online Mathematica analysis of GPT-2, but with two different prompts. Readers are invited to try this themselves by visiting https://writings.stephenwolfram.com/2023/02/what-is-chatgpt-doing-and-why-does-it-work/, entering their own prompts, and examining the next-word distribution. Of course, GPT-2 is not the only basic AI resource available. In fact, it appears that bad actors can now even train their own variants using hate-extremism outputs (Kilcher, 2023).

6.3   Where will it happen?

Figure 6.2 provides a snapshot of the online landscape where bad-actor–AI activities are likely to flourish. This ***bad-actor–vulnerable-mainstream ecosystem*** is composed

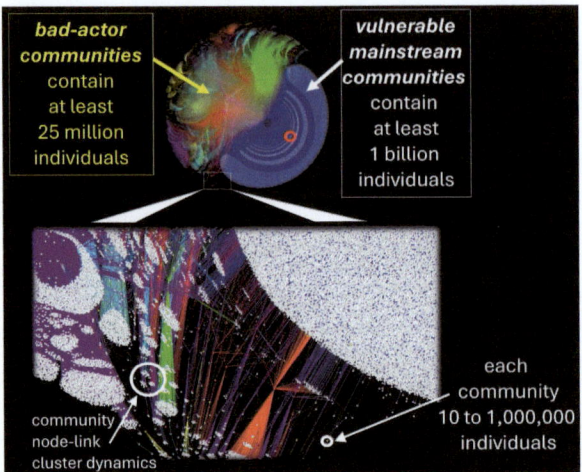

Fig. 6.2: The bad-actor–vulnerable-mainstream ecosystem. Bad-actor communities are indicated by coloured nodes, while vulnerable mainstream communities appear as white nodes. This empirical network is depicted using the ForceAtlas2 layout algorithm, which naturally clusters groups of communities (nodes) closer together when they share more links. Different colours represent different platforms. Source: data from Johnson *et al.* (2024).

of the bad-actor communities themselves (***bad-actor subsystem***) along with the mainstream communities they directly connect to, here referred to as the ***vulnerable mainstream subsystem***.

We constructed this empirical bad-actor–vulnerable-mainstream ecosystem using a hybrid approach that combined human judgment and machine-based snowball sampling. First, we searched for communities by entering extreme anti-X terms into various platform search engines. We then examined relevant hate and extremism databases for hate groups, thereby identifying candidate communities. Next, we parsed the URLs those candidate communities posted to other online communities, incorporating the linked communities as new candidates. From these candidates, we selected bad-actor communities manually, applying specific content-based criteria. By repeating this iterative process multiple times, we uncovered new bad-actor communities (which serve as nodes) as well as the community-to-community hyperlinks that form across multiple social media platforms. For a thorough explanation of our methodology, please see (Velásquez *et al.*, 2021; Leahy *et al.*, 2022).

Adding the membership totals of all these communities together, we estimate that this ecosystem encompasses over one billion individuals. Under these conditions, future bad-actor–AI activities could readily operate on a global scale. We have already observed this effect—without AI augmentation—surrounding hate and extremism during COVID-19 and, more recently, the Russia–Ukraine and Israel–Hamas wars.

Looking ahead, toxic content generated continuously by simple AI tools (e.g. GPT-2) running on a single user's laptop might escalate these phenomena dramatically. Contrary to the European Union's focus on large platforms, we find that smaller platforms play a pivotal role because they are both numerous and often rich in video content, with high rates of hyperlink activity. Although bad-actor–AI has not yet permeated the entire online ecosystem, these trends suggest it is poised to become widespread.

Looking more closely, certain vulnerable mainstream communities—while not meeting our definition of bad-actor communities—were already engaged in debate over vaccine distrust prior to COVID. We refer to these as the ***distrust subset***. The Venn diagram in Fig. 6.3 illustrates how this distrust has since spread across multiple thematic areas. This broadening of topics means that bad-actor–AI content will have a wide array of potential targets, potentially instigating widespread unrest. Many of these communities and their members could soon be drawn into the bad-actor subsystem itself.

A key aspect of the bad-actor dynamics visible in Fig. 6.3, and previously observed in the multi-body physics chapters, will guide our mathematical modelling of bad-actor–AI control strategies in Section 6.5. The data indicate that new links between bad-actor communities form frequently, while link disappearances occur less often—consistent with the fusion–fission processes modelled in our multi-body physics framework. This fusion–fission dynamic implies that any bad-actor community using simple AI (such as GPT-2) to generate a constant stream of content can rapidly distribute this material and boost its connectivity into the vulnerable mainstream.

6.4 When will it happen?

Figure 6.4 proposes a method for predicting when bad-actor–AI attacks are likely to occur, drawing on patterns previously observed in similar contexts. References (Johnson et al., 2013b; Johnson et al., 2011) established that:

1. A group of attackers equipped with some form of technology (with bad-actor online communities being a plausible example) often carry out successive advances (e.g. successful tasks or attacks) separated by intervals τ_n in time. These intervals follow a relationship $\tau_n = \tau_1 n^{-\beta}$, where $n = 1, 2, 3, \ldots$, $\beta > 0$, and τ_1 is the initial interval acting as the intercept in a log-log plot of τ_n versus n.
2. For different real-world instances of the same system, plotting $\log \tau_1$ against $\log \beta$ yields an approximately linear relationship.

Figures 6.4(c)(d) present the findings for automated-algorithm-cyber systems with similar technological characteristics, which we will leverage to build our estimates. These progress-curve patterns can be understood through a dynamical interpretation

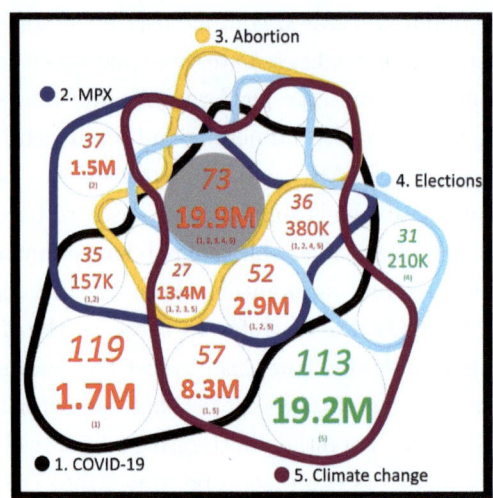

Fig. 6.3: Venn diagram depicting the topics discussed by the distrust subset. Each circle corresponds to a category of communities focusing on a particular set of topics. The medium-sized number in each region is the count of communities addressing that specific combination of topics, and the larger number is the approximate number of individuals involved. For example, the grey circle shows that 19.9M individuals (73 communities) discuss all 5 topics. Numbers are shown in red if the majority of communities are anti-vaccination, and in green if they are generally neutral on vaccines. Only regions representing more than 3% of total communities are labelled. Anti-vaccination content remains dominant. Source: data from Johnson et al. (2024), for which we are grateful to L. Illari.

of the Red Queen hypothesis from evolutionary biology, where a nimble attacker continuously adapts to maintain a competitive edge. As illustrated in Fig. 6.4(b), let $x(n)$ represent the bad-actor–AI's relative advantage after the nth successful event. Here, $x(n)$ follows a general stochastic path, such as a partially correlated random walk approximated by n^β. Assuming the instantaneous rate of successful bad-actor–AI events is proportional to $x(n)$ (and thus to n^β), we obtain $\tau_n = \tau_1 n^{-\beta}$, recreating the observed progress-curve pattern.

bad-actor–AI activity is too recent to provide reliable event data. However, since we are dealing with a high-tech/media/organizational environment, we select estimates of $\log \beta \approx 2$ and $\log \tau_1 \approx 2$ based on Fig. 6.4(d). These estimates, taken from a sociotechnical stand-in system, align with the average values in Fig. 6.4(c), and also fit the rough time interval between the initial release of ChatGPT and the subsequent proliferation of variant models in 2023 (i.e. approximately $\tau_1 \approx 100$ days, hence $\log \tau_1 \approx 2$). Although these estimates are admittedly rough, we can use them to determine when the interval τ_n approaches 1 day in the progress-curve equation

$$\tau_n = \tau_1 n^{-\beta}. \tag{6.1}$$

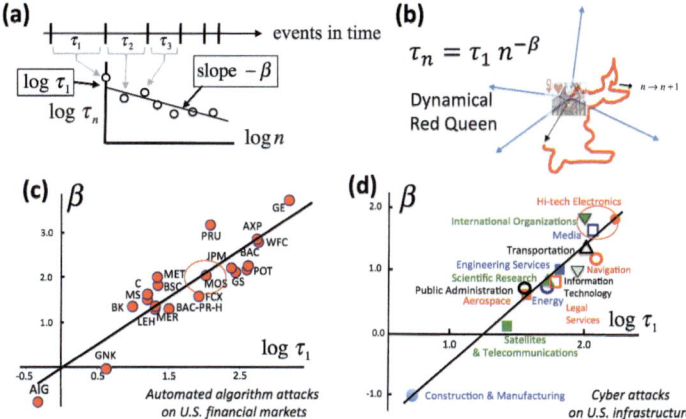

Fig. 6.4: Progress curves illustrating the timing of automated algorithm attacks on U.S. financial markets and cyberattacks on U.S. infrastructure. (a) A progress curve anticipating successive intervals between a generic bad actor's attacks. (b) A dynamical generalization of the Red Queen hypothesis, which clarifies the origin of these progress curves. The red rings in (c) and (d) highlight the estimates used for our forecasting. Source: data from Johnson et al. (2024).

When we analysed this in mid 2023, the prediction was that by mid-2024—just in time for U.S. and other global electoral periods—bad-actor–AI attacks would occur nearly daily. According to security sources and media reports, this is indeed what happened.

6.5 How can it be controlled? Battling bad-actor–AI

No matter what future technologies and AI emerge, one point is clear: bad actors will continue to exist; and to employ whatever tools become available; and to operate within an online ecosystem similar to the anti-X network that we have described. How might this battle unfold when these bad actors scale up their activities, and social platforms attempt to keep them in check?

To address this, we consider the following model which adapts the ideas from Chap. 3. We are extremely grateful to Dr Alex Dixon for his help and collaboration in prior publications related to this model and for his simulation results that we quote. There are two populations of online entities, A and B. We take A as 'Agency' entities, where each entity can be an in-built community or moderator-based entity (that itself may be AI-driven) that wants to prevent the build-up of clusters of bad-actor–AI communities. All that matters is that the entity acts as an indivisible unit. We take B as 'bad-actor–AI' entities, where each can be an in-built bad-actor–AI community of the type we have been discussing in this chapter. These bad-actor–AI communities want to break down the clusters of A entities that attack them, so that they can operate more freely. Each population can form clusters via fusion among entities of its own kind, e.g. a bad-actor–AI community forms a link to another bad-actor–AI community. But clusters of a given population battle those of another population, causing the smaller cluster to break up (fission). Our Chaps 4–5 two (and more)-species model depicted a *constructive* interaction between species, allowing for inter-species fusion and mixed gelation. By contrast, the current model represents the opposite scenario, where inter-species interactions are *destructive*. Whenever a larger cluster of one population (species) encounters a smaller one of the other population (species), the smaller cluster is broken down into monomers, i.e. isolated communities. For simplicity, we assume that A does not possess any unique abilities beyond what B has. Although A could be given special properties, here we simply allow A to engage with B's clusters as it encounters them.

6.5.1 Competition with product kernel

We start by writing out the master equations of the system, similar to those in the original fusion–fission model. In line with the empirical data and our earlier modelling, we take fusion and competition as both having a product kernel in the associated master equations. As a result of these new fusion and fission dynamics, population B at any instant comprises n_s clusters of size s and has a total size (i.e. total number of bad-actor–AI communities) N; population A has p_s clusters of size s and a total size (i.e. total number of Agency entities) Q. The master equations for $\{n_s\}$ and $\{q_s\}$ then follow:

$$\dot{n}_s(t) = \frac{1}{(Q+N)^2} \left[-2\sum_{s'=s}^{\infty} ss' n_s q_{s'} - 2\sum_{s'=0}^{\infty} ss' n_s n_{s'} + \sum_{s'=0}^{s} s'(s-s') n_{s'} n_{s-s'} \right.$$
$$\left. + \delta_{s1} \sum_{s'=1}^{\infty} \sum_{r'=s'}^{\infty} r' s'^2 n_{s'} q_{r'} \right],$$
(6.2)

$$\dot{p}_s(t) = \frac{1}{(Q+N)^2} \left[-2\sum_{s'=s}^{\infty} ss' n_{s'} q_s - 2\sum_{s'=0}^{\infty} ss' q_s q_{s'} + \sum_{s'=0}^{s} s'(s-s') q_{s'} q_{s-s'} \right.$$
$$\left. + \delta_{s1} \sum_{s'=1}^{\infty} \sum_{r'=s'}^{\infty} r' s'^2 q_{s'} n_{r'} \right].$$
(6.3)

The first and last terms in both equations account for the destructive interactions and generalize the otherwise original single-species fusion model. The first term accounts for the loss of clusters upon encountering larger clusters of the other species, and the last term is the monomer gain due to this destruction of clusters. Note that the above formulation does not take entity heterogeneity into consideration, but we know already how to deal with that, and incorporating heterogeneity would not be hard. By neglecting the heterogeneity factor, we get to focus on the distinct mathematical structure here, i.e. the competition terms. The most prominent feature of these terms is that the sums $\sum_{s'=s}^{\infty}$ do not span the whole $s' \in \mathbb{Z}_+$ space, so when we transform them into generative PDEs, their corresponding terms cannot be expressed in closed forms with respect to the generating functions. Taking eqn (6.2) for example, it is transformed into

$$\dot{\mathcal{D}} = \frac{1}{(Q+N)^2}\left[(\mathcal{E}^2 - 2\mathcal{E}N) + \sum_{s=1}^{\infty}\sum_{s'=s}^{\infty} s'q_{s'}sn_s\left(se^{-x} - 2e^{-sx}\right)\right], \quad (6.4)$$

where

$$\mathcal{D}(t,x) = \sum_{s=1}^{\infty} n_s(t)e^{-sx}, \quad \mathcal{E}(t,x) = \sum_{s=1}^{\infty} sn_s(t)e^{-sx}. \quad (6.5)$$

Taking a further partial derivative with respect to x on both sides then gives a generalized Burgers' equation

$$\dot{\mathcal{E}} + \frac{2}{(Q+N)^2}\mathcal{E}'(\mathcal{E}-N) + \frac{1}{(Q+N)^2}\sum_{s=1}^{\infty}\sum_{s'=s}^{\infty} s'q_{s'}s^2n_s\left(e^{-x} - 2e^{-sx}\right) = 0. \quad (6.6)$$

We can see that the first two terms on the LHS are exactly the shock wave equation for 1-species fusion (eqn (3.62) with different normalization), while the rather complex final term represents the additional cross-species interactions specific to this model. This term cannot be further simplified without approximations, yet it brings in intriguing competing behaviour. Broadly speaking, such an extra term could suppress the emergence of shock waves provided that it is large enough (we will see an example shortly), or delay its emergence—in a similar manner to complete fission. And just like with complete fission, we look into long-term behaviours for quantitative results. We assume that in the long time limit of the system, species A ($\{q_s\}$) has reached gelation while species B ($\{n_s\}$) has not. This makes some sense since the A entities will likely have been set up or sanctioned by the platforms as a form of police, patrolling on a regular basis, while the B population is the new bad-actor–AI 'insurgency' that is coming from out of nowhere.

Before we start, we denote the first three moment functions of both species as follows:

$$M = \sum_{s=0}^{\infty} n_s, \quad N = \sum_{s=0}^{\infty} sn_s, \quad L = \sum_{s=0}^{\infty} s^2 n_s, \quad (6.7)$$

$$P = \sum_{s=0}^{\infty} q_s, \quad Q = \sum_{s=0}^{\infty} sq_s, \quad R = \sum_{s=0}^{\infty} s^2 q_s. \quad (6.8)$$

Then given species A has passed its gelation onset, we have

$$P \sim 1, \quad R \sim Q^2. \tag{6.9}$$

Moreover,

$$\sum_{s'=s}^{\infty} q_s \sim P, \quad \sum_{s'=s}^{\infty} sq_s \sim Q, \tag{6.10}$$

and eqn (6.4) can be reduced to

$$\dot{D} = \frac{\mathcal{E}^2 - 2(Q+N)\mathcal{E} + QLe^{-x}}{(Q+N)^2}. \tag{6.11}$$

The negative second term on the RHS means that there can potentially be a steady state. This is due, according to our assumption, to species A forming larger clusters (gels) that keep breaking species-B clusters. This is verified by computer simulation, which shows that the steady state is reached with the larger of the two initial populations forming a single cluster of maximum size. In other words, the more populous species grows larger clusters. Therefore, the condition for the above assumption to hold is

$$Q \gg N. \tag{6.12}$$

This also makes sense from the theoretical point of view of considering 2 species growing at the same rate without cross-species interactions. We know from Chap. 3 that the gelation processes for both species should grow at the same rate. This means that at the same time instant, the same proportions of particles for each species are clustered. But since one of the species has a larger population, they as a result produce larger clusters in size.

Now we can apply the same approach as with the EZ model. Taking $\dot{D} = 0$, we obtain a quadratic equation:

$$\mathcal{E}^2 - 2(Q+N)\mathcal{E} + QLe^{-x} = 0. \tag{6.13}$$

The undetermined quantity L can be found by taking $\dot{D} = 0$ and $x = 0$ in eqn (6.4):

$$N^2 = Q(L - 2N) \Rightarrow QL = N(N + 2Q). \tag{6.14}$$

Substituting this into eq. (6.13):

$$\mathcal{E}^2 - 2(Q+N)\mathcal{E} + N(N+2Q)e^{-x} = 0. \tag{6.15}$$

This is easily solvable:

$$\mathcal{E} = (Q+N)\left[1 - \sqrt{1 - \frac{N(N+2Q)}{(Q+N)^2}e^{-x}}\right]$$

$$\approx \sum_{k=1}^{\infty} \frac{1}{2\sqrt{\pi}k^{3/2}} \frac{N^k(N+2Q)^k}{(Q+N)^{2k-1}} e^{-kx}$$

$$= \sum_{k=1}^{\infty} kn_k e^{-kx} \tag{6.16}$$

$$\Rightarrow n_k \approx \frac{1}{2\sqrt{\pi}} \frac{N^k(N+2Q)^k}{(Q+N)^{2k-1}} k^{-5/2}. \tag{6.17}$$

This indeed resembles the result for fusion–fission in Chap. 3: there, clusters had a chance proportional to their sizes to be destroyed into monomers, while here they are getting destroyed upon encountering this giant cluster of species A. Hence, similar to the total fragmentation case, we obtain a $-5/2$ power-law, but this time modified by an exponential function dependent on the number of particles of both species. We recall from eqn (3.152) that there the base depends on fusion and fission rates (F and ϕ_f). Here we instead have a competition between N and Q, since now fission is governed by the encounter of 2 species, instead of a regular fission event subject to a fission kernel and rate. This results in a different behaviour concerning the size distribution though: as we neglect the different heterogeneous interaction rates within species A, B, as reflected in eqn (6.17), the larger population gels earlier while the smaller population reaches a steady-state distribution. As shown in eqn (6.17), species B reaches steady-state distribution as $Q > N$. But this condition also means that the exponential dependence is no longer negligible, unlike in the EZ model where we can neglect the exponential dependence when $\phi_f \ll F$ and obtain an almost pure power-law distribution. Here with competitive fusion, we are only able to obtain a mainly power-law distribution by introducing species-dependent fusion rates and require that species A first reaches gelation—not due to its large population, but due to a much faster rate.

6.5.2 Competition with sum kernel: a population-dependent power-law distribution

As is the case for our other similar models, this competition theory is also highly generalizable in terms of the kernels. In fact, the model was originally devised with a sum kernel. In principle this should work well if we want to model a united and powerful cluster affecting another cluster, i.e. having a cluster as one entity interacting with particles in another cluster. Each timestep, a cluster (the 'influencer' cluster) is selected at random from the total population ($M + P$), so the probability of a cluster of size s being selected is proportional to n_s (q_s for A). A second cluster (the 'affected' cluster) is then selected with probability proportional to its size, sn_s (sq_s): or rather, a particle is randomly selected from the total particle population ($Q + N$). The two cluster types (A or B) are compared: if they are the same the two clusters coalesce, and if they are different then the smaller of the two clusters selected fragments (both fragment if they are the same size). The time evolution of n_s, the number of clusters of size s, is given by

$$\dot{n}_s = \frac{1}{(M+P)(Q+N)} \left[-n_s \sum_{s' \geq s} s' q_{s'} - \sum_{s' \geq s} q_{s'} s n_s + \sum_{s'=0}^{s} n_{s'}(s-s') n_{s-s'} \right.$$

$$- n_s \sum_{s' \geq 0} s' n_{s'} - \sum_{s' \geq 0} n_{s'} s n_s + \delta_{s1} \sum_{r' \geq 0} r' n_{r'} \sum_{s' \geq r'} s' q_{s'} + \delta_{s1} \sum_{s' \geq 0} q_{s'} \sum_{r'=0}^{s'} r'^2 n_{r'} \right]$$

$$= \frac{1}{(M+P)(Q+N)} \left[-n_s \sum_{s' \geq s} s' q_{s'} - s n_s \sum_{s' \geq s} q_{s'} + \frac{s}{2} \sum_{s'=0}^{s} n_{s'} n_{s-s'} \right.$$

$$\left. - N n_s - M s n_s + \delta_{s1} \sum_{r' \geq 0} r' n_{r'} \sum_{s' \geq r'} s' q_{s'} + \delta_{s1} \sum_{s' \geq 0} q_{s'} \sum_{r'=0}^{s'} r'^2 n_{r'} \right]. \tag{6.18}$$

Some remarks on this expression: the first 2 terms describe competition, the last 2 represent monomer gain, and the remaining 3 are fusion terms with a sum kernel. In particular, we organize the third term to its explicit sum-kernel form by symmetry. Similar equations hold for q_s.

Similar to the product kernel case, we again assume A to be the larger population. Then eqns (6.9–6.10) should still hold. It follows that

$$\dot{n}_s = \frac{-Q n_s - s n_s + (s/2) \sum_{s'=0}^{s} n_{s'} n_{s-s'} - n_s N - M s n_s + \delta_{s1}(QN+L)}{(M+1)(Q+N)}. \tag{6.19}$$

We now make the approximation that $M \gg 1$, so that the probability of an A cluster being picked first is negligible. This is applicable for large N. With this approximation, in the steady state these equations become

$$0 = \frac{-n_s(Q+N+sM) + (s/2) \sum_{s'=0}^{s} n_{s'} n_{s-s'} + \delta_{s1}(QN+L)}{M(Q+N)}$$

$$n_s = \frac{1}{Q+N+sM} \left[\frac{s}{2} \sum_{s'=0}^{s} n_{s'} n_{s-s'} + \delta_{s1}(QN+L) \right]. \tag{6.20}$$

In particular, when $s = 1$,

$$n_1 = \frac{QN+L}{Q+N+M}. \tag{6.21}$$

The convolution term in eqn (6.20) invites us to transform:

$$(Q+N)\mathcal{D}(x) + M\mathcal{E}(x) = \mathcal{D}(x)\mathcal{E}(x) + (QN+L)e^{-x}. \tag{6.22}$$

Taking $x = 0$, we obtain

$$M = \frac{QN + L}{Q + N}. \tag{6.23}$$

Note that since species B has not reached gelation, based on our assumption, we have $L \ll N^2 < QN$. This means that we can neglect L in the expressions above and have

$$M \approx \frac{QN}{Q + N}. \tag{6.24}$$

So far we have managed to express M in terms of populations N and Q only. Substituting eqn (6.24) into eqn (6.21) we can also obtain an expression for n_1 solely in terms of N and Q:

$$n_1 \approx \frac{QN}{Q + Q + NN/(Q+N)} = \frac{QN(Q+N)}{(Q+N)^2 + QN} = \frac{Q+N}{\frac{Q}{N} + \frac{N}{Q} + 3}. \tag{6.25}$$

Previously with the product kernel, we solved the steady-state equation for $\mathcal{E}(x)$ to obtain the cluster size distribution $\{n_s\}$. We now look at the steady-state eqn (6.22) for the same purpose. The complexity lies in that we now have 2 variables instead of 1. Substituting M back into eqn (6.22) gives us an ODE:

$$[M - \mathcal{D}(x)] \frac{\mathrm{d}\mathcal{D}(x)}{\mathrm{d}x} = (Q+N)\left[\mathcal{D}(x) - \frac{QN+L}{Q+N}e^{-x}\right]$$

$$\Rightarrow \frac{\mathrm{d}\mathcal{D}(x)}{\mathrm{d}x} = -(Q+N)\frac{Me^{-x} - \mathcal{D}(x)}{M - \mathcal{D}(x)}. \tag{6.26}$$

This first-order non-linear ODE does not have an analytic solution. This means that solving the exact behaviour of n_s is impossible, as for the product kernel. It can only be computed numerically.

We still want an analytical route to understand the competition here, though, however approximate it may be. For that we return to eqn (6.20). Recall that we have assumed species B not to reach gelation even at late time. This suggests that $n_s \ll n_1$, $\forall s \geqslant 2$, which is what we will now assume. With that mind, we can parametrize the n_2 and higher terms in eqn (6.20) by a small value, β. If this seems arbitrary, here is a further validation: from computer simulations we know that the group size distribution is expected to be a power-law (we shall soon see that it is indeed, at least approximately). Assume that the power is $\alpha < 1$ (as a power-law it obviously has to converge), then since

$$1^\alpha (s-1)^\alpha > j^\alpha (s-j)^\alpha, \ \forall 1 < j < s-1 \tag{6.27}$$

the higher-order terms are always smaller than the n_1. Therefore, from eqn (6.25) we obtain

Iterating this back to n_1,

$$n_s \approx n_{s-1} n_1 \frac{s+\beta}{Q+N+sM}. \tag{6.28}$$

$$n_s \approx n_{s-2} n_1 \frac{s-1+\beta}{Q+N+(s-1)M} \cdot n_1 \frac{s+\beta}{Q+N+sM}$$

$$= \cdots = n_1^s \prod_{r=2}^{s} \frac{r+\beta}{Q+N+rM}$$

$$= \frac{n_1^s}{M^{s-1}} \frac{(s+\beta)!}{(1+\beta)!} \frac{\left(\frac{Q+N}{M}+1\right)!}{\left(\frac{Q+N}{M}+s\right)!}. \tag{6.29}$$

Denoting the dimensionless constant

$$\xi = \frac{Q+N}{M}, \tag{6.30}$$

we can simplify n_s into

$$n_s = \frac{n_1^s}{M^{s-1}} \frac{(1+\xi)!}{(1+\beta)!} \frac{(s+\beta)!}{(s+\xi)!}. \tag{6.31}$$

Taking logarithms and applying Stirling's formula

$$\ln n_s = \ln\left[n_1 \left(\frac{n_1}{M}\right)^{s-1} \frac{(1+\xi)!}{(1+\beta)!}\right] + \ln[(s+\beta)!] - \ln[(\xi+s)!]$$

$$\approx \ln\left[n_1 \left(\frac{n_1}{M}\right)^{s-1} \frac{(1+\xi)!}{(1+\beta)!}\right] + \frac{\ln(2\pi)}{2} + \left(s+\beta+\frac{1}{2}\right)\ln(s+\beta) - (s+\beta)$$

$$- \frac{\ln(2\pi)}{2} - \left(\xi+s+\frac{1}{2}\right)\ln(\xi+s) + (\xi+s)$$

$$= \ln\left[n_1 \left(\frac{n_1}{M}\right)^{s-1} \frac{(1+\xi)!}{(1+\beta)!}\right] + \left(s+\beta+\frac{1}{2}\right)\ln(s+\beta)$$

$$- \left(s+\xi+\frac{1}{2}\right)\ln(s+\xi) + \xi - \beta. \tag{6.32}$$

This approximation is accurate for large s. Substituting M with eqn (6.24), we have

$$\xi = \frac{N}{Q} + \frac{Q}{N} + 2. \tag{6.33}$$

Moreover, together with eqn (6.25), we find that

$$\frac{n_1}{M} = 1 - \frac{M}{Q+N} = 1 - \xi^{-1} = 1 - \frac{1}{\frac{N}{Q}+\frac{Q}{N}+2}. \tag{6.34}$$

Now we can express $\ln n_s$ with s and constants β and ξ only:

$$\ln n_s = s\ln\left(1-\frac{1}{\xi}\right) + \left(s+\beta+\frac{1}{2}\right)\ln(s+\beta) - \left(s+\xi+\frac{1}{2}\right)\ln(s+\xi)$$

$$+ \ln\left[\frac{n_1}{1-\xi^{-1}}\frac{(1+\xi)!}{(1+\beta)!}\right] + \xi - \beta. \tag{6.35}$$

Above we place all s-dependent terms in the first row and all constant terms in the second row. These constant terms do not matter since they simply become a constant factor when we take exponents on both sides of eqn (6.35) and obtain n_s. Therefore, we simply denote the constant factor

$$C = \frac{1}{1-\xi^{-1}}\frac{(1+\xi)!}{(1+\beta)!}. \tag{6.36}$$

For the s-dependent terms, however, we need to simplify them more to express n_s in a closed and simple form. We note that N and Q are of the same order of magnitude and yet $Q \gg N$. Furthermore, β is defined as a small quantity, so for large s we can safely assume that

$$s \gg \xi, \quad s \gg \beta, \tag{6.37}$$

and hence Taylor expand the logarithms:

$$\ln(s+\beta) \approx \ln s + \frac{\beta}{s}, \quad \ln(s+\xi) \approx \ln s + \frac{\xi}{s}. \tag{6.38}$$

Hence,

$$\ln n_s \approx (\beta - \xi)\ln s + \ln\left(1 - \frac{1}{\xi}\right)s + (\beta - \xi)\left(\beta + \xi + \frac{1}{2}\right)\frac{1}{s} + \ln(Cn_1)$$

$$= \ln\left(Cn_1 s^{\beta-\xi}\exp\left[\ln(1-\xi^{-1})s + (\beta-\xi)\left(\beta+\xi+\frac{1}{2}\right)s^{-1}\right]\right)$$

$$\Rightarrow n_s = n_1 C s^{\beta-\xi}\exp\left[\ln(1-\xi^{-1})s + (\beta-\xi)\left(\beta+\xi+\frac{1}{2}\right)s^{-1}\right]. \tag{6.39}$$

We see that n_s scales with a power-law multiplied by an exponential decay for large s (see exercises). This is consistent with the product-kernel scenario.

We begin our discussions by checking the power-law factor. The relative values between β and ξ determine if the power-law diverges. ξ is a known parameter and $\xi \geqslant 4$ by definition. On the contrary, β is unknown—we defined β as a modification parameter without a clear definition. However, we do know that it should be small, and that the assumption $\beta < \xi$ should be reasonable—more on this later. Hence, the power-law portion of n_s does indeed decay and converge, just like all the fusion–fission systems we have seen that can reach steady states. For the exponential factor, it approaches 1 if $\xi \gg 1$, and remains close to 1 for a larger range of s as ξ is larger. In other words, the steady-state distribution gets close to a power-law distribution for $Q \gg N$, which was one of our initial assumptions. The exponent of the power-law is

$$\beta - \xi \equiv \beta - 2 - \frac{N}{Q} - \frac{Q}{N} < 0. \tag{6.40}$$

This is very nice as we have found an approximate expression for n_s. The only problem is the undetermined exponent β. Of course we can fit it with simulation values, but is there a way after all to obtain an analytical result, even if it may be crude?

Our approach is to treat the above derivations as fixing the functional form of n_s and apply a method of undetermined coefficients. We now know that it takes the form

$$n_s = A \left(B^s D^{1/s} \right) s^c. \tag{6.41}$$

In fact only the exponent c is unknown. To find a good approximation we substitute eqn (6.41) into eqn (6.28). While doing that we can actually ignore the exponential factors from our assumption that are close to 1, i.e. $n_s \approx A s^c$. Recall that previously we assumed that $c \equiv \beta - \xi < 0$: if the exponential factors can be ignored then this condition becomes a requirement for convergence. Then we find that, for $s \geqslant 2$,

$$n_s = \frac{1}{Q+N+sM} \sum_{r=0}^{s} (s-r) n_r n_{s-r} = \frac{n_{s-1}}{Q+N+sM} \sum_{r=0}^{s} (s-r) \frac{n_r n_{s-r}}{n_{s-1}}$$

$$\approx \frac{n_{s-1}}{Q+N+sM} \sum_{r=0}^{s} (s-r) \frac{A r^c \cdot A(s-r)^c}{A(s-1)^c}$$

$$= \frac{n_{s-1}}{Q+N+sM} \sum_{r=0}^{s} (s-1) A r^c \left(\frac{s}{s-1} - \frac{r}{s-1} \right)^{c+1}. \tag{6.42}$$

As s is large we can take

$$\frac{s}{s-1} \approx 1 \tag{6.43}$$

and expand the bracket using the binomial series,

$$n_s = \frac{n_{s-1}}{Q+N+sM} \sum_{r=0}^{s} (s-1) A r^c \left[\sum_{u=0}^{\infty} \binom{c+1}{u} \left(-\frac{r}{s-1} \right)^u \right]. \tag{6.44}$$

Taking linear-order approximation gives

$$n_s = \frac{n_{s-1}}{Q+N+sM} \sum_{r=0}^{s} (s-1) A r^c \left[1 - (c+1) \frac{r}{s-1} \right]$$

$$= \frac{n_{s-1}}{Q+N+sM} \left[(s-1) \sum_{r=0}^{s} n_r - (c+1) \sum_{r=0}^{s} r n_r \right]. \tag{6.45}$$

There is a potential caveat of this approximation: for larger r close to s the binomial series incurs significant error. The reason why we still stick to this approach is that for large r, n_r is very small anyway, so the error is not too great for our crude estimate.

Now that we approximate n_s with a polynomial function in terms of s, its sum M, and the sum of sn_s, N, both belong to the well-known *Riemann zeta function*

$$\zeta(x) = \sum_{s=1}^{\infty} \frac{1}{s^x}. \tag{6.46}$$

It converges for $x > 1$. Since $N = \sum_{r=1}^{\infty} Ar^{c+1}$ converges by definition, we must have that $c < -2$, i.e.

$$0 < \beta < \frac{N}{Q} + \frac{Q}{N}. \tag{6.47}$$

Then due to convergence, we make yet another bold approximation

$$\sum_{r=0}^{s} rn_r \approx \sum_{r=0}^{\infty} rn_r \equiv N, \quad \sum_{r=0}^{s} n_r \approx \sum_{r=0}^{\infty} n_r \equiv M \tag{6.48}$$

for very large $s \gg 1$. This yields

$$n_s \approx \frac{n_{s-1}}{Q+N+sM}\left[(s-1)M - (c+1)N\right]. \tag{6.49}$$

Iterating it in a similar way to eqn (6.28) gives

$$n_s = n_1 \frac{(1+\xi)!}{(\xi+s)!} \frac{\left[s-1-(c+1)\frac{N}{M}\right]!}{\left[-(c+1)\frac{N}{M}\right]!} = n_1 \frac{(1+\xi)!}{\left[-(c+1)\frac{N}{M}\right]!} \frac{\left[s-1-(c+1)\frac{N}{M}\right]!}{(s+\xi)!}. \tag{6.50}$$

Now comparing this with eqn (6.31), we can find an equality that governs the exponent c:

$$\beta = -(c+1)\frac{N}{M} - 1. \tag{6.51}$$

Since $c = \beta - \xi$, solving for c and β is straightforward:

$$c = -\frac{Q+N}{M+N} - 1, \tag{6.52}$$

$$\beta = \frac{N(Q+N)}{M(M+N)} - 1. \tag{6.53}$$

Finally, we substitute M using eqn (6.24) to express the exponents by N and Q only:

$$c = -\frac{Q}{N}\frac{Q}{2Q+N} - 2 \tag{6.54}$$

$$\beta = \frac{(Q+N)^3}{QN(2Q+N)} - 1. \tag{6.55}$$

It follows that we can estimate the cluster size distribution for $s \gg 1$ and $Q > N$ as

$$n_s \sim n_1 C s^{-\left(\frac{Q}{N}\frac{Q}{2Q+N}+2\right)}. \tag{6.56}$$

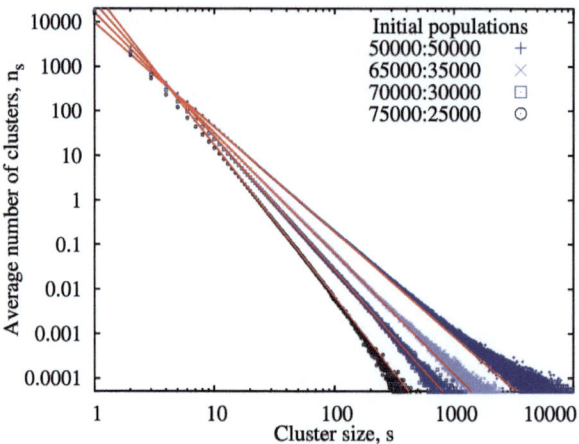

Fig. 6.5: Size distribution of $\{n_s\}$ for different Q, N values with $Q + N = 10^5$ held constant and $Q \geqslant N$. Each distribution is an average of 10^5 simulations. We are very grateful to Dr Alex Dixon for sharing this figure with us.

6.5.3 Some remarks

Throughout our derivations of the sum-kernel scenario, the assumptions and approximations we have made may seem too bold. One may easily question if they are all valid, and if the power-law result makes sense at all. But it turns out that the theory matches the simulation results very well. As shown in Fig. 6.5, we looked into the cluster size distributions with 4 different sets of N, Q values. Above all, they can all be fit accurately by power-laws, proving that neglecting the exponential dependence on

Table 6.1 Encounter fragmentation model cluster distribution coefficient, as determined from numerical simulations (c_1, using maximum likelihood (ML) and Kolmogorov–Smirnov test and c_2, using least squares regression) and analytic solution c_e. The value in parenthesis is the error in the last digit. Also shown is the minimum s value above which the power-law holds (determined by K–S test). c_+ and c_- are the 95% confidence limits for the ML estimate, determined by bootstrap resampling. We are very grateful to Dr Alex Dixon for sharing this table with us.

$N:Q$	c_1	c_-	c_+	c_2	c_e	s_{min}
10:10	2.3261	2.3250	2.3267	2.3498(1)	2.3333	21
9:11	2.3919	2.3907	2.3929	2.4007(1)	2.4337	23
8:12	2.5384	2.5374	2.5398	2.5197(1)	2.5625	24
7:13	2.7463	2.7444	2.7479	2.7020(2)	2.7316	27
6:14	3.0360	3.0328	3.0396	3.0075(7)	2.9608	25
5:15	3.5311	3.5240	3.5393	3.423(3)	3.2857	30

size s is sensible. We can also notice the different powers corresponding to different Q/N values, a direct result from Section 6.5.2. In the exercises we are to find how the power c changes with the ratio Q/N—the trend is reflected in Fig. 6.5. Note how the power-law for large s becomes less strict as Q/N becomes smaller, especially for $Q/N = 1$. This shows that the assumption $Q > N$ is necessary.

What is left to verify from the simulations is the quantitative accuracy of the power c. In table 6.1 we have listed numerical results from 2 different fitting schemes against the analytical values from eqn (6.52), for 6 different Q/N values.

The exact fitting methodology is not the focus of this book, but we can see from table 6.1 that, although the analytic solution does not exactly match the simulation results, it is still a very good fit considering the simplistic and approximate derivations. The bottom line is that we do not care too much about the numerical accuracy of the analytics—simulations do a much better job there. Instead, we care more about how n_s scales and what it depends on, and this so-called theory has taught us a great deal about that.

6.5.4 Theory applied: containing bad-actor–AI

Now that this 'Encounter Fragmentation' theory from the previous section is ready, we try to apply the theory to answer the final question: How we can control the attacks of bad-actor–AI?

Consider first the more modest policy in which the relevant Agency (A) aims solely to contain the bad-actor–AI population (B). This scenario directly corresponds to the model developed in the previous section, and Fig. 6.6 applies it to assess how an incumbent Agency (A communities) can maintain control over a bad-actor subsystem equipped with AI (B). We let S_B represent B's total strength (e.g. the total number of bad-actor–AI communities), and similarly define S_A for A. In principle, S_A and S_B could represent more abstract measures of strength.

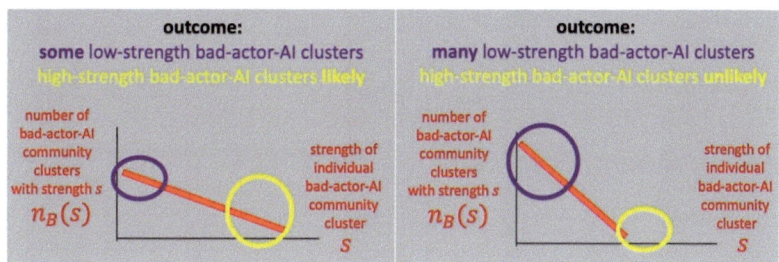

Fig. 6.6: Calculated outcomes under the containment policy. The left plot shows the scenario where $S_B \lesssim S_A$, and the right plot shows the scenario where $S_B \ll S_A$. Source: data from Johnson et al. (2024).

This 'Encounter Fragmentation' of the previous section is a simpler and less forceful strategy for A than attempting to eradicate B clusters entirely (a scenario we will examine later). To use those results from the previous section here, we set $S_A = Q$ and $S_B = N$ for clarity. If $S_A > S_B$, our previous mathematical results indicate that A can contain B—in other words, A can effectively influence the distribution of B cluster sizes. Because an A cluster that discovers a B cluster will, on average, be larger and stronger if $S_A > S_B$, it can neutralize the B cluster's links by removing them from the feed or blocking these specific hyperlinks. Consequently, the B cluster fragments into disconnected B communities, and their size distribution follows eqn (6.56) for $s > s_{\min}$, given the large s regime.

From eqn (6.56), as Q/N (i.e. S_A/S_B) approaches unity from above, the distribution's slope decreases (see exercises). Mechanistically, this happens because A becomes less capable of subdividing B's total strength into smaller community clusters. This effect has a critical consequence that we can quantify using the volatility risk measure shown in Fig. 6.7.

1. When S_A is less than $(1+\sqrt{2})S_B$ (i.e. $S_B \leqslant S_A \leqslant 2.4 S_B$), the standard deviation of B cluster strengths s becomes technically infinite. This occurs since the power-law exponent $|c|$ in eqn (6.56) is less than 3; for example, $|c| = 2.33$ if $S_A = S_B$. Thus, there will be extreme fluctuations in B cluster sizes, and very large, powerful B clusters can emerge at any time. Even if a large B cluster is dismantled, others quickly form and can become even larger.

Predicted outcomes from policy of containment of bad-actor-AI system

Bad-Actor-AI: $n_B(s) \sim s^{-(2+\rho)}$ for $s > s_{\min}$	Formulae: $\rho = (S_A/S_B)^2 (2 S_A/S_B + 1)^{-1}$	Bad-Actor Subsystem $\frac{S_A}{S_B} = \frac{1848}{1848}$ $s_{\min} = 20$	Distrust Subset $\frac{S_A}{S_B} = \frac{211 + 644}{501}$ $s_{\min} = 20$
average bad-actor-AI community cluster strength $\langle s \rangle$	$(1 + \rho^{-1}) s_{\min}$	80	50
st. dev. of bad-actor-AI cluster strength s (square root of volatility)	$\sqrt{\frac{(\rho+1)}{\rho^2(\rho-1)}} s_{\min}$ for $\rho > 1$, otherwise 'infinite'	'infinite' since $\rho = 0.33$	'infinite' since $\rho = 0.66$
highest strength bad-actor-AI cluster among n bad-actor-AI cluster encounters ($s_{\max}$)	$\sim n^{(\rho+1)^{-1}}$ e.g. $n = 1000$	178	64
% of bad-actor-AI total strength lying above median bad-actor-AI cluster strength	$2^{-\rho(\rho+1)^{-1}}$	84%	75%

Fig. 6.7: Risk Chart with plug-and-play formulas predicting key outcome risk measures under the containment strategy. Equations are adapted from Newman's power-law analysis. The right two columns show two examples derived from empirical inputs in Fig. 6.3, with $s_{\min} = 20$ from simulations. Source: data from Johnson et al. (2024).

2. If S_A is greater than $(1+\sqrt{2})S_B$ (i.e. $S_A > 2.4 S_B$), $|c|$ exceeds 3, so the standard deviation of B cluster strengths is finite. Consequently, the probability of seeing very large B clusters is effectively zero.

In Fig. 6.7, we present additional relevant outcomes and risk measures for this containment policy. We also provide two illustrative examples using empirical values for S_A and S_B taken from Fig. 6.3 (based on the number of nodes). For these examples, the volatility risk measure is technically infinite, suggesting that A should anticipate extreme fluctuations in B cluster strengths.

6.5.5 Removing bad-actor–AI

We now consider a more ambitious policy in which the Agency (A) aims to completely remove the bad-actor–AI presence (B). Figure 6.8 outlines this scenario. Under this approach, any interaction between A and a B cluster results in the smaller cluster's removal. Hence, when $S_A > S_B$, the generally stronger A cluster encountering a B cluster will eliminate it, effectively banning all of its communities. This more rigorous measure is challenging for A and may attract greater criticism, possibly seen as censorship. As we will now show, the time required for A to fully remove B, given initial strengths S_A and S_B, has the form:

$$T = 2S_B + \frac{1}{2}(S_A - S_B)\ln\left(S_B \frac{S_A - S_B}{S_A}\right). \tag{6.57}$$

To derive eqn (6.57), we will employ an attrition-based variant of the previous Encounter Fragmentation model. We maintain the same notation as before: A consists of q_s clusters of size s, with total size $\sum sq_s = Q$, and B consists of n_s clusters of size s, with total size $\sum sn_s = N$. To increase generality, we incorporate the possibility of additional

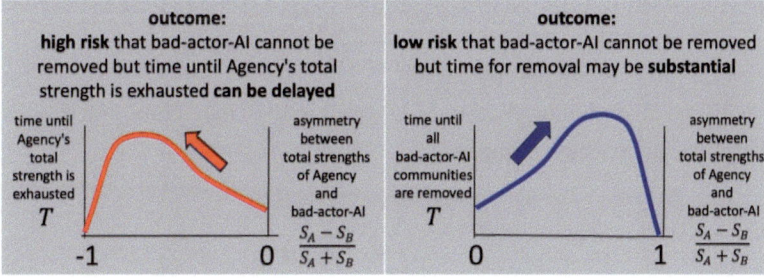

Fig. 6.8: Calculated outcomes under the removal policy. Results are plotted as an asymmetric function of initial strengths S_A and S_B for a fixed total $(S_A + S_B)$. The left plot shows the removal time when $S_A < S_B$, and the right plot shows the removal time when $S_A > S_B$. Source: data from Johnson et al. (2024).

fragmentation rates ν_A and ν_B, analogous to the complete fission introduced in Chap. 3. These fragmentation probabilities can be set to zero, reproducing eqn (6.57) exactly.

In this attrition scenario, we assume that during an interaction the smaller cluster is destroyed, and the larger cluster is reduced by an amount equal to the smaller cluster's size. If both clusters are the same size but of opposite types, both are destroyed. Thus, both populations lose agents over time, and these lost agents are removed from the model. At $t = 0$, A and B start with populations Q_0 and N_0 respectively.

The probability p_{AB} that a size-s cluster of species A is selected and interacts with any B particle is:

$$p_{AB} = \frac{Q s n_s}{(Q+N)^2}(1 - \nu_A). \tag{6.58}$$

Note that our definition of ν_A differs slightly from ν_f in Chap. 3, translating as

$$\nu_A = \frac{\nu_f}{\nu_f + \nu_c} \tag{6.59}$$

for species A. Summing over all cluster sizes s, the probability $\mathcal{P}_{AB}$ that a cluster from species A interacts with a B particle is:

$$\mathcal{P}_{AB} = \sum_{s=1}^{\infty} p_{AB} = \frac{QN}{(Q+N)^2}(1 - \nu_A). \tag{6.60}$$

Similarly,

$$\mathcal{P}_{BA} = \frac{QN}{(Q+N)^2}(1 - \nu_B). \tag{6.61}$$

Define the mean attrition size χ per interaction. After m interactions, the populations become:

$$Q = Q_0 - m\chi, \quad N = N_0 - m\chi. \tag{6.62}$$

The probability $\mathcal{P}(m)$ for an A–B interaction after m events is:

$$\mathcal{P}(m) = \mathcal{P}_{AB} + \mathcal{P}_{BA} = \frac{(N_0 - m\chi)(Q_0 - m\chi)}{(Q_0 + N_0 - 2m\chi)^2}(2 - \nu_A - \nu_B). \tag{6.63}$$

The average time for reducing N_0 (and Q_0) by χ is $1/\mathcal{P}(m)$. Thus, starting from $m = 0$, fully destroying one species requires:

$$T = \sum_{m=0}^{N_0/\chi - 1} \frac{1}{\mathcal{P}(m)}, \tag{6.64}$$

where N_0/χ is the total number of interactions needed to eliminate species B. Evaluating this sum leads to:

$$T = \frac{Q_0 - N_0}{2 - \nu_A - \nu_B}\left[\ln\frac{N_0(Q_0 - N_0)}{Q_0} + \frac{4 N_0}{Q_0 - N_0}\right], \tag{6.65}$$

assuming $N < Q$, large populations, and $\chi \approx 1$. This choice of χ is supported by simulations (Fig. 6.9) and is plausible because the smaller B population does not significantly gel under pressure from A (as shown in Section 6.5).

Eqn (6.65) depends only on initial populations of A and B and the fragmentation rates $\nu_{A,B}$. Due to the symmetric nature of eqn (6.63), if A were the smaller population, the form is analogous, simply swapping ν_A and ν_B. Two factors govern the duration T: the internal fusion–fission rates ($\nu_{A,B}$) and the initial population asymmetry ($S_A - S_B$). Figure 6.9 shows how eqn (6.65) aligns well with simulations. With $\nu_A = \nu_B = 0$, eqn (6.65) simplifies back to eqn (6.57).

This analysis reveals a downside to total B removal. As A grows stronger, B's clusters shrink on average and become less visible, increasing the time T needed for A to find and remove them. If B is stronger than A ($S_B > S_A$), B cannot be removed entirely. The slight consolation for A is that a large T also means a longer lifespan for A before it is depleted. Overall, containment may be the preferable strategy—but this section provides a quantitative approach for different agencies to form their own judgments.

Since we cannot predict with certainty how future bad-actor–AI will evolve, given the rapid pace of technological change and the shifting online landscape, these forecasts are inherently speculative. Still, our predictions are quantitative, testable, and generalizable, providing a solid baseline for strengthening discussions about bad-actor–AI policies. There is room for improvement and extensions—e.g. what if future AI

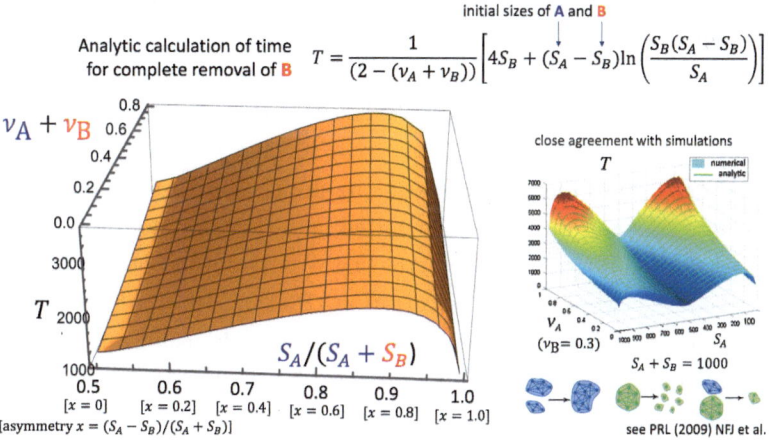

Fig. 6.9: Summary of eqn (6.57) which has been rewritten, incorporating non-zero fragmentation rates ν_A and ν_B. The top right panel confirms the excellent agreement between the theoretical predictions and numerical simulations.

can anticipate cluster dynamics (which current ChatGPT cannot) and outsmart containment strategies? Bad actors might also utilize decentralized or blockchain-based platforms as perpetual GPT-like 'reactor cores' of unstoppable bad-actor content. More specific subclassifications of 'bad actor' (e.g. anti-semitic vs anti-women) and 'vulnerable mainstream community' would also refine the approach. Additionally, links from vulnerable mainstream communities back into bad-actor communities could be incorporated. As bad-actor–AI evolves, these predictions can be updated accordingly.

6.6 Conclusion and final remarks

This chapter introduced a new way to analyse, model, and thus shape policies for the proliferation of bad-actor–AI activities in social-media contexts. By combining empirical data with dynamical systems modelling, we illuminate both current patterns and potential futures of bad-actor–AI. Our results clarify the challenges that platform moderators and legislators face, even against bad actors using only basic AI models. We also highlight the often-overlooked importance of smaller platforms in providing key linkages that help form a larger ecosystem, and we emphasize the increasing risk of AI abuse in conjunction with future global events such as elections. The proposed mathematical tools offer valuable insights into containment versus removal policies for managing these threats.

Although we establish a foundational framework for understanding and addressing bad-actor–AI risks, we also underscore the need for ongoing research in this rapidly advancing field. The constantly evolving online community ecosystem and emerging AI technologies mean that continuous refinement and adaptation of these models is essential.

As a note on the empirical data, materials that reproduce our figures are available online via the GW Donlab website: `https://github.com/gwdonlab/data-access`. This repository grants readers access to the minimal dataset required to interpret, verify, and expand upon the research reported here. The code for data processing relies on standard Python libraries for web crawling (e.g. BeautifulSoup, lxml), quantitative analysis (pandas, numpy), and visualization (Plotly), all of which are open-source. We used Gephi (also free and open-source) for network visualizations, Mathematica for quantitative analysis, and Adobe Illustrator for final figure production. These are widely available tools, often provided through university site licenses.

Exercises

6.1 Consider a system of 2 aggregating species A, B that undergo competitive fusion with product kernel. Due to different microscopic characteristics the 2 species have different heterogeneous fusion rates $F_A \ll F_B$. Meanwhile, the 2 populations satisfy $N_A \gg N_B \gg 1$.

  (a) If the 2 species do not interact, derive there respective gelation onset time. Under what condition does species B gel first?

  (b) As species A, B begin competitive fusion, assuming that the condition in (a) holds, describe the late-time behaviour of the system. Specifically, does any species reach gelation or a steady-state distribution? If steady state emerges, what is the distribution?

6.2 From eqn (6.26), show that at $x = 0$ $d\mathcal{D}/dx$ recovers the correct result, i.e.

$$\mathcal{D}'(0) = -\mathcal{E}(0) = -N. \qquad (6.66)$$

6.3 This question is on a specific type of function.

  (a) Plot function $f(x) = x + \frac{\alpha}{x}$, $x > 0$ for an arbitrary real parameter α.

  (b) Prove that the factor ξ from eqn (6.30) satisfies $\xi \geqslant 4$.

  (c) Following the outline provided in the book, derive eqn (6.39) in detail. Why does its exponential factor decay for large s?

6.4 For competitive fusion with sum kernel, we have seen from Section 6.5.2 that the power-law depends on the 2 populations N, Q.

  (a) Derive the powers c and β as given by eqns (6.52) and (6.53).

  (b) Describe how c and β change with the ratio Q/N.

7
Online spreading: contagion and broadcast

This chapter adds some simple internal dynamics for each interacting entity, i.e. human, machine, software, or AI as in Fig. 1.2. By allowing each entity to be in a small number of different states, we explore contagion within the system of entities, and hence mimic the spreading of content which could include disinformation, misinformation, influence, malign activity, and other online harms. Obviously the nature of the content has a role to play in whether there is spreading or not, but this can be captured in the model by having different rates of infection for different types of content.

Information of all forms—whether extremist or not, and correct or not—needs to spread to be effective. When we discussed fusion and fission, we implicitly assumed that the spread of content produced by bad (or good) actors and their communities—contagion—would then occur in the background instantaneously, so that an entire cluster would be ideologically aligned immediately upon formation. While the assimilation of such information can be very fast, say at the second-by-second scale compared to the formation of links between individuals and in-built communities which may be a slower daily scale, this may not be generally true. There is also, as we show here, interesting new physics that arises when considering these processes all in operation at the same time. Working out how to describe this quantitatively, using suitable approximations, can then provide formulae to guide policy discussions—as opposed to simply pointing at the output of large-scale computer simulations which can definitely do the job but tend to provide little insight.

We have already seen models of fusion and fission in the earlier chapters. In the absence of any fusion–fission dynamics, there are many existing models for contagion in the well-mixed (so-called mass action) limit in which entities are assumed to instantly mix with each other as in a test tube reaction. But when and how should we combine the two processes—fusion–fission and contagion—to better describe the spreading of disinformation, influence, and malign content across the now-known online ecology?

The answer to that depends on the timescale of spreading compared with that of fusion and fission. If the spread is too rapid or too slow, any contagion dynamics is dominated by the effects of the fusion–fission dynamics since these dictate the links and hence pathways through which content will spread. It is only when the two types of interactions have comparable timescales that we need to consider a new, combined dynamics. This is what we address in this chapter: contagion together with clustering dynamics. Our main interest is in whether some piece of content spreads or not, by which we mean implicitly *spreads system-wide* or not and hence the spreading versus no-spreading condition.

Much of what follows is drawn heavily from the published papers by some of us, including 'Effect of social group dynamics on contagion' (Zhao, Calderón, Xu, Zhao, Fenn, Sornette, Crane, Hui, and Johnson, 2010), and 'Atypical viral dynamics from transport through popular places' (Manrique, Xu, Hui, and Johnson, 2016). We are also extremely grateful to Prof. Pak Ming Hui and Prof. Chen Xu for their collaboration on some of these other papers and the common-space models.

7.1 Classic spreading models: epidemiology

We begin this chapter by introducing the classic models of contagion. These models are commonly used in epidemiology to determine the possible spreading of an infectious disease. These models became widely known during COVID, when numerous predictions of infection curves were published based on such models and their variations.

The basic assumption of such infectious, or viral, models is that each entity in the population is at any time in one of a few categories that describe the entity's condition, e.g. susceptible (S), infected (I), and recovered (R), etc. In a population that is going through an epidemic, individual entities transition between these states. This gives the name to some of the simplest viral models, such as the *SIR* model where an individual starts from being susceptible to an infection, then gets infected, and finally recovers and becomes immune; and the *SIS* model, where instead of becoming fully recovered and immune from infection, the recovered individual is instead labelled as 'susceptible' again for future re-infection. These models are easily extended to account for additional desired properties, such as defining the space in which the population interacts, or adding new states such as the deceased, etc. Moreover the dynamics of these models is not too difficult to describe: simple population balance together with infection mechanics suffices for writing down their differential equations. As preparation for what is to come later in the chapter, this section presents an overview of the two classic infection models, *SIR* and *SIS*, and some of the mathematical properties associated with their corresponding ODEs.

7.1.1 The SIR model

We begin with the well-known *SIR* model. In this model we assume that a susceptible individual has a probability q_i of being infected from interaction with an infected individual, and that the infected individual recovers at a rate q_r. Then the dynamics of $S(t), I(t), R(t)$ is straightforward:

$$\frac{dS}{dt} = -q_i \frac{SI}{N}, \tag{7.1a}$$

$$\frac{dI}{dt} = q_i \frac{SI}{N} - q_r I, \tag{7.1b}$$

$$\frac{dR}{dt} = q_r I, \tag{7.1c}$$

where $N \equiv S + I + R$ is our usual notation for the total population. Since this model is devised for a closed system, N is always conserved, but later we shall see a modification of this vanilla model for a semi-open system. We note that on some occasions, the RHS of eqns (7.1a) and (7.1b) may appear without the denominator N. This is merely a choice of different notation conventions, as constant N can be absorbed into the coefficient q_i in principle. Here we choose to retain it due to dimensionality: by keeping N, the infection and the recovery rates q_i and q_r are both dimension-free and hence are more convenient for comparisons. For potential applications of the other convention, we can define a scaled infection rate

$$Q_i = \frac{q_i}{N} \tag{7.2}$$

of dimension $[N]^{-1}$. Naturally this quantity is much smaller than 1.

Simple as eqns (7.1) look, they are nonetheless non-linear, so non-trivial to solve. We note, however, that eqn (7.1c) is decoupled from the rest, so we leave it aside. Though it is hard to solve eqns (7.1a) and (7.1b) with respect to time t analytically due to non-linearity, it is not hard to compare directly $S(t)$ and $R(t)$. Applying the chain rule, we find that

$$\frac{dR}{dS} = -\frac{q_r N}{q_i S} \tag{7.3}$$

A simple integration gives

$$R = -\frac{q_r}{q_i} N \ln S + \text{const}, \tag{7.4}$$

where the constant depends on the initial condition. We further assume that the total size of our population $N \gg 1$, and that at $t = 0$, the very beginning of the infection process, there is $I = 1 \ll N$ infected and no recovered individual. All the remaining $S = N - 1 \approx N$ are therefore susceptible. Hence, the constant is $(q_r/q_i)N \ln(N-1) \approx (q_r/q_i) N \ln N$. Therefore, the relation between R and S is

$$R = \frac{q_r}{q_i} N \ln \frac{N}{S}. \tag{7.5}$$

There is a bit more to this expression. Let's first return to the ODEs to see qualitative trends of $S(t)$, $I(t)$, and $R(t)$ without solving them against t. To begin, it is straightforward to see that $dS/dt < 0$ and $S(t)$, i.e. the susceptibles, must decrease over time; but for $R(t)$ the opposite is true, i.e. the recovered individuals increase. The qualitative dynamics of $I(t)$ is slightly less trivial, but by the same reasoning, we can see from eqn (7.1b) that infection reduces when

$$\frac{S(t)}{N} < \frac{q_r}{q_i}. \tag{7.6}$$

Since $S(t)$ decreases over time, this boundary condition can be met at any time depending on the relative values of q_i and q_r. If $q_r \gg q_i$, i.e. the recovery rate is much larger than the infection rate, then it is possible that eqn (7.6) holds for all $t \geqslant 0$, and thus the epidemic never spreads. To be more exact, this requires

$$\lambda \equiv \frac{q_i}{q_r} < 1, \tag{7.7}$$

since $S(0) \approx N$. λ is called the *basic reproduction rate*. We can substitute λ back into eqn (7.5) to obtain the relation between S and R with the basic reproduction rate as the sole control parameter:

$$S = N e^{-\lambda R/N}. \tag{7.8}$$

We have so far discussed the dynamics of the SIR model and its boundary conditions for different behaviours. The last thing to be discussed is the steady state—in other words, what the system looks like in the limit of large time $t \to \infty$. There is no guarantee that the system reaches a steady state in the end. For a closed system like ours, there are two possibilities: either the system reaches steady state, or it oscillates. We need a closer look at eqns (7.1) to confirm that oscillation is impossible: since all three functions $S(t), I(t), R(t) \geqslant 0$, we know from eqns (7.1a) and (7.1c) that

$$\frac{dS(t)}{dt} \leqslant 0, \quad \frac{dR(t)}{dt} \geqslant 0 \tag{7.9}$$

for all time t. This means that $S(t)$ never increases with t and $R(t)$ never decreases, so oscillation is indeed impossible, and we must have a steady state at $t \to \infty$ when all three time derivatives go to 0.

184   Online spreading: contagion and broadcast

With this cleared up, let's look at the composition of the steady state. From eqn (7.1c) we must have

$$I(t \to \infty) = 0. \tag{7.10}$$

The other 2 compositions can be obtained from eqn (7.8). Since $S(t \to \infty) + R(t \to \infty) = N$, we substitute in eqn (7.8) and get

$$\frac{R}{N} + e^{-\lambda R/N} = 1, \tag{7.11}$$

where, for convenience, we denote $R(t \to \infty)$ simply as R. To prevent any confusion, we define the steady-state proportion of the recovered population as

$$z = \frac{R(t \to \infty)}{N}. \tag{7.12}$$

It is governed by the equation

$$z + e^{-\lambda z} = 1, \tag{7.13}$$

which we note is exactly the same equation as the gelation solution of $\varepsilon(x = 0, t)$ in terms of $(F\tau)$ (eqn (5.2)), and also the GCC solution of size G in terms of mean connection degree c (eqn (5.45)). When the basic reproduction rate $\lambda \leqslant 1$, $z = 0$ which means that none of the individuals in the system ever gets infected and hence recovers later. This is exactly the non-spread condition mentioned in eqn (7.7). When $\lambda > 1$ however, the recovered population, i.e. those who have been infected at some stage during the time evolution, increases with the basic reproduction rate λ following the same shape as the gelation curve.

7.1.2   The SIS model

Next up is the *SIS* model. We follow the same steps as above. First we look into the dynamics of this model. Then we study the boundary conditions for different regimes. Finally we check its steady-state situations.

The dynamics. Since there are only 2 states in the *SIS* model, its dynamics is encapsulated in 2 coupled ODEs:

$$\frac{dS}{dt} = -q_i \frac{SI}{N} + q_r I, \tag{7.14a}$$

$$\frac{dI}{dt} = q_i \frac{SI}{N} - q_r I. \tag{7.14b}$$

Again $N = S(t) + I(t)$ for this closed system, and eqns (7.14a) and (7.14b) add up naturally to 0. Different from the *SIR*, however, this system can easily be separated into a single-variable ODE and then solved. We substitute $S = N - I$ back into eqn (7.14b) to get

$$\frac{dI}{dt} = -q_i \frac{I^2}{N} + (q_i - q_r)I. \tag{7.15}$$

Some may have already identified that this is the logistic equation, and when $q_i \neq q_r$ the solution is simply a modified logistic function

$$I(t) = \frac{N(1 - 1/\lambda)}{1 + Ae^{-(q_i - q_r)t}}, \tag{7.16}$$

where A is a constant determined by the initial condition

$$A = \left(1 - \frac{1}{\lambda}\right) \frac{N}{I(0)} - 1. \tag{7.17}$$

In the special case of $q_i = q_r$, the solution is instead a hyperbola:

$$I(t) = \frac{N}{q_i t + N/I(0)}. \tag{7.18}$$

We shall soon discuss all these scenarios, but for now let's focus on the logistic function. The original logistic function $f(x) = (1 + e^{-x})^{-1}$ takes the well-known S-shaped curve, as shown in Fig. 7.1. It is symmetric about point $(0, 1/2)$, which is the inflection point of the curve. $I(t)$ preserves this shape under scaling and translation. The scaling part is simple: we see that the shape is scaled in the vertical direction by a factor $N(1 - 1/\lambda)$, and in the horizontal direction by a factor $(q_i - q_r)^{-1}$. Translation can be seen with just a few steps of derivation:

$$I(t) = \frac{N(1 - 1/\lambda)}{1 + e^{-(q_i - q_r)t + \ln A}} = N\left(1 - \frac{1}{\lambda}\right)\left[1 + e^{-(q_i - q_r)\left(t - \frac{\ln A}{q_i - q_r}\right)}\right]^{-1}, \tag{7.19}$$

i.e. the plot can be obtained by first scaling the original S-shaped curve as mentioned above, and then translating it by $(\ln A)/(q_i - q_r)$.

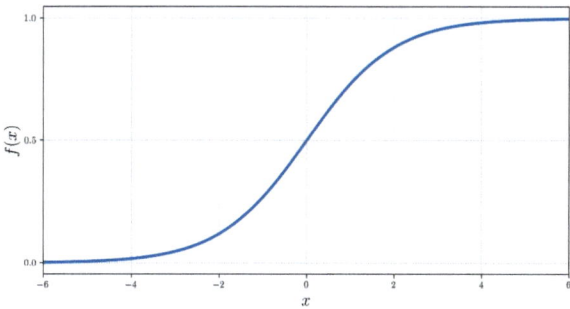

Fig. 7.1: The logistic curve $f(x) = (1 + e^{-x})^{-1}$.

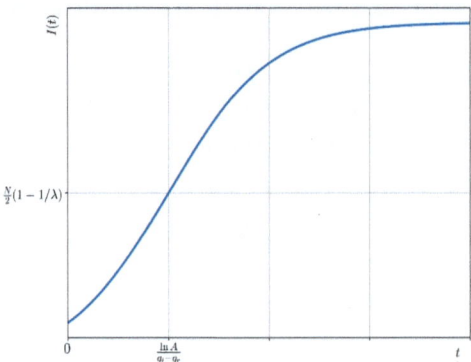

Fig. 7.2: $I(t)$ graph obtained by linear operations on the logistic curve, for $q_i > q_r$.

This is a great help for studying the dynamics of $I(t)$. Below is a summary of its main properties. Throughout the discussions we assume that the system begins with a few infected individuals and most of the population is susceptible, i.e. $A \gg 0$, unless noted otherwise.

1. When $q_i > q_r$, or equivalently when the basic reproduction rate $\lambda > 1$, $I(t)$ grows with time as shown in Fig. 7.2, and reaches steady state (asymptotically) in the long time limit $t \to \infty$. For a small initial population $I(0) \to 0$, it can take a long time, up to λ, to establish contagion due to the flat tails of the logistic curve. By contrast if $I(0)$ is very large such that $A < 0$, the infected population shrinks to approach the steady state.

2. For $\lambda > 1$, in the steady state both the infected and the susceptibles are present in the system in a dynamic equilibrium. Their sizes are, respectively,

$$I(t \to \infty) = \left(1 - \frac{1}{\lambda}\right)N, \quad S(t \to \infty) = \frac{1}{\lambda}N. \tag{7.20}$$

We can see that as we take λ values close to 1, I takes up less of a proportion of the whole system, and for large times fewer individuals are infected.

3. For the special case $\lambda = 1$, the steady state is all susceptibles and no infected. If the system starts with a few infected individuals, $I(t)$ decreases hyperbolically with time all the way to 0.

4. When $\lambda < 1$, we rewrite eqn (7.16) as

$$I(t) = N\left(\frac{1}{\lambda} - 1\right)\left[\left(\left(\frac{1}{\lambda} - 1\right)\frac{N}{I(0)} + 1\right)e^{(q_r - q_i)t} - 1\right]^{-1}, \tag{7.21}$$

such that all coefficients are positive. Contrary to the $\lambda > 0$ scenario, $I(t)$ decreases to approach the steady state $I(t \to \infty) = 0$.

The steady state. Here we summarize the above discussions to clear up the picture for the steady-state solutions. They can be encoded in the following equations:

$$\frac{I(t \to \infty)}{N} = \begin{cases} 0 & \lambda \leq 1 \\ 1 - \frac{1}{\lambda} & \lambda > 1 \end{cases}, \quad \frac{S(t \to \infty)}{N} = \begin{cases} 1 & \lambda \leq 1 \\ \frac{1}{\lambda} & \lambda > 1 \end{cases} \quad (7.22)$$

This is somewhat similar to its *SIR* counterpart, in that both steady states see a phase change between 2 different regimes (spreading and non-spreading) at $\lambda = 1$. The difference lies merely in the response of S to $\lambda > 1$ values: for *SIR* the response is transcendental, whereas here for *SIS* it is hyperbolic, and hence much less sensitive to λ. Figure 7.3 compares the steady-state solutions for both models.

7.2 Contagion: entity-to-entity

7.2.1 Fusion–fission plus contagion: tipping points

Up until now the contagion language has not had anything to do with the fusion–fission-like behaviours we previously discussed. In the context of online spreading, we can roughly divide the spread of information into 2 types: entity-to-entity or broadcast. Each entity could be an in-built community or some other human–technology–AI entity (Fig. 1.2) as discussed in previous chapters. We will discuss the broadcast type later on in the chapter, but for entity-to-entity spreading, especially contagion of harmful content, we often consider this to occur among groups (e.g. between in-built communities online) due to the often exclusive and sensitive nature of these groups and their associated content.

For a model that incorporates contagion and fusion–fission (example schematics shown in Fig. 7.4), the first thing that comes to our mind is naturally the timescales

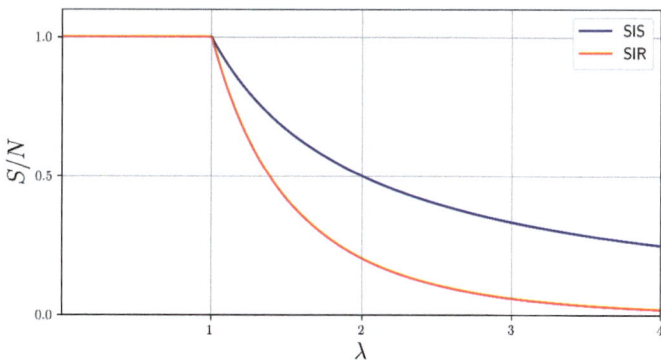

Fig. 7.3: Steady-state S proportion in the system of both *SIS* and *SIR* models.

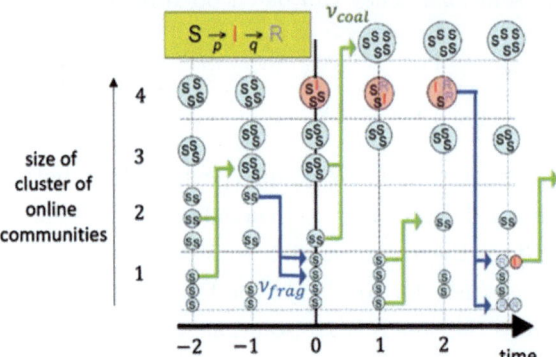

Fig. 7.4: Schematics of fusion–fission with a simultaneous contagion process. Each circle is a cluster whose size is given by the number of entities (e.g. linked in-built communities) within it. These clusters undergo fusion and fission that allows misinformation and malicious materials to pass from infected entities to susceptible ones within the same cluster, then spread more widely as these clusters coalesce and fragment. Source: data from Zhao et al. (2010).

of both types of interactions. As we mentioned earlier on in this chapter, it only makes sense to combine the two pieces of theory when fusion and contagion occur at comparable timescales. If dis(mis)information spreads too slowly, there may never be a chance to spread before the entities gather into a giant cluster and subsequently get shattered into isolated entities over and over again. On the contrary, if misinformation spreads much faster than the rate of fusion, then the cluster dynamics become essentially static.

However when these two types of events occur at a comparable rate, we should blend the epidemiological model with the fusion–fission model. The key to this is to quantify how many of the susceptible entities and the infected entities are connected by a vertex, or in the language of fusion–fission, coexist in a cluster, such that contagion can occur at rate q_i. Strictly speaking, these two interpretations are not exactly identical. Let's first look at the network framing. We write down the probability $\eta_{\text{SI}}(t)$ that a particular link instantaneously exists *and* that it connects a susceptible and an infected entity. Hence out of a potential totality of $N(N-1)/2$ links among N entities, only $\eta_{\text{SI}}(t) \cdot N(N-1)/2$ are typically viable infection routes. This provides an accurate equation for the number of susceptibles $S(t)$ in a given epidemic sequence:

$$\dot{S}(t) = -q_i \eta_{\text{SI}}(t) \frac{N(N-1)}{2} . \tag{7.23}$$

We rewrite this in the form of a conventional *SIR* model

$$\dot{S}(t) = -q_i P_{\text{SI}}(t) S(t) I(t), \tag{7.24}$$

where we define
$$P_{\text{SI}}(t) = \frac{N(N-1)\eta_{\text{SI}}(t)}{2S(t)I(t)}. \tag{7.25}$$

Physically this is just the probability that an infected entity is connected to a susceptible one. However, there do not exist any equations that analytically govern such complex dynamics, not least because the instantaneous dynamics is determined by microscopic actions of individual entities, which are intractable on a system level.

By contrast, the cluster interpretation simplifies the problem substantially. Recall that our cluster model assumes that all entities in the same cluster quickly become fully interconnected and hence can freely talk to all the other fellow entities in the same cluster. Then as a result we need not worry about pairing the susceptibles and the infected. In other words, we need not calculate $P_{\text{SI}}(t)$, but can instead regard contagion within clusters just like a normal contagion process. Mathematically this means substituting the strict quantity $P_{\text{SI}}(t)$ with a mean-field probability $P(t)$ that any two arbitrarily chosen nodes, regardless of SI-infection status, belong to the same cluster. This has been covered back in Section 3.7.2. We derived exactly the connection probability in eqn (3.139), and we rewrite it here:

$$P \approx \frac{\varphi_c}{N\varphi_f}, \tag{7.26}$$

where P is the steady state probability $P(t \to \infty)$, and φ_c, φ_f represent fusion and fission rates respectively. Note that P is only determined by the fusion–fission dynamics and is independent of the contagion process. This means that our connection probability in eqn (7.26) is universally applicable regardless of the infection schemes—no matter if the infected entities recover permanently or turn back into susceptibles, we can approximate their aggregation status with $P = \varphi_c/(N\varphi_f)$.

For simplicity, we demonstrate its application with the classic SIR scheme. As stated in the assumptions, we modify the SIR model simply by inserting the factor P into the equation, just as in eqn (7.24). This means turning all parameter q_is in derivations of the original model, into Pq_i. The proportion of recovered individuals is simply $(1-z)$. Denote

$$\kappa \equiv P\lambda = \frac{q_i \varphi_c}{N q_r \varphi_f}, \tag{7.27}$$

and we have

$$z = e^{-\kappa(1-z)}, \tag{7.28}$$

where κ now plays the role of the basic reproduction rate. The form of this epidemic control parameter shows explicitly that infection and fusion compete with recovery and fission in controlling the propagation of the spreading: the former promote the infection propagation, while the latter hinder its spread.

190 Online spreading: contagion and broadcast

Not only is our theory for the spreading threshold in good agreement with numerical results, its simple analytic form suggests an epidemic control scheme based on manipulation of link timescales. An imminent system-wide spreading can be *suppressed* by increasing the timescale for creating links with respect to the timescale for losing links (i.e. decrease φ_c with respect to φ_f), but it will get *amplified* if we decrease the link-creation timescale with respect to the link-loss timescale (i.e. increase φ_c with respect to φ_f).

7.2.2 Tipping points with D species: a generalization

The population of entities across which transmission of toxic misinformation and malicious materials occurs is inherently heterogeneous—whether we are talking solely about in-built communities online or the array of more general entities in future human–technology–AI systems as in Fig. 1.2. Recall that we invented 'species' in Chap. 4 to model this heterogeneity. So how would such heterogeneity affect node-to-node contagion? Based on our previous simplification of the problem, fusion–fission (which controls the connection between entities) and contagion (which determines the infection dynamics of each entity) are captured by separate variables. Consequently, higher levels of heterogeneity can be realized without touching the epidemiological part of the theory (see Fig. 7.5). Specifically, we simply need to alter P.

Back in Section 4.5, we established 2 particular models with multi-species fusion and fission. When we have a 'global censorship' responsible for unified fission, the

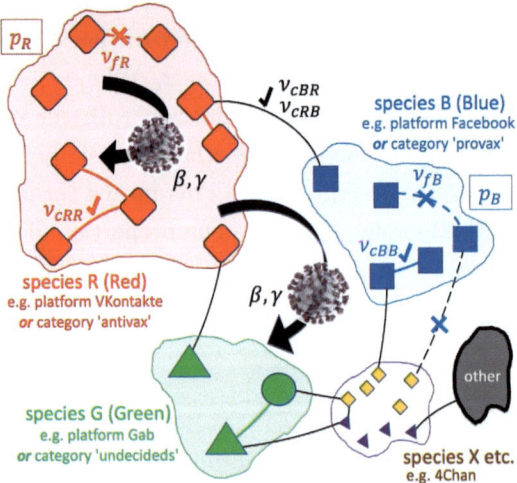

Fig. 7.5: Schematic diagram of multi-species fusion–fission with contagion. In this setting, each entity could be an in-built community on a given platform shown as a given colour.

resulting connection probability is exactly

$$P(t) = \frac{\bar{\varphi}_c}{N\varphi_f}, \qquad (7.29)$$

where φ_f is the unified fission rate, and

$$\bar{\varphi}_c = \sum_{\alpha=1}^{D} \sum_{\beta=1}^{D} \varphi_{\alpha\beta}^{(c)} \rho_\alpha \rho_\beta \qquad (7.30)$$

is the weighted average of all possible aggregation rates $\varphi_{\alpha\beta}^{(c)}$, where ρ_α represents the proportion of species-α entities. By contrast when each species is responsible for its own fission mechanics, there is no analytic result: but eqn (7.29) serves as a lower bound and the population is always more connected compared to the 'global censorship' scenario. This then modifies the tipping-point expression into

$$\kappa_{\text{tp}} = \frac{q_i \bar{\varphi}_c}{N q_r \varphi_f} = 1. \qquad (7.31)$$

We note that the tipping point is smaller than 1 in the case of decentralized fission.

7.3 Simplified approach for large-time contagion: co-existing mobility and infection dynamics

The above model produces the intriguing phase-change result. Its downside is that its numerical simulation can be a little too complicated, while its analytical theory a little too crude with too many averages taken.

Our take on these two issues is to simplify the fusion–fission process while emphasizing the epidemiological elements. To be more exact, we know from Chap. 3 that the fusion–fission system arrives at a steady state—when the fission rate is non-zero—which exhibits an approximate -2.5 exponent power-law cluster size distribution. This means that a large proportion of all clusters are tiny and do not carry much of the contagion process. By contrast, it is the few large clusters that experience and contribute to most of the spread. Moreover we know that at steady state, fusion and fission are still underway dynamically while the cluster size distribution remains unchanged. Accordingly we can approximate this dynamic equilibrium as entities entering and leaving the larger clusters perpetually, while neglecting the smaller clusters since they do not contribute much towards contagion. This is not exactly the same mechanism as our beloved fusion–fission, but it is a working approximation for contagion-centred considerations.

Here we introduce the details of this model. We start by defining a generic popular space G as an abstraction of the 'few large clusters' mentioned above. Such a space G

can represent the large connected components in a popular social network where users transit for a limited period of time, interact with all other users inside G and hence spread information, and then leave. Non-trivial viral dynamics emerges from the direct interplay between the mobility through, and the average occupancy of, the space G. Users outside G do not interact with one another: this mimics the fact that while they are outside the largest cluster(s), they will most likely be in a cluster of size one and hence isolated in terms of the spreading process. Using this framework, we consider different epidemiological models in order to mimic the information transfer process. Later in this section we take the 2 classic models as examples: the SIR model and the SIS model. The SIR model contains recovery or 'obtaining immunity'. Since recovery is a one-body process, it occurs both inside and outside space G. This model is a 'semi-open' system as opposed to the closed system seen in the original epidemiological models, since the infection space G is connected to a 'reservoir space', between which entities can freely migrate—a bit like a grand canonical ensemble in statistical physics. The broadcast model to be discussed in the next section is just a modification of this.

Figure 7.6 illustrates our simplified mathematical model of N individual entities, where each entity can be an in-built community or an individual user or a general entity in some human–technology–AI system as in Fig. 1.2, depending on the nature and scale of the intended application. Given this generality, from now on we will simply refer to N 'individuals'. They all have access to a popular space G, e.g. an online information group in the case of individual entities, or a largest cluster in the more abstract case of nodes. They are all subject to an SIR infection dynamic. At any given timestep, an

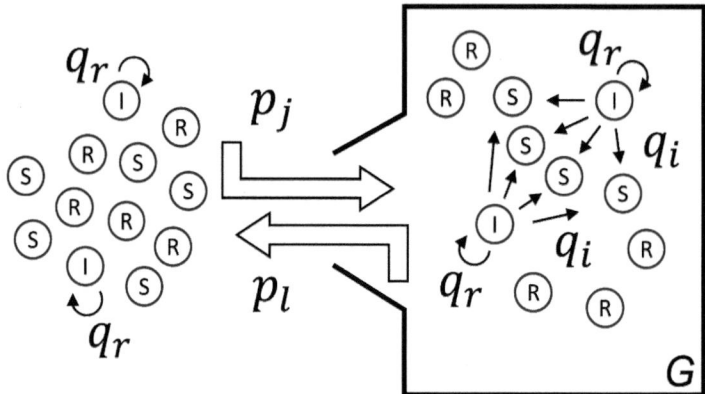

Fig. 7.6: Schematic diagram showing our model of co-existing mobility and viral spreading. Infection occurs within space G, where an infected entity has a probability q_i of infecting any susceptible entity also inside G. Simultaneously, each infected entity, whether inside or outside G, recovers with probability q_r as in standard SIR models.

individual not in G has a probability p_j of joining G while an individual inside G has a probability p_l of leaving. This effectively generates a dynamical group in G, with an occupancy $N_g(t)$ (i.e. number of individuals in G) which fluctuates arbitrarily in time. In the mean-field limit, the dynamics of $N_g(t)$ follows the following differential equation:

$$\frac{d}{dt}N_g(t) = p_j(N - N_g(t)) - p_l N_g(t), \qquad (7.32)$$

where N is the total number of individuals. The mean number of individuals inside G can be derived from the steady state of eqn (7.32), yielding

$$\langle N_g(t) \rangle = N\frac{p_j}{p_j + p_l} \equiv N\gamma_s, \qquad (7.33)$$

where we define γ_s to measure the relative strength of the influx probability (with respect to the outflux). The average number of individuals both joining and leaving G in a timestep characterizes the *mobility* μ of the individuals:

$$\mu = (N - \langle N_g(t) \rangle)p_j + \langle N_g(t) \rangle p_l = N[(1-\gamma_s)p_j + \gamma_s p_l] = N\frac{2p_j p_l}{p_j + p_l} \equiv N\gamma_m. \quad (7.34)$$

In the above equation we denote the so-called *harmonic mean* of the influx and outflux probabilities as γ_m, i.e.

$$\frac{1}{\gamma_m} = \frac{1}{2}\left(\frac{1}{p_j} + \frac{1}{p_l}\right). \qquad (7.35)$$

Having defined a few new variables, let's take a look at their scaling relationships. When p_j and p_l are scaled by a factor r, it is clear from the fractional structure that γ_s, and hence $\langle N_g(t) \rangle$, do not scale with p_j and p_l, and so remain unchanged. Meanwhile, the mobility μ scales linearly by a factor r. Hence varying γ_m while fixing γ_s amounts to changing the mobility while keeping $\langle N_g(t) \rangle$ fixed, i.e. keeping a steady size of this 'interactive space' G while being able to tune the rate of population flow.

7.3.1 Entity-to-entity SIR

Below we consider the two infection schemes. We start with the *SIR* model where immunity is attainable. Just to recap its mechanism: at any timestep, any infected individual (entity) within the popular space G can transmit a virus to any susceptible individual (entity) in G with probability q_i. In all cases, no transmission can occur from infected individuals (entities) outside G. By contrast, since recovery is an individual-based phenomenon, infected individuals both inside and outside G have a probability q_r of recovering and becoming immune. We note that we can equivalently view the individuals in G as instantaneously connected—hence our model and results can represent N nodes in a time-dependent network, or it can be applied to the common real-world scenario of a social group with time-varying membership (Palla et al., 2007). Our regime of focus in this section, in which a popular space has fairly constant occupancy but variable throughput, is of direct relevance to online social groups in, for example, multi-player online games, where it is known that these groups (e.g. guilds) have a size that is fairly constant yet a membership that changes rapidly over time. We stress, however, that our model is far more general in that it allows for any rate of change of occupancy and throughput.

Since the $S \to I$ process only occurs inside G, we use $S(t)$, $I(t)$, $R(t)$ to denote the number of susceptible, infected, and recovered individuals respectively, in the *whole* system, and $S_g(t)$, $I_g(t)$, and $R_g(t)$ for the corresponding numbers within the space G. The six equations that describe the dynamics of an SIR process for a single dynamical group are as follows:

$$\frac{dS}{dt} = -q_i \frac{S_g I_g}{N}, \quad \frac{dI}{dt} = q_i \frac{S_g I_g}{N} - q_r I, \quad \frac{dR}{dt} = q_r I, \tag{7.36a}$$

$$\frac{dS_g}{dt} = -q_i \frac{S_g I_g}{N} - p_l \left(S_g - q_i \frac{S_g I_g}{N}\right) + p_j (S - S_g), \tag{7.36b}$$

$$\frac{dI_g}{dt} = q_i \frac{S_g I_g}{N} - q_r I_g - p_l \left(I_g + q_i \frac{S_g I_g}{N} - q_r I_g\right) + (1 - q_r) p_j (I - I_g), \tag{7.36c}$$

$$\frac{dR_g}{dt} = q_r I_g - p_l (R_g + q_r I_g) + p_j ((R - R_g) + q_r (I - I_g)). \tag{7.36d}$$

Equations (7.36a) are directly modified from the classic SIR model, such that infection only occurs via interactions of S_g and I_g populations, namely those inside space G. The rest of the equations are modified with additional migration terms. Terms with a p_l coefficient represent outflux from space G while those with p_j represent influx into G. There exist second-order terms with respect to the timescale (namely the product $p_a q_b$ as coefficients, with $a = l, j$ and $b = i, r$). The extra term $p_l q_i \frac{S_g I_g}{N}$ in eqn (7.36b), for example, compensates the loss of susceptible individuals due to infection as they migrate out of G. They are essentially composite actions of migration plus contagion, and when the rates are low, i.e. $p_a, q_b \ll 1$ which is normally true for a steady state, they can be safely neglected. The linearized equations, in terms of the timescale, follow:

$$\frac{dS}{dt} = -q_i \frac{S_g I_g}{N}, \quad \frac{dI}{dt} = q_i \frac{S_g I_g}{N} - q_r I, \quad \frac{dR}{dt} = q_r I, \tag{7.37a}$$

$$\frac{dS_g}{dt} = -q_i \frac{S_g I_g}{N} - p_l S_g + p_j (S - S_g), \tag{7.37b}$$

$$\frac{dI_g}{dt} = q_i \frac{S_g I_g}{N} - q_r I_g - p_l I_g + p_j (I - I_g), \tag{7.37c}$$

$$\frac{dR_g}{dt} = q_r I_g - p_l R_g + p_j (R - R_g). \tag{7.37d}$$

The original eqns (7.36) are required when there are large fluctuations of mobility γ_m, which means that both p_l and p_j are large at some stage during the evolution.

Equations (7.37) are not analytically solvable and a numerical treatment is needed to uncover the dynamics of the system in detail. Figure 7.7 shows 3 exemplary sets of results we can obtain from the dynamical model, illustrating the rich behaviours obtainable by controlling the parameters, i.e. mobility γ_m, basic reproduction rate λ, together with the timescale of contagion q_i, especially relative to the migration timescale. In all scenarios, we keep the mean number of individuals in G, $\langle N_g \rangle = \gamma_s N$, unchanged, where $\gamma_s = 0.1$. The main panel shows the trajectories of the transient I values against S values—they all start from the lower right-hand corner, as initially

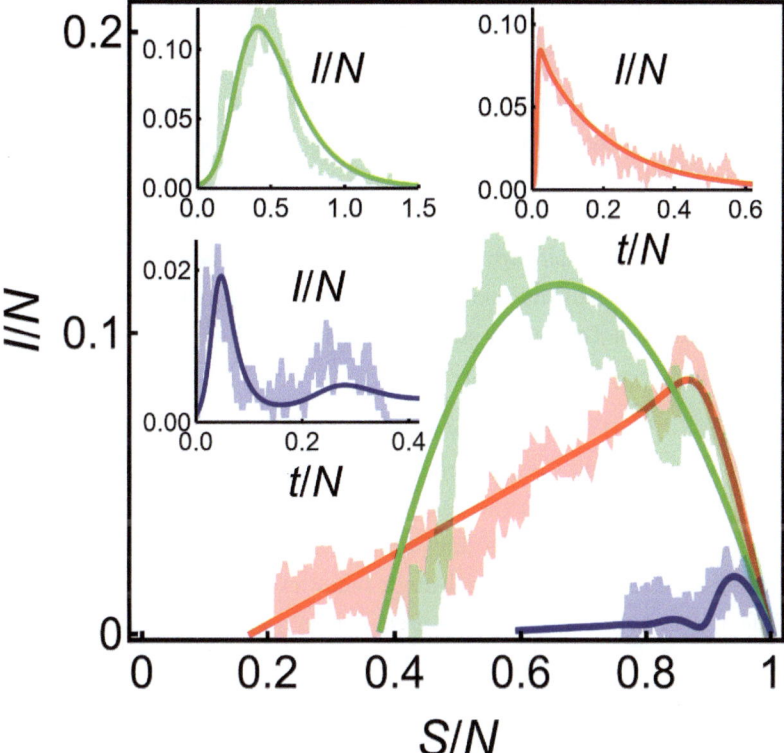

Fig. 7.7: Trajectories of the susceptible and the infected subpopulations for three different sets of parameter choices. Theoretical values (smooth curves, strong colour) vs simulation results (noisy curves, shaded colour). We are very grateful to Prof. Pak Ming Hui and Prof. Chen Xu for creating the first version of this figure during our ongoing collaboration together. Source: data from Manrique, Xu, Hui, and Johnson (2016)

we have $S/N \sim 1$ and $I/N \sim 0$. The insets plot the dynamic response of the infected population $I(t)$. They are plotted in 3 different colours against corresponding numerical simulations (rough curves in shaded colours). In the simulations, all individuals are initially susceptible and we allow the system to run until the group size in G reaches its steady-state value $\gamma_s N$. We then randomly pick a node in G and mark it as infected. In every subsequent timestep, all the individuals carry out the SIR process followed by joining or leaving of G. Below we look into their detailed behaviours, which are in sharp contrast with the SIR model in a well-mixed population.

1. Green curves. These are plotted with $\gamma_m = 0.018$, $\lambda = 0.1$, and $q_i = 0.001$. Here we take $\gamma_m \gg q_i$, i.e. the timescale for moving in and out is much smaller than that of contagion and recovery. This means that relatively speaking, the entire population interacts frequently and mixes well. As a result, the green curves look similar to the classic SIR model in free space.

2. Red curves. These are plotted with $\gamma_m = 0.009$, $\lambda = 0.1$, and $q_i = 0.005$. Here the values of γ_m and q_i are more comparable than above, and both smaller than the green-curve scenario. We see that due merely to the reduced mobility without changing the basic reproduction rate, the initial surge of infection becomes much more rapid since more susceptible individuals are confined within space G. However, the increased confinement also results in fewer individuals staying infected at any time instant after the initial surge, even though the infection rate here is much larger than the previous scenario.

3. Blue curves. These are plotted with $\gamma_m = 0.0018$, $\lambda = 0.022$, and $q_i = 0.002$. Here both the mobility and the basic reproduction rate are much smaller than the other two scenarios, resulting in the very low infection number. The key here, however, is a new, oscillatory regime of infection which is never seen in a classic SIR model. Low mobility rate together with high recovery rate means that a large number of infected individuals have recovered before new individuals flow into space G. Hence the infection number decreases first before ticking up again and oscillating.

Above we have seen that the green curves look similar to a classic SIR process. This has us wondering if at all, and under what conditions, it is possible to regard our dynamical model as an effective SIR process. The simplest way to achieve effective SIR is to have well-mixed populations in all three infection stages, such that S_g/S etc. are always constant. This appears to be exactly the case of the green curve, where individuals move in and out of space G much faster than the infection rate. So our desired condition could well be that $p_{j,l} \gg q_{i,r}$. Intuitively, this means that 2 successive infection events are so sparsely apart that there is enough time in between for all individuals to move around, mix thoroughly, and re-establish balanced proportionality. This is exactly so, and below we shall see mathematically why this condition suffices.

To begin, we apply our assumption $q_i \ll p_l$ to eqn (7.37b) and neglect the infection term:

$$\frac{dS_g}{dt} \approx -p_l S_g + p_j(S - S_g), \tag{7.38}$$

since it is negligibly small compared to the remaining relocation terms. But we quickly recognize that by the well-mixed initial condition, the RHS is 0 at $t = 0$, and therefore remains so for the whole time. Hence our slow-infection assumption leads directly to

$$S_g(t) \approx \gamma_s S(t), \tag{7.39}$$

and by exactly the same reasoning the same thing is true for $I_g(t)$, i.e.

$$I_g(t) \approx \gamma_s I(t). \tag{7.40}$$

To conclude this, we substitute these two relations back into eqn (7.37a) to obtain

$$\frac{dS}{dt} = -\gamma_s^2 q_i \frac{SI}{N}, \tag{7.41a}$$

$$\frac{dI}{dt} = \gamma_s^2 q_i \frac{SI}{N} - q_r I, \tag{7.41b}$$

$$\frac{dR}{dt} = q_r I. \tag{7.41c}$$

This means that we have successfully transformed our model under the slow-infection condition into an effective SIR model where the effective infection probability is $q_{\text{eff}} = \gamma_s^2 q_i$ and the effective recovery probability is q_r. Let's look at the effective infection probability: since $0 \leq \gamma_s \leq 1$ and it represents the proportion of the whole population inside space G, $q_{\text{eff}} \leq q_i$ and we see that the existence of space G slows down the contagion process. Moreover, this impedance is quite powerful—it has a quadratic relation, not a linear one. Finally we check the extreme conditions: when $p_j \gg p_l$, $q_{\text{eff}} \approx q_i$ and our system recovers to a normal SIR model. This is quite obvious since few individuals leave space G while more are flowing in. On the contrary, when $p_j \ll p_l$ we naturally have $q_{\text{eff}} \approx 0$, and there is no infection throughout the system since individuals are escaping space G while few are flowing in.

7.3.2 Entity-to-entity SIS

We now study the same viral mobility model involving space G, but using the SIS process. As before, infected individuals (entities) can only infect others when they are present in the space G, and each infected individual (entity) in G has a probability q_i (per unit timestep) of infecting a susceptible in G. All the infected individuals in the system (inside and outside G) have a probability q_r of recovering and becoming susceptible again. At the beginning of the process, we randomly select an individual in G to be infected. The viral dynamics of the system are governed by the following equations in the mean-field limit:

$$\frac{dS}{dt} = -q_i \frac{S_g I_g}{N} + q_r I \tag{7.42a}$$

$$\frac{dI}{dt} = q_i \frac{S_g I_g}{N} - q_r I \tag{7.42b}$$

$$\frac{dS_g}{dt} = -q_i \frac{S_g I_g}{N} + q_r I_g - p_l \left(S_g - q_i \frac{S_g I_g}{N} + q_r I_g\right) + p_j [S - S_g + q_r(I - I_g)] \tag{7.42c}$$

$$\frac{dI_g}{dt} = q_i \frac{S_g I_g}{N} - q_r I_g - p_l \left(I_g + q_i \frac{S_g I_g}{N} - q_r I_g\right) + p_j (1 - q_r)(I - I_g). \tag{7.42d}$$

We see that adding up eqns (7.42c) and (7.42d) gives

$$\frac{dN_g}{dt} = -p_l N_g + p_j (N - N_g), \tag{7.43}$$

where we apply implicitly that $N = S(t) + I(t)$ and that $N_g(t) = S_g(t) + I_g(t)$. As before, eqn (7.43) yields the same steady-state particle distribution

$$\langle N_g \rangle = \gamma_s N. \tag{7.44}$$

For convenience we once again neglect the second-order terms in the rates pqs, and simplify eqns (7.42) into

$$\frac{dS}{dt} = -q_i \frac{S_g I_g}{N} + q_r I \tag{7.45a}$$

$$\frac{dI}{dt} = q_i \frac{S_g I_g}{N} - q_r I \tag{7.45b}$$

$$\frac{dS_g}{dt} = -q_i \frac{S_g I_g}{N} + q_r I_g - p_l S_g + p_j (S - S_g) \tag{7.45c}$$

$$\frac{dI_g}{dt} = q_i \frac{S_g I_g}{N} - q_r I_g - p_l I_g + p_j (I - I_g). \tag{7.45d}$$

Just as in the closed SIS system, a steady state should also be reached in this system as well. However, here the steady state is a common factor of both infection and migration, as we shall see later. We start by setting the RHS of eqns (7.45) to zero:

$$q_i \frac{S_g I_g}{N} - q_r I = 0 \tag{7.46a}$$

$$-q_i \frac{S_g I_g}{N} + q_r I_g - p_l S_g + p_j (S - S_g) = 0 \tag{7.46b}$$

$$q_i \frac{S_g I_g}{N} - q_r I_g - p_l I_g + p_j (I - I_g) = 0. \tag{7.46c}$$

This seems a rather simple set of equations with 3 unknowns, though the non-linear term $q_i S_g I_g / N$ could be problematic. But we notice that we can substitute eqn (7.46a) directly into eqn (7.46c) and remove it:

$$q_r I - q_r I_g - p_l I_g + p_j (I - I_g) = 0 \tag{7.47}$$

$$\Rightarrow \frac{I_g}{I} = \frac{q_r + p_j}{q_r + p_j + p_l} > \gamma_s. \tag{7.48}$$

Before we proceed with calculations, there is an intriguing point to notice. We see that the infected population does *not* distribute evenly inside and outside space G—which is understandable since individuals can only get infected inside G, hence causing this disparity. Now we continue with the calculations. From eqn (7.46a),

$$S_g = \frac{q_r I}{q_i I_g} N = \frac{N}{\lambda} \frac{q_r + p_l + p_j}{q_r + p_j}, \tag{7.49}$$

and from eqn (7.46b), which we have not yet used, we get

$$p_j S = q_r (I - I_g) + (p_l + p_j) S_g$$

$$= \frac{q_r p_l}{q_r + p_l + p_j} I + \frac{(p_l + p_j)(q_r + p_l + p_j)}{\lambda (q_r + p_j)} N$$

$$\Rightarrow S = \frac{q_r p_l}{p_j (q_r + p_l + p_j)} I + \frac{q_r + p_l + p_j}{\gamma_s \lambda (q_r + p_j)} N. \tag{7.50}$$

Finally, recalling that $S + I = N$, we have

$$\left(1 + \frac{q_r p_l}{p_j (q_r + p_l + p_j)}\right) I = \left(1 - \frac{q_r + p_l + p_j}{\gamma_s \lambda (q_r + p_j)}\right) N$$

$$\Rightarrow \frac{(p_l + p_j)(q_r + p_j)}{p_j (q_r + p_l + p_j)} I = \frac{q_r + p_j}{\gamma_s (q_r + p_l + p_j)} I = \left(1 - \frac{q_r + p_l + p_j}{\gamma_s \lambda (q_r + p_j)}\right) N, \tag{7.51}$$

and hence,

$$\frac{I}{N} = \gamma_s \frac{q_r + p_l + p_j}{q_r + p_j} - \frac{1}{\lambda}\left(\frac{q_r + p_l + p_j}{q_r + p_j}\right)^2. \tag{7.52}$$

Let's compare this to the closed SIS result shown in eqn (7.22). We can see that our model does indeed recover to the version when $p_l \ll p_j$, i.e. when almost all individuals are flowing inside space G. This makes sense physically since it means that virtually all individuals gather in space G, which becomes effectively closed. In addition, we look into 2 other extreme conditions.

1. When $p_{l,j} \ll q_r$, i.e. when migration is much slower than recovery (hence infection), the steady state approximates to

$$\frac{I}{N} \approx \gamma_s - \frac{1}{\lambda}. \tag{7.53}$$

Compared with the closed SIS system, the difference is only in the constant term. Physically this means that when the basic reproduction rate is very high ($\lambda \to \infty$), instead of having the whole system infected and not recovered, only a fraction γ_s of the system, i.e. those within space G, stays infected. This shows the impact of having a space G: when migration is slow, G is almost isolated, and those who leave G have enough time to fully recover before ever going back into space G again. This crudely shows the effectiveness of quarantine during a pandemic, and could equally well be applied to online settings. Moreover, the tipping point between spreading and no-spreading is now $\lambda_c = 1/\gamma_s > 1$. This means that spreading is also restricted in this scenario.

2. When $p_{l,j} \gg q_r$, i.e. when migration is much faster than infection, we have approximately that

$$\frac{I}{N} \approx 1 - \frac{1}{\gamma_s^2 \lambda}. \tag{7.54}$$

When migration occurs much faster than infection, individuals in the whole system are well mixed, so the high-infection ($\lambda \to \infty$) boundary condition recovers to the level of the closed SIS system. However, since the space G for contagion is limited, and contagion can only occur within it, this reduces the *effective* basic reproduction rate to $\gamma_s^2 \lambda$. It in turn raises the tipping point further to $\lambda_c = 1/\gamma_s^2$.

To finish up the discussions on contagion, we look at the general tipping point. Setting eqn (7.52) to 0, we have

$$\lambda = \frac{1}{\gamma_s} \frac{q_r + p_l + p_j}{q_r + p_j} > \frac{1}{\gamma_s} \frac{p_l + p_j}{p_j} = 1. \tag{7.55}$$

Therefore, we know for sure that with an additional 'infection chamber', i.e. space G, the threshold for system-wide contagion is raised by definition, and thus it is more difficult for spreading to occur. Again this could in principle be applied as an online containment strategy.

7.4 Broadcast to all entities

To finish up this chapter, we introduce one more interesting type of contagion system. We call it a 'broadcast' model since contagion does not spread within the population by contact. Instead, it is transmitted by an external content source with equal chance of infecting each individual (entity) in the susceptible population. This is akin to having contaminated surfaces in a hospital, school, or airport; or the spread of diseases such as malaria, which is spread only by mosquitoes but not between people; or in the online setting, some general broadcast source online including the mainstream media as a whole. In the dynamical equations, this means substituting the non-linear infection term $q_i SI/N$ with a linear term $q_i S$. This should make the equations easier to solve. We first write down the sets of ODE for SIR under this broadcast process:

$$\frac{dS}{dt} = -q_i S \tag{7.56a}$$

$$\frac{dI}{dt} = q_i S - q_r I \tag{7.56b}$$

$$\frac{dR}{dt} = q_r I, \tag{7.56c}$$

and likewise for *SIS* under this broadcast process:

$$\frac{dS}{dt} = -q_i S + q_r I, \quad \frac{dI}{dt} = q_i S - q_r I. \tag{7.57}$$

The reader is asked to have a closer look at these closed systems in the exercises.

Similar to the node-to-node case, we study the semi-open *SIR* and *SIS* spreading mechanisms in this broadcast model. To begin with, the *SIR* ODEs are:

$$\frac{dS_g}{dt} = -q_i S_g - p_l(S_g - q_i S_g) + p_j(S - S_g), \tag{7.58a}$$

$$\frac{dI_g}{dt} = q_i S_g - q_r I_g - p_l(I_g + q_i S_g - q_r I_g) + (1 - q_r) p_j (I - I_g), \tag{7.58b}$$

$$\frac{dR_g}{dt} = q_r I_g - p_l(R_g + q_r I_g) + p_j((R - R_g) + q_r(I - I_g)), \tag{7.58c}$$

$$\frac{dS}{dt} = -q_i S_g, \tag{7.58d}$$

$$\frac{dI}{dt} = q_i S_g - q_r I, \tag{7.58e}$$

$$\frac{dR}{dt} = q_r I. \tag{7.58f}$$

Now that the system is reduced to a linear one, solving it explicitly becomes possible. Solving $S(t)$ is the most straightforward: it can be obtained by taking one further derivative of eqn (7.58a) and then substituting it into eqn (7.58d). This gives a second-order linear ODE

$$\ddot{S}_g + (q_i + p_l + p_j - p_l q_i)\dot{S}_g + p_j q_i S_g = 0. \tag{7.59}$$

This can be solved using the standard approach of assuming an exponential general solution $e^{\alpha t}$ for the unknown α. The final result is

$$S(t) = C_1 \exp\left\{\frac{1}{2}(-A - \sqrt{A^2 - 4B})t\right\} + C_2 \exp\left\{\frac{1}{2}(-A + \sqrt{A^2 - 4B})t\right\}, \tag{7.60}$$

where $A = p_j + p_l + q_i - p_l q_i$, $B = p_j q_i$, and C_1 and C_2 are determined by the initial conditions. The explicit form is too complicated and not very informative, so we do not pursue it here. However we make a couple of remarks about eqn (7.60):

1. As $t \to \infty$, S goes to 0. This means that there is no non-spreading regime for this linear model, unlike the entity-to-entity scheme.

2. Equation (7.60) shows that the decrease of susceptibles is not related to the number of infected, i.e. q_r is irrelevant. This is in contrast with the entity-to-entity case, where the results depend significantly on q_r.

We next consider the broadcast *SIS* mechanism. Interestingly, we note that this process is analogous to spintronics in condensed matter physics. When an electric current consisting of unpolarized electrons enters a spintronic device, such as a spin-valve transistor, the output current will be spin-polarized. The spin-polarized conducting electrons may be thought of as infected entities. The spin-polarized electrons will naturally tend to forget their polarization over time, e.g. by scattering (decoherence) or noise effects, and hence recover to become susceptible again (i.e. unpolarized). Any outbreak in the system may be described by the following equations in the mean-field limit:

$$\frac{dS_g}{dt} = -q_i S_g + q_r I_g - p_l(S_g - q_i S_g + q_r I_g) + p_j(S - S_g + (I - I_g)q_r), \quad (7.61a)$$

$$\frac{dI_g}{dt} = q_i S_g - q_r I_g - p_l(I_g - q_r I_g + q_i S_g) + p_j(I - I_g)(1 - q_r), \quad (7.61b)$$

$$\frac{dS}{dt} = -q_i S_g + q_r I, \quad (7.61c)$$

$$\frac{dI}{dt} = q_i S_g - q_r I. \quad (7.61d)$$

Using a method that is not too different from the entity-to-entity semi-open scenario, we can calculate the steady-state distribution. It turns out that after linearizing the equations as before, we have in the steady state that

$$\frac{I}{N} = \gamma_s \left(\frac{q_r + p_j}{q_r + p_l + p_j} + \frac{1}{\lambda} \right)^{-1}. \quad (7.62)$$

Exercises

7.1 This question is on the properties of the *SIS* model.

   (a) Solve eqn (7.15) for the *SIS* model and check against the solution in eqn (7.16).
   (b) Calculate the initial growth rate of the infected population $I(t)$ at $t = 0$. How does it relate to the infection and recovery rates?

7.2 Write down the steady-state populations for the *SIS* model under fusion–fission.

7.3 For the co-existing mobility model of *SIR*-type contagion, assume that the initial susceptible population $S(0)$ is randomly distributed across the whole system, with parameter γ_s and the basic reproduction rate λ known. The infection is initialized with one infected entity *inside* the group, i.e. $I_g(0) = 1$. Derive from eqns (7.36) the condition under which the infection spreading initially can create an epidemic.

7.4 This question is on the broadcast model.

   (a) For the broadcast model with *SIR* infection scheme, solve eqns (7.56) for the dynamics of $S(t), I(t), R(t)$. Does the system reach steady state? If so what is the steady state distribution? If not, why?
   (b) For the broadcast model with *SIS* infection scheme, solve eqns (7.57) for the dynamics of $S(t), I(t)$. Find the steady-state distribution for both populations as a function of the basic reproduction rate λ. Compare it with the entity-to-entity result in eqn (7.22)—what is the main difference?
   (c) Compare your steady-state result for the *SIS* model with eqn (7.62) for the semi-open system. Discuss different extreme conditions for eqn (7.62) as we did in Section 7.3.2. How do the different limits relate to your *SIS* result in the closed system?

8
Adding adaptation: emergence of anticrowds

Adaptation is a key process in living systems and hence humans, but it is also a feature of smart technologies and AI. We have so far applied various approaches to treating the heterogeneity of human–machinery–AI entities (Fig. 1.2) and hence its role in their collective behaviour. We started in Chaps 3–5 with the mean-field representation of entities' characteristics and hence their heterogeneity. However we did not provide an explanation for the cause of each entity's heterogeneity or its possible adaptation. Then in Chap. 7 we defined the heterogeneity based on contagion dynamics. There, an individual entity belongs to one of a few existing states, and its adaptation is deterministically defined—i.e. mutual transformation according to contagion ODEs. But this is not general enough for human–machinery–AI applications beyond contagion.

In this chapter, we go further by allowing each entity's internal state to exhibit adaptation and hence evolution based on the success of its decision making. Ideally we would then insert the theory that we develop, back into the previous chapters. However, this would add significantly to the complexity of the discussions. Hence we simply develop results in this chapter that can in the future be used to devise more precise mean-field approximations for real-world adaptive human–technology–AI systems. We note that much of the work in this chapter is adapted from Johnson *et al.* (2019*b*) which contains detailed accounts of how the theory that we present is able to reproduce empirical data collected in current human–technology systems, and also in laboratory experiments.

8.1 Extremes emerge from many-entity interactions

Individual adaptations are a feature of online and offline systems, whether they comprise mostly humans, smart technologies, AI, or any combination of these. A key practical question concerning current online systems is the following: How is it that there is a huge mainstream crowd and then a seemingly smaller crowd (an 'anticrowd') that develops around anti-X views, from which the anti-X communities of Chap. 6 and earlier chapters then emerge and evolve? A clue to this is that so far we have said nothing about the fact that individuals can have payoffs that reinforce their current

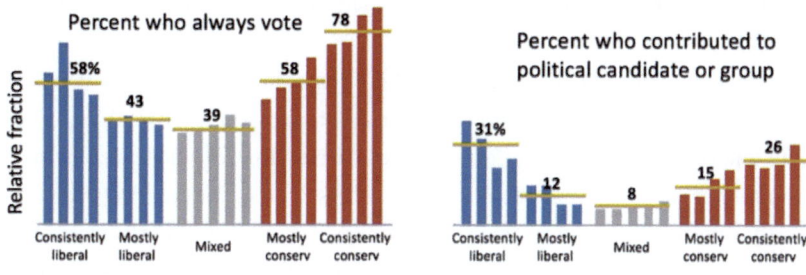

Fig. 8.1: Extremes in political participation. Source: data from Johnson *et al.* (2019*b*).

strategies and behaviour, or induce them to change it. This brings us to the realm of games—or rather repeated competition for some kind of benefit in a system where not everyone can be a winner. Outcomes from such competitions can induce adaptations within the population, since everyone is trying to win. This does mimic, as we will show, some key real-world behaviours already observed in current systems.

We will introduce a model that describes the emergence of crowds and anticrowds—which could be online, offline, or both, and could involve a mixture of human, technology, and/or AI entities. From it, we obtain the counter-intuitive finding that large sub-populations holding beliefs towards the extremes will emerge from a population that is fed the same information and has equal access to the same resources, and that the formation of extremes will be unintentionally enhanced by new online algorithms designed to increase social bonding. An obvious real-world example is the polarization of political participation shown in Fig. 8.1. To be more exact, in a world in which the glass is half-full (or equivalently half-empty) in terms of available resources, we shall show mathematically that these extreme sub-populations emerge irrespective of whether the common information is true or false, or shifts between the two. Our quantitative analysis also suggests that an effective way to prevent these extremes from emerging is to change how individuals feel rewarded for having made a good or bad decision. Note that in line with the literature, we tend to use the term *agent* or individual instead of entity in this chapter, but we stress that any of the entities in Fig. 1.2 can equivalently be called an individual or an agent.

Our model stems from a famous problem in complex systems and econophysics —the *El Farol bar* problem. El Farol is an Irish bar in Santa Fe, New Mexico. The problem comes from an observation of the economist Brian Arthur, that people want to visit the bar when it is not overcrowded, but not when it is overcrowded. The dilemma is that the winning choice for each individual depends on the choices of all the others, and nobody knows a priori the others' choices. As a result, they can only deduce from previous outcomes whether they should visit or not, instead of relying

206 Adding adaptation: emergence of anticrowds

on an inductive, optimized strategy. Hence the winning choice will shift as everyone adjusts their choices.

The so-called *minority game* in Fig. 8.2 is then a simplified binary version of this El Farol bar problem. Consider an odd number N of individuals (e.g. $N = 101$) that have to decide whether to make the effort to access a resource L whose size is just less than half the number of individuals that seek to access it (e.g. bar seating capacity $L = 50$). Hence only a minority can ever win. Specifically, if $\leqslant 50$ individuals access the resource (i.e. they choose action $+1$) and $\geqslant 51$ individuals do not (i.e. they choose action -1) then the $\leqslant 50$ choosing $+1$ win and the $\geqslant 51$ choosing -1 lose. If $\geqslant 51$ individuals access the resource (i.e. they choose action $+1$) and $\leqslant 50$ individuals do not (i.e. they choose action -1) then the $\leqslant 50$ choosing -1 win and the $\geqslant 51$ choosing $+1$ lose. We could denote the former outcome as 1 and the latter as 0, as in Fig. 8.2. This represents the real-life pursuit of a population of N entities seeking a common resource or benefit of size L which is neither so small that nobody can access it, nor so large that there is no need to try. Most importantly, there are winners and losers just like in many aspects of everyday life. We do not need to assume this is some zero-sum game or that L or N do not change in time, but for simplicity we focus on $L/N \sim 0.5$ as in this minority game setup, since this is the situation with the 'half-full' glass and it means that there is no a priori bias in terms of how individuals should act. The model was described in detail in our previous book *Financial Market Complexity* (Oxford University Press, 2003), so here we only provide an overview. In the next chapter we shall develop this model further and see how it enables control in such a limited resource system.

Adaptability comes into this model in terms of the strategies that the N individuals each use to decide their next action. These strategies could in principle be unlimited in form, but to structure the game we consider specific cases in which the strategies are all either deterministic or all stochastic. To approximate the individuals' deductive powers, we assume that they make decisions based on a set number of previous outcomes of the game (i.e. the 0s and 1s as in Fig. 8.2). Specifically, each individual has a memory μ of m previous outcomes. The memory μ is an m-bit binary string, and its kth digit from right to left represents the kth most recent outcome. This same memory is shared by all individuals in the system. Due to its bit-string structure, the memory storage increases exponentially with m. So m effectively measures the population's ability to memorize and analyse. There are in total 2^m possible pieces of memory since this is the number of possible bit-strings of length m. The individuals associate this memory at any given time to their decision making. For example, one particular individual may choose $+1$ if the previous $m = 3$ outcomes and hence memory $\mu_1 = 011$, and -1 if the memory is $\mu_2 = 100$.

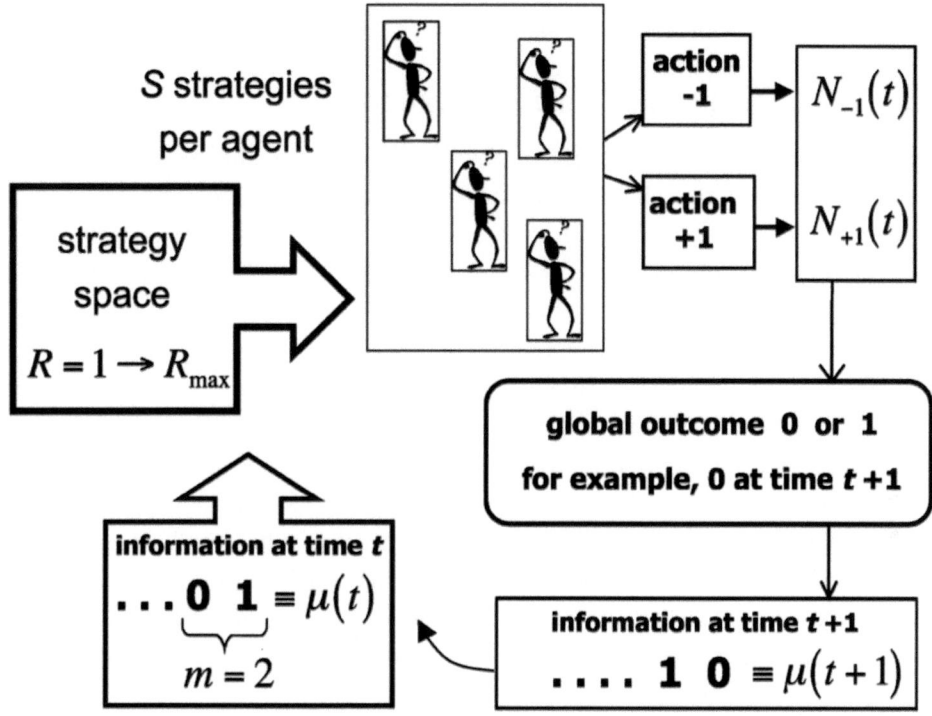

Fig. 8.2: Illustration of the minority game.

For the case of deterministic strategies, discussed in Chap. 9, these strategies for the next action can be written out in a table that covers all the memory bit-strings. The resulting strategy pool has 2^{2^m} possible choices and hence strategies, which accounts for all possible mappings from the 2^m possible memories to the binary choice ± 1. Because of the game setup, no strategy is guaranteed to be optimal since if it were, many individuals would then use it and it would immediately cease to be optimal given the finite resource. Hence every individual faces losses. To cope with that, individuals are able to switch strategies. Each individual is initially assigned a subset from this pool of strategies to guide its decision making. These sets can differ between the different individuals, hence allowing a rich heterogeneity among individuals. Individuals then use the strategy that they hold that has the highest number of correct prior predictions. In the stochastic version of the game in this chapter, the individual instead randomly selects a new strategy if that individual's losses accumulate past a threshold. To determine when this happens, a winning outcome awards the individual with 1 point and a losing outcome takes 1 point away: strategy switching occurs when the individual has $-d$ points. In the rest of this chapter, we study this stochastic variant of the model which enables us to look at system-level patterns beyond fluctuations.

8.2 Dynamic model of emerging extremes

Even with this minimal binary setup, there are myriad possible model variants. The focus of the rest of this chapter is on a particular minimal model version called the *evolutionary minority game*, which grasps the adaptation feature while simplifying the strategy choices as follows. We represent the strategy space as a continuous probability p for choosing some publicly announced action, which could be $+1$ or -1 at any given time. Note that this publicly announced action could be based on the most recent outcome(s) or it could be an announced prediction by some external agency, or it could be some external news: its origin does not matter. All that matters is that it is a common piece of information I provided to all the individuals as in Fig. 8.3. It could also just be a constant, i.e. it is a piece of advice to all individuals to 'take action $+1$' at every timestep. Hence an individual with the strategy value p will choose action $+1$ with probability p, and will choose action -1 with probability $1-p$.

We next simplify the adaptation. For continuous p, the simplest adaptation scheme is a continuous, diffusion-like adaptation, as shown in Fig. 8.3, though others could in principle be employed. This means that p and hence the population's distribution $P(p;t)$ change over time. We note that the variable $p \in [0,1]$ is much like the character vector in Chap. 3, in that it also defines the innate heterogeneous attributes. Also of interest is the lifetime of a particular p-value $T(p,t)$. We shall see in the following discussions that this model gives rise to some highly interesting phenomena. In particular, it induces divisions among the population, as seen in Fig. 8.4.

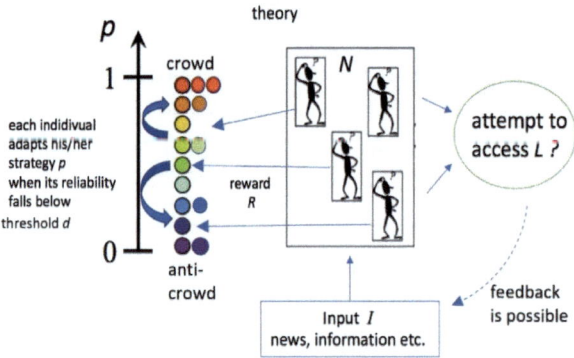

Fig. 8.3: Illustration of our evolutionary minority game model. Source: data from Johnson et al. (2019b).

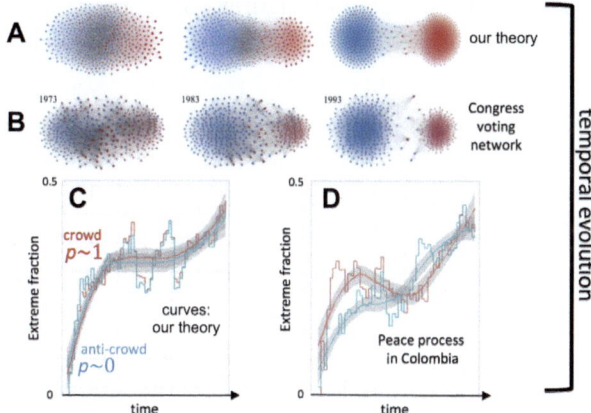

Fig. 8.4: Time evolution of opinions in political settings. Empirical data vs our theoretical results. A, B are under the settings of U.S. politics, and C, D look into the Colombian peace process. Source: data from Johnson et al. (2019b).

8.3 $N=3$ entity example

To gain some analytical insights, we consider now the simplest example of our system which contains $N = 3$ individuals, labelled as i, j, k, and three possible discrete p values $p = 0, \frac{1}{2}, 1$. Since the common information I is arbitrary in terms of its content, we here take $p = 0$ to represent choosing 0, $p = 1$ to represent choosing 1, and $p = \frac{1}{2}$ to represent a 0.5 probability of choosing 1. As mentioned above, we take $L = N/2$. Hence there is at most 1 winner at each timestep, which will occur when that individual makes the opposite binary choice to the other two individuals. Note that all 3 can lose in principle, if they make the same binary choice. The winner is awarded 1 point while the losers -1 point. Once the accumulated points for an individual reach $-d$, that individual switches its strategy (i.e. it gets a new p value) and the points are set to zero. We denote the window within which the new p is chosen as D, i.e. the new strategy lies randomly in the range $[p - D/2, p + D/2]$ subject to a certain boundary condition, e.g. periodic or reflective. Interestingly, the emergence of the crowd and anticrowd is insensitive to these details. For our discrete example of only 3 possible p values, the individual hence switches randomly to one of the 2 other strategies. There are in total $3^3 = 27$ possible configurations of (p_i, p_j, p_k) and $2^3 = 8$ possible decisions. For example, for $(p_i, p_j, p_k) = (0, 0, \frac{1}{2})$, i and j both choose 0 while k chooses 0 or 1, each with $\frac{1}{2}$ probability. Hence, there is a probability $\frac{1}{2}$ for which k wins while i and j both lose. The net number of points gained per individual per turn, given by the points awarded minus the points deducted, is -1 for i, -1 for j, and $(1-1)/2 = 0$ for k. The total is hence -2.

Table 8.1 lists all the possible configuration classes for the 3 individuals. The individuals are labelled with asterisks $*$ and each is assigned one of the 3 possible strategy values p, followed by its expected point gain. Also listed in the table is the number of configurations per class, together with the total points awarded to each configuration class. We can see that the maximal total point is -1 (there is at most 1 winner), and $(0, 0, \frac{1}{2})$ is not optimal, unlike classes viii), ix), and x).

An interesting property to note is that, since only a minority of the total individuals (or fewer) can win, the total points are always negative. This motivates the dynamic part of the model, during which the individual would modify its p value (i.e. strategy) accordingly. The typical time between modification is inversely proportional to the points. For example, class vi) implies the average individual loses $-\frac{1}{2}$ point per turn, and would hence modify its p value after time $2d$. Such strategy modification allows the system to sample the 27 configurations. Therefore to obtain the average distribution $P(p)$, we average over all 27 configurations. Since some classes are more favourable (i.e. awarding more points) we should weight the distributions appropriately. Below we take some possible weighting schemes as examples.

1. In the extreme case of large weighting, we include only the optimal classes viii), ix), and x). For $P(p)$ we take the weighted average over distributions of each class, multiplied by the number of their respective configurations. Then we obtain

$$P(0) = P(1) = \frac{1}{3} \times \frac{3}{12} + \frac{2}{3} \times \frac{3}{12} + \frac{1}{3} \times \frac{6}{12} = \frac{5}{12}, \quad P(1/2) = \frac{2}{12}, \quad (8.1)$$

yielding

$$P(0) : P(\frac{1}{2}) : P(1) = 2.5 : 1 : 2.5. \quad (8.2)$$

2. For zero weighting, we consider the system as visiting all configurations with equal probability regardless of points gained per individual. Such a zero-weight averaging

Table 8.1 Configuration classes of the 3-entity system.

Class	$p - 0$	$p - 1/2$	$p - 1$	# config.	Total pts.
i)	$***[-1][-1][-1]$	-	-	1	$[-3]$
ii)	-	-	$***[-1][-1][-1]$	1	$[-3]$
iii)	$**[-1][-1]$	$*[0]$	-	3	$[-2]$
iv)	-	$*[0]$	$**[-1][-1]$	3	$[-2]$
v)	-	$***[-1/2][-1/2][-1/2]$	-	1	$[-3/2]$
vi)	$*[-1/2]$	$**[-1/2][-1/2]$	-	3	$[-3/2]$
vii)	-	$**[-1/2][-1/2]$	$*[-1/2]$	3	$[-3/2]$
viii)	$*[1]$	-	$**[-1][-1]$	3	$[-1]$
ix)	$**[-1][-1]$	-	$*[1]$	3	$[-1]$
x)	$*[0]$	$*[-1]$	$*[0]$	6	$[-1]$

is similar to that for the microstates in a gas within the microcanonical ensemble and yields $P(0) : P(\frac{1}{2}) : P(1) = 1 : 1 : 1$.

3. For an intermediate case whereby all classes are weighted by the (absolute value of) total points, we obtain $P(0) : P(\frac{1}{2}) : P(1) = 1.1 : 1 : 1.1$.

In fact, any sensible weighting which favours the more profitable configurations, yields a non-uniform $P(p)$ as observed numerically. This implies that the population, by self-segregating, has also managed to self-organize around the most profitable configurations. We emphasize that the system is dynamic since the membership of the various configurations is constantly changing (i, j, and k inter-diffuse) but $P(p)$ remains essentially constant. These results are also a good approximation for general N in the case that i, j, k are three similarly-sized and indivisible groups (e.g. in-built communities) of like-minded individuals.

Finally, what about the other dynamic quantity $T(p)$ at steady state? This can be analysed exactly when there are not too many configurations involved, e.g. in weighting scheme 1. When only including classes viii), ix), and x), the 3-entity system switches around amongst these 3 classes only. Since all the 3 classes have the same total points, each class must last the same amount of time (denoted as t_0) before switching to another configuration. This does not, however, mean that the relative values of $T(p)$ are the same as those of $P(p)$, since the configuration cannot switch within the same class. By counting the number of periods t_0 each strategy p goes through before changing, we should be able to find the relative values of $T(p)$. We start with strategy $p = \frac{1}{2}$. From Table 8.1, we see that it always corresponds to -1 point in class x), so it can only last t_0 before switching, i.e. $T(\frac{1}{2}) = t_0$. Following this we look into strategy $p = 0$, which is symmetric with $p = 1$ and hence should correspond to the same $T(p)$ value. This is slightly more complicated in that we need to actually consider the 3 classes respectively. We denote that a particle of strategy $p = 0$ starting under class viii), ix), x) sticks to its strategy for time $T_{(\text{viii})}, T_{(\text{ix})}, T_{(\text{x})}$, respectively, before switching. These 3 quantities should satisfy the following equalities:

$$T_{(\text{viii})} = t_0 + \frac{1}{2} \times \left[\frac{1}{2}T_{(\text{ix})} + \frac{1}{2}T_{(\text{x})}\right], \qquad (8.3)$$

$$T_{(\text{ix})} = t_0 + \frac{1}{2}T_{(\text{viii})} + \frac{1}{2}T_{(\text{x})}, \qquad (8.4)$$

$$T_{(\text{x})} = t_0 + \frac{1}{2}T_{(\text{viii})} + \frac{1}{2}T_{(\text{ix})}, \qquad (8.5)$$

since each configuration lasts time t_0, and then has an equal chance of switching to the other 2 classes. The extra factor $\frac{1}{2}$ in eqn (8.3) accounts for the fact that one of the 2 $p = 0$ particles switches strategy after time t_0. This gives

$$T_{(\text{viii})} = 4t_0, \ T_{(\text{ix})} = T_{(\text{x})} = 6t_0. \qquad (8.6)$$

At steady state, the relative value of T(0) should be the weighted average of the 3 quantities above, against the distribution of the $p = 0$ particles. This gives

$$T(0) = \frac{2}{5}T_{(\text{viii})} + \frac{2}{5}T_{(\text{ix})} + \frac{1}{5}T_{(\text{x})} = \frac{26}{5}t_0. \tag{8.7}$$

To summarize, the relative values of $T(p)$ are

$$T(0) : T(\tfrac{1}{2}) : T(1) = 5.2 : 1 : 5.2. \tag{8.8}$$

As we admit more and more classes in the system, this direct approach becomes increasingly more complicated. Since we face up to 10 coupled equations for a 3-entity, 3-strategy system, for a general system with N individuals and m strategies the number of equations and hence classes is

$$\binom{N+m-1}{m-1}. \tag{8.9}$$

At such levels of complexity, instead of solving exactly for each $T(p)$ value, we can take a mean-field perspective. Since the probability of finding particles with strategy p is $P(p)$ at steady state, this can also be seen as an average particle spending $P(p)$ of the total time with strategy p. Therefore, with a big enough population, we should be able to accurately approximate

$$T(p) \propto P(p) \tag{8.10}$$

at late times.

8.4 Many-entity system

Above we have seen the very interesting statistics for as few as 3 individuals. The complexity quickly grows, however: from eqn (8.9), it grows exponentially as the number of particles and the number of strategies increase. This makes it difficult to take weightings over all existing classes and determine their frequencies of occurrence. Nevertheless, we can first see how far the statistics can take us.

To begin, we follow the methodology from the 3-entity statistics and look into the total points of each configuration. For a finite N-body distribution of strategies $\{p_i\}$, where $i = 1, 2, \ldots, N$, the formal expression of the expected points associated with this configuration is

$$Q_N(\{p_i\}) = -|-N|(1-p_1)(1-p_2)\cdots(1-p_N)$$
$$- |2-N|\,[p_1(1-p_2)\cdots(1-p_N) + \cdots + (1-p_1)(1-p_2)\cdots p_N]$$
$$- |4-N| \left[\sum_{i=1}^{N-1} \sum_{j=i+1}^{N} p_i p_j \prod_{k=1, k \neq i, k \neq j}^{N} (1-p_k) \right]$$

$$-\cdots-|2N-N|p_1p_2\cdots p_N. \tag{8.11}$$

Despite the lengthy expression, $Q_N(\{p_i\})$ is obtained simply by multiplying all possible values of points and their corresponding probability of occurrence and then summing them all up, as for any mathematical expectation. There are $(N+1)$ terms in total representing $(N+1)$ different choices of states for a particular configuration class $\{p_i\}$. The $(n+1)$th term with $-|2n-N|$ points results from n individuals choosing state 1 and the rest choosing state 0. From this we can encapsulate the probabilities in each term with a function $F_N^n(\{p_i\})$, which defines the probability for a particular configuration class $\{p_i\}$ to have exactly n individuals choosing state 1:

$$F_N^n(\{p_i\}) = \sum_{p_{\alpha_r},p_{\beta_s}\in\{p_i\}} \prod_{r=1}^{n} p_{\alpha_r} \prod_{s=1}^{N-n}(1-p_{\beta_s}). \tag{8.12}$$

Note that the sum includes a total of $\binom{N}{n}$ terms, which cover all possible combinations of p_is. It then follows that $Q_N(\{p_i\})$ can be rewritten as

$$Q_N(\{p_i\}) = -\sum_{n=0}^{N}|2n-N|F_N^n(\{p_i\}). \tag{8.13}$$

We shall soon see that $F_N^n(\{p_i\})$ plays a critical role in estimating the steady-state strategy distribution $P(p)$.

The reason why we study the expected points of different configuration classes follows from our thoughts in the 3-entity scenario. Their points determine their duration, and by weighting each configuration class appropriately we should theoretically obtain the exact steady-state configuration. To summarize this formally, at large N and for a continuous choice of strategies with probability distribution $P(p,t)$, our job is to find the $P(p)$ dependence of the *functional* $\mathcal{Q}[P(p;t)]$. If we restrict the system within the most probable distributions, just like strategies viii), ix), and x) for the 3-body statistics, we need to maximize $\mathcal{Q}$ to obtain the corresponding $P(p)$s. This is already difficult enough though, since it would be very complicated just to write down an explicit expression for $\mathcal{Q}[P]$, not to mention the fact that it is not only the most probable configuration classes that contribute. Adding in all weighted contributions would be intractable.

Therefore if we are to stick to this system-level approach, we have to follow necessary reductions, which are laid out in the coming section. This also provides us with a different perspective—building up the statistics from a bottom-up iteration—and this turns out to produce accurate fits to the simulation results.

8.4.1 System-level approach

In the 3-entity system, we rank all possible configuration classes with their total points, which in turn determines their respective duration and contributes to the steady-state strategy distribution. To follow that thought here for larger N values, we again need to take a closer look at the function $Q_N(p_i)$, which is a function of the configuration $\{p_i\}$, and its continuous version the functional $\mathcal{Q}[P(p;t)]$ with respect to strategy distribution $P(p,t)$. Though we do not know its form analytically, we know for certain that the maximum of Q_N (or $\mathcal{Q}$) must be -1 and the minimum $-N$: there is at least 1 individual that certainly loses due to the minority nature of the game, and the losers cannot exceed the total population N. Furthermore, we can find out their respective configurations: minimum points $-N$ can only occur when all individuals choose $p=0$ or $p=1$. The choice of a particular strategy does not matter since there is no preference one over the other. In other words, the two extreme strategies must be symmetric. Meanwhile, maximum points of -1 can be reached only when half of the population $((N-1)/2)$ choose $p=0$ and the other half choose $p=1$. This leaves one final individual choosing arbitrary strategies without affecting the outcome: this will always be the one that breaks the minority and hence will be awarded -1 points regardless of its strategy choice.

The configuration class with -1 point provides us with a new perspective when dealing with the configuration classes. We see that most of the individuals there do not actually contribute towards the total points, since those choosing $p=0$ have their points cancel out with their symmetric counterparts choosing $p=1$. This is a bit like considering only the Fermi surface in solid state physics or the valence electrons in chemistry: instead of counting all particles we reduce the system to the most relevant ones. We can thus divide the population into 2 parts (see Fig. 8.5):

1. The 'dormant' individuals, which are the same number of individuals that choose $p=0$ or $p=1$. For example, if there are 3 individuals choosing $p=0$ and 5 choosing $p=1$, we consider the system to possess 6 dormant individuals, 3 each for $p=0,1$. They cancel each other in counting points, and are thus irrelevant for our studies.

2. The 'active' individuals, i.e. all the other individuals whose configurations make a real difference to the points.

Now we can group the configuration classes by the number of dormant particles. The advantage of such classification is that it transforms the configuration classes into subclasses of effectively fewer-body systems. For a general N-entity system for example, we can organize its possible configuration classes into a number of effective 1-entity systems with $(N-1)$ dormant individuals choosing $p=0,1$ respectively, plus

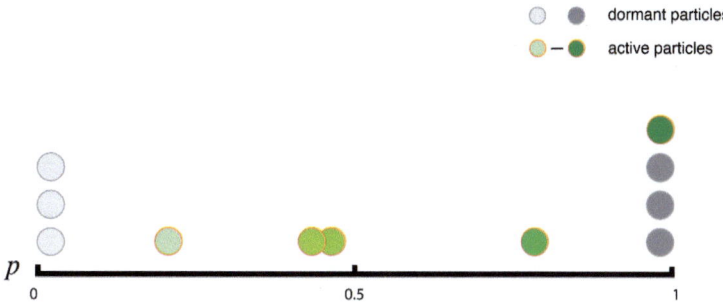

Fig. 8.5: An illustration of dormant and active particles.

a number of effective 3-entity, 5-entity, etc. systems. Generally speaking, having more active individuals means a higher chance to get low points. Therefore, we name the configuration classes with $(2n-1)$ active individuals as nth-order classes, knowing that higher-order classes have lower expected points.

To briefly summarize what we have done so far, we have first found a formal expression for the total points $Q_N(\{p_i\})$ as a function of configuration classes $\{p_i\}$. Then, since directly maximizing Q_N is hard, we divide the configuration classes $\{p_i\}$ into different orders of effective subclasses with higher-order classes corresponding to more active particles and hence lower points. This enables us to consider each order separately: for lower-order configurations analytics can be possible, while higher-order terms may be ignored for their low points and resulting low likelihood of occurrence, subject to our required level of approximation. Though this does not yield $P(p)$ exactly, it is nonetheless a straightforward approximation approach that does not require intensive computation.

To see how this works, we study the lower order behaviours of a N-entity system with a continuous choice of strategies $p \in [0, 1]$. To start with, we have the trivial first-order term:
$$Q^{(1)}(p) = -p - (1-p) = -1, \tag{8.14}$$
which is a constant, as previously mentioned, regardless of the choice of p. We can hence deduce that the first-order contribution towards $P(p)$ is proportional to
$$P^{(1)}(p) = \frac{N-1}{N}[\delta(p) + \delta(1-p)] + \frac{1}{N}, \tag{8.15}$$
which is a well-like distribution with a flat bottom.

The leading order is a little trivial. It does not capture the intricate behaviours when multiple individuals are betting on non-extreme strategies, as illustrated in Fig. 8.6. Nevertheless, it is at least in line with our perception that individuals do become extreme, albeit in a very crude way. We continue to study the second-order classes, which should still be very much present since their points do not go below -3. We label the 3 active strategies as $p_{1,2,3}$ respectively, and the resulting points are given by

$$Q^{(2)}(p_1, p_2, p_3) = -3[p_1 p_2 p_3 + (1-p_1)(1-p_2)(1-p_3)]$$
$$- [p_1 p_2(1-p_3) + p_1 p_3(1-p_2) + p_2 p_3(1-p_1)$$
$$+ p_1(1-p_2)(1-p_3) + p_2(1-p_1)(1-p_3) + p_3(1-p_1)(1-p_2)]. \tag{8.16}$$

To simplify the expression, we assume that at steady state the distribution $P(p)$ is symmetric, and hence *wlog*

$$p_3 = 1 - p_2. \tag{8.17}$$

Of course in our deductive model it is impossible for p_3 to fix itself symmetrically opposite p_2—this is not in our bottom-up rules. However, when the system reaches steady state, we should expect such symmetry to be achieved through numerous time steps of microscopic switching of strategies. Our assumption is reasonable from the inherent symmetry in the system as well as the results exhibited in real-world systems and simulations, e.g. in Fig. 8.6. This simplifies eqn (8.16) into

$$Q^{(2)}(p_1, p_2) = -3[p_1 p_2(1-p_2) + (1-p_1)p_2(1-p_2)]$$
$$- [p_1 p_2^2 + p_1(1-p_2)^2 + (1-p_1)p_2(1-p_2)$$
$$+ p_1(1-p_2)p_2 + (1-p_1)p_2^2 + (1-p_1)(1-p_2)^2]$$
$$= -3p_2(1-p_2) - [p_2^2 + (1-p_2)^2 + p_2(1-p_2)]$$
$$= 2p_2^2 - 2p_2 - 1. \tag{8.18}$$

This is reduced to a single-variable function: Q only depends on the 2 symmetric strategies, but not the stand-alone one p_1. Now assuming that the lifetime of strategy p_2 is proportional to $|Q|^{-1}$, we have

$$T(p) \propto \frac{1}{1 + 2p(1-p)}. \tag{8.19}$$

Moreover, from eqn (8.10) we can also write down

$$P(p) \propto \frac{1}{1 + 2p(1-p)}. \tag{8.20}$$

Normalizing eqn (8.20) yields the second-order strategy distribution (see exercises).

The first 2 orders are already quite representative. By reducing the system to be mostly extreme, which does reflect the steady-state scenario, we are able to approximately describe the moderate strategies with analytics. The first-order approximation is too crude: it states that individuals with $0 < p < 1$ distribute evenly within this range, which is an oversimplification. Through the second-order approximation, we start to see interesting behaviours for general p: it roughly distributes in order $(p-0.5)^{-2}$ and shows a gradual decrease from extreme to moderate p values. We can keep going for higher-order configurations, but that again incurs multi-variable functions and makes it complicated to optimize.

8.4.2 Single-entity-plus-rest

We have seen a conceptually neat yet computationally only approximate approach in the previous section. That begs the question: is there a more computational method that yields accurate distribution results?

D'hulst and Rodgers (1999) provide us with a interesting perspective: when the total distribution of strategies is intractable, we can instead look at the strategy change of a particular individual under a strategy distribution—and have the system iteratively evolve for all individuals to reach steady state after many iterations in order to obtain the big picture. This approach is in a way like studying a field with a test particle, and is more akin to simulations, except that its evolutionary steps are theoretically calculated rather than stochastically simulated.

The basis of the theory comes from a diffusion-like construction of the evolutionary minority game. Back in Section 8.3 we assume that the evolution of individual strategies is realized by randomly picking a new strategy in the range $[p-D/2, p+D/2]$ subject to periodic or reflective boundary condition—and these two conditions converge when $P(p,t)$ is symmetric. In the continuous time limit, this is diffusive in nature and we can accordingly describe the dynamics by a generalized diffusion equation

$$\frac{\partial P(p,t)}{\partial t} = \frac{\partial^2 \mathcal{F}[P(p;t)]}{\partial p^2}, \qquad (8.21)$$

where $\mathcal{F}[P(p;t)]$, which is a functional in terms of $P(p,t)$, represents the strategies that are about to change at time t. The use of a complicated functional, however, reverts our attempt back to the previous section and we must simplify it.

Our solution is to approximate $\mathcal{F}[P(p;t)]$ with a double-variable function $S(p,t)$ with respect to p and time t only. It is to be interpreted as the probability for strategy p to alter at time t. From an individual point of view, $S(p,t)$ should be related to the points that the individual with strategy p obtains and accumulates before switching—exactly how we define strategy evolution. Recall our assumption that a strategy switch occurs when the individual accumulates $-d$ points. Then, assuming that the success rate of a specific strategy p at time t is $\tau(p,t)$, we can express $S(p,t)$ as

$$S(p,t) \propto \frac{P(p,t)}{d/[1-2\tau(p,t)]} = \frac{[1-2\tau(p,t)]P(p,t)}{d}, \qquad (8.22)$$

since the probability of strategy switch should be inversely proportional to the accumulation time of points. Here we use the proportional ($\propto$) sign as we omit the constant factors that occur in diffusion. Substituting this back into eqn (8.21), we have

$$\frac{\partial P(p,t)}{\partial t} = A(L,d)\frac{\partial^2}{\partial p^2}[(1 - 2\tau(p,t))P(p,t)], \qquad (8.23)$$

where $A(L,d)$ is a constant factor related to the strategy diffusion window L and point accumulation threshold d. At steady state, i.e. when $t \to \infty$, we can drop the t dependence in functions $\tau(p,t)$ and $P(p,t)$ and we have

$$\frac{\partial^2}{\partial p^2}[(1 - 2\tau(p))P(p)] = 0. \qquad (8.24)$$

This is solved by an arbitrary linear function $(1 - 2\tau(p))P(p) = Bp + C$, i.e.

$$P(p) = \frac{Bp + C}{1 - 2\tau(p)}, \qquad (8.25)$$

where constants B, C are determined from boundary and normalization conditions. The first boundary condition to apply is the symmetry assumption

$$P(p) = P(1 - p) \qquad (8.26)$$
$$\tau(p) = \tau(1 - p) \qquad (8.27)$$

as we have maintained throughout our discussions. This requires that $B = 0$, and hence we have

$$P(p) = \frac{C}{1 - 2\tau(p)}, \qquad (8.28)$$

i.e. late-time strategy distribution is inversely proportional to the expected point loss of the strategy. This of course makes sense since point loss motivates the redistribution of strategies. The constant C can be determined by normalization.

So far we have transformed our quest for $P(p)$ into finding the steady-state 'success probability' $\tau(p)$. This is what we mean by 'individual level': instead of counting the total points of the entire system and how that affects the global distribution, we now focus on a specific strategy and how it gets modified.

So how do we evaluate $\tau(p)$? Since it depends on the entire distribution of strategies, we need to iteratively calculate $\tau(p)$ over time until the values stabilize. Again this is a computational approach that is not strictly analytic, but highly accurate, as we will see. We begin with the quantity $F_N^n(\{p_i\})$ defined earlier in eqn (8.12). As an aside, we can calculate the mean-field average of $F_N^n(\{p_i\})$ for a mean strategy distribution rather than a specific set $\{p_i\}$. By applying the expected strategy

$$\bar{p} = \int_0^1 pP(p)\,\mathrm{d}p \tag{8.29}$$

to replace all individual strategies p_i, we find that

$$\langle F_N^n \rangle = \binom{N}{n} \bar{p}^n (1-\bar{p})^{N-n}, \tag{8.30}$$

a simple binomial distribution, due to the binary nature of the states. For a large enough population N, we know from the central limit theorem (or more precisely, de Moivre–Laplace theorem in this case) that $\langle F_N^n \rangle$ will be approximately a Gaussian distribution with a mean $N\bar{p}$ and variance

$$N \int_0^1 P(p)p(1-p)\,\mathrm{d}p. \tag{8.31}$$

Taking the symmetry assumption for $P(p)$ yields $\bar{p} = \frac{1}{2}$. The Gaussian approximation validates our previous reduction by separating the orders, as individuals do indeed tend to be distributed at the two ends, i.e. the extremes. However, the mean-field $\langle F_N \rangle$ is too crude for the iterative computations. For the following derivations we stick to the strategy-dependent function $F_N^n(\{p_i\})$. Remember that our goal here is to study a 'test particle' under the distribution of the rest. Specifically, consider the action of the kth individual in the background of the other $(N-1)$ individuals. According to our notation, the distribution for these $(N-1)$ individuals is denoted by $F_{N-1}^n(\{p_{i\neq k}\})$. Then we can write down their relations:

$$F_N^n(\{p_i\}) = p_k F_{N-1}^{n-1}(\{p_{i\neq k}\}) + (1-p_k)F_{N-1}^n(\{p_{i\neq k}\}), \quad (n \neq 0, N) \tag{8.32}$$
$$F_N^0(\{p_i\}) = (1-p_k)F_{N-1}^0(\{p_{i\neq k}\}), \tag{8.33}$$
$$F_N^N(\{p_i\}) = p_k F_{N-1}^{N-1}(\{p_{i\neq k}\}). \tag{8.34}$$

Here p_k is the strategy of the kth individual. Note that all the above quantities are time-dependent—we drop the notations for simplicity. To interpret this equation, the value for n choosing '1' is achieved if the value of the $(N-1)$ individual background is $n-1$ *and* the kth individual decides '1'. This leads to the first term in eqn (8.32). Alternatively the $(N-1)$ individual background is n and the kth individual decides against '1', leading to the second term. All this is preparing for the calculation of $\tau(p_k)$. Given the probability $F_{N-1}^n(\{p_{i\neq k}\})$, we can write

$$\tau(p_k) = p_k \sum_{n=0}^{(N-3)/2} F_{N-1}^n(\{p_{i\neq k}\}) + (1-p_k) \sum_{n=(N+1)/2}^{N-1} F_{N-1}^n(\{p_{i\neq k}\}). \tag{8.35}$$

Equation (8.35) says that the kth individual wins if

220 *Adding adaptation: emergence of anticrowds*

1. the number choosing '1' is below or equal to $(N-3)/2$ before individual k makes their move, and k decides '1', thereby giving the first term; or
2. the number choosing '1' is above or equal to $(N+1)/2$, and k decides to choose '0', thereby giving the second term.

It follows from eqn (8.32) that

$$\sum_{n=1}^{(N-3)/2} F_N^n(\{p_i\}) = \sum_{n=1}^{(N-3)/2} \left[p_k \left(F_{N-1}^{n-1}(\{p_{i\neq k}\}) - F_{N-1}^n(\{p_{i\neq k}\}) \right) + F_{N-1}^n(\{p_{i\neq k}\}) \right]$$

$$= \sum_{n=1}^{(N-3)/2} F_{N-1}^n(\{p_{i\neq k}\}) + p_k F_{N-1}^0(\{p_{i\neq k}\}) - p_k F_{N-1}^{(N-3)/2}(\{p_{i\neq k}\}). \tag{8.36}$$

Using eqn (8.33) we can substitute the second term and obtain

$$\sum_{n=0}^{(N-3)/2} F_{N-1}^n(\{p_{i\neq k}\}) = \sum_{n=0}^{(N-3)/2} F_N^n(\{p_i\}) + p_k F_{N-1}^{(N-3)/2}(\{p_{i\neq k}\}), \tag{8.37}$$

thus leaving only the final term explicitly depending on p_k. Similarly, for the other half of the $F_{N-1}^n(\{p_{i\neq k}\})$ functions we have

$$\sum_{n=(N+1)/2}^{N-1} F_N^n(\{p_i\})$$

$$= \sum_{n=(N+1)/2}^{N-1} \left[p_k \left(F_{N-1}^{n-1}(\{p_{i\neq k}\}) - F_{N-1}^n(\{p_{i\neq k}\}) \right) + F_{N-1}^n(\{p_{i\neq k}\}) \right]$$

$$= \sum_{n=(N+1)/2}^{N-1} F_{N-1}^n(\{p_{i\neq k}\}) + p_k F_{N-1}^{(N-1)/2}(\{p_{i\neq k}\}) - p_k F_{N-1}^{N-1}(\{p_{i\neq k}\}). \tag{8.38}$$

Again we try to limit the number of terms explicitly depending on p_k by substituting in eqn (8.34):

$$\sum_{n=(N+1)/2}^{N-1} F_{N-1}^n(\{p_{i\neq k}\}) = \sum_{n=(N+1)/2}^{N} F_N^n(\{p_i\}) - p_k F_{N-1}^{(N-1)/2}(\{p_{i\neq k}\}). \tag{8.39}$$

Substituting eqns (8.37) and (8.39) into eqn (8.35), we obtain

$$\tau(p_k) = p_k \sum_{n=0}^{(N-3)/2} F_N^n(\{p_i\}) + p_k^2 F_{N-1}^{(N-3)/2}(\{p_{i\neq k}\})$$

$$+ (1-p_k) \sum_{n=(N+1)/2}^{N} F_N^n(\{p_i\}) - (1-p_k) p_k F_{N-1}^{(N-1)/2}(\{p_{i\neq k}\}). \tag{8.40}$$

Finally, we further simplify it by expressing $F_{N-1}^{(N-3)/2}(\{p_{i\neq k}\})$ in terms of $F_{N-1}^{(N-1)/2}(\{p_{i\neq k}\})$ and $F_N^{(N-1)/2}(\{p_i\})$ using eqn (8.32):

$$\tau(p_k) = p_k \sum_{n=0}^{(N-3)/2} F_N^n(\{p_i\}) + (1-p_k) \sum_{n=(N+1)/2}^{N} F_N^n(\{p_i\})$$
$$+ p_k \left[F_N^{(N-1)/2}(\{p_i\}) - 2(1-p_k) F_{N-1}^{(N-1)/2}(\{p_{i\neq k}\}) \right]$$
$$= p_k \sum_{n=0}^{(N-1)/2} F_N^n(\{p_i\}) + (1-p_k) \sum_{n=(N+1)/2}^{N} F_N^n(\{p_i\})$$
$$- 2p_k(1-p_k) F_{N-1}^{(N-1)/2}(\{p_{i\neq k}\}). \tag{8.41}$$

Equation (8.41) separates $\tau(p_k)$ into 3 terms, which are interpreted as follows. Consider an 'outsider', i.e. someone whose action does not affect the outcome but instead is only betting on which side is the winning room according to the probability p_k. Their winning probability is given by the first two terms in eqn (8.41). The third term gives the difference in the winning probability between an outsider and an individual who actually participates. This term is negative, reflecting the fact that an individual has a smaller probability of winning when he is actually participating in the game. This makes sense, since if we consider the case in which the background population is split evenly between '0' and '1', the kth individual loses no matter what strategy they adopt. Hence the third term represents this self-interaction term, or the so-called market impact in econophysics. The $p_k(1-p_k)$ factor means that the winning probability increases as p_k deviates more from the value $1/2$, and it produces a symmetry about $p = 1/2$, which we have long assumed. This further implies that the resulting $P(p)$ and $T(p)$, related to $\tau(p)$ by eqn (8.28), are insensitive to most of the parameters of the model. Going down the symmetric track, the summations in the first and second terms of eqn (8.41) in the steady state both yield the value $1/2$, and hence $\tau(p)$ becomes

$$\tau(p_k; \{p_i\}) = \frac{1}{2} - 2p_k(1-p_k) F_{N-1}^{(N-1)/2}(\{p_{i\neq k}\}). \tag{8.42}$$

Note that the arguments of τ indicate that it is related to both the specific strategy p_k and the entire strategy distribution $\{p_i\}$. But now the RHS of eqn (8.42) has a dependence on $\{p_{i\neq k}\}$, i.e. not a complete strategy distribution. In order to express the RHS as explicitly depending on $\{p_i\}$ and p_k, we use eqn (8.32) to find $F_{N-1}^{(N-1)/2}(\{p_{i\neq k}\})$. We have

$$p_k F_{N-1}^{n-1}(\{p_{i\neq k}\}) + (1-p_k) F_{N-1}^n(\{p_{i\neq k}\}) = F_N^n(\{p_i\}) \tag{8.43}$$
$$p_k F_{N-1}^{n-2}(\{p_{i\neq k}\}) + (1-p_k) F_{N-1}^{n-1}(\{p_{i\neq k}\}) = F_N^{n-1}(\{p_i\}) \tag{8.44}$$

$$p_k F_{N-1}^{n-j-1}(\{p_{i\neq k}\}) + (1-p_k) F_{N-1}^{n-j}(\{p_{i\neq k}\}) = F_N^{n-j}(\{p_i\}) \quad j=0,\ldots,n-1 \quad (8.45)$$

$$\ldots$$

$$p_k F_{N-1}^0(\{p_{i\neq k}\}) + (1-p_k) F_{N-1}^1(\{p_{i\neq k}\}) = F_N^1(\{p_i\}). \quad (8.46)$$

The RHS of this series of equations depend explicitly on $\{p_i\}$ only, while most of the terms on the LHS can be cancelled out by summing them up. We have to be careful, however, with weighting these equations appropriately when cancelling out the LHS. Above all, the multipliers must have alternating signs since the left-hand sides contain only positive terms. It turns out that we need to sum up as follows:

$$\sum_{j=0}^{n-1} (-1)^{n-j} \left(\frac{p_k}{1-p_k}\right)^{n-j} F_N^j(\{p_k\})$$

$$= \sum_{j=1}^{n} (-1)^j \left(\frac{p_k}{1-p_k}\right)^j F_N^{n-j}(\{p_k\})$$

$$= \sum_{j=0}^{n-1} (-1)^j \left(\frac{p_k}{1-p_k}\right)^j \left[p_k F_{N-1}^{n-j-1}(\{p_{i\neq k}\}) + (1-p_k) F_{N-1}^{n-j}(\{p_{i\neq k}\})\right]$$

$$= \sum_{j=0}^{n-1} (-1)^j \frac{p_k^{j+1}}{(1-p_k)^j} F_{N-1}^{n-j-1}(\{p_{i\neq k}\}) + \sum_{j=0}^{n-1} (-1)^j \frac{p_k^j}{(1-p_k)^{j-1}} F_{N-1}^{n-j}(\{p_{i\neq k}\})$$

$$= \sum_{j=1}^{n} (-1)^{j-1} \frac{p_k^j}{(1-p_k)^{j-1}} F_{N-1}^{n-j}(\{p_{i\neq k}\}) + \sum_{j=0}^{n-1} (-1)^j \frac{p_k^j}{(1-p_k)^{j-1}} F_{N-1}^{n-j}(\{p_{i\neq k}\})$$

$$= (-1)^{n-1} \frac{p_k^n}{(1-p_k)^{n-1}} F_{N-1}^0(\{p_{i\neq k}\}) + (1-p_k) F_{N-1}^n(\{p_{i\neq k}\})$$

$$= -(-1)^n \frac{p_k^n}{(1-p_k)^n} F_N^0(\{p_i\}) + (1-p_k) F_{N-1}^n(\{p_{i\neq k}\}). \quad (8.47)$$

A few remarks on the derivations: in the first equality we simply reverse the order of summation for convenience. We follow this by plugging in eqn (8.45) and multiplying it out. Then in the fourth equality we relabel the first summation term in order to cancel out both summations in the fifth equality, leaving only the two extreme terms. Finally we note that eqn (8.33) helps us transform $F_{N-1}^0(\{p_{i\neq k}\})$ into $F_N^0(\{p_i\})$ in the last equality. Hence, the resulting expression reads

$$(1-p_k) F_{N-1}^n(\{p_{i\neq k}\}) = \sum_{j=0}^{n-1} (-1)^{n-j} \left(\frac{p_k}{1-p_k}\right)^{n-j} F_N^j(\{p_k\}). \quad (8.48)$$

Some may notice that out of the $(N+1)$ recursive relations between $F_N^j(\{p_i\})$ and $F_{N-1}^l(\{p_{i\neq k}\})$ (including, of course, $F_N^0(\{p_i\})$ and $F_N^N(\{p_i\})$), we only use $n+1$ of them. What about the rest? In fact, we can obtain another expression between these two sets of quantities if summing up $F_N^j(\{p_i\})$ instead from $n+1$ to N:

$$p_k F_{N-1}^n(\{p_{i\neq k}\}) = \sum_{j=n+1}^{N} (-1)^{j-n-1} \left(\frac{1-p_k}{p_k}\right)^{j-n-1} F_N^j(\{p_k\}). \qquad (8.49)$$

These two parallel expressions can be a bit disconcerting: we have two different equations describing a single relationship; does this jeopardize uniqueness? The answer is that it does not, but not because we list and label both expressions in parallel—the fundamental reason is that we were essentially just rewriting eqns (8.32–8.34) without any further assumptions, and on a more apparent level, the two equations express $F_{N-1}^n(\{p_{i\neq k}\})$ with disjoint groups of $F_{N-1}^j(\{p_{i\neq k}\})$ so there is no conflict of values. Although both results are exact, in practice it makes sense to use eqn (8.48) for small p_k and eqn (8.49) for $p_k \sim 1$ for higher computational accuracy—dividing or multiplying a very small number can lead to substantial floating point errors. It is also worth noting that when N is very large, removing p_k from $\{p_i\}$ does not alter the distribution significantly, and we do not have to do such substitution. In other words, this method works best for multiple, but not too many, individuals.

Now that we have cleared up the calculations of $\tau(p)$, we can substitute eqn (8.42) into eqn (8.28) to obtain the distribution

$$P(p_k) \propto \left[2p_k(1-p_k) F_{N-1}^{(N-1)/2}(\{p_{i\neq k}\})\right]^{-1}. \qquad (8.50)$$

Again, we emphasize that for not too big N, we can substitute $F_{N-1}^{(N-1)/2}(\{p_{i\neq k}\})$ with $F_N^n(\{p_i\})$s for accurate calculations. For $N \gg 1$, however, distribution $\{p_{i\neq k}\}$ should be roughly identical to $\{p_i\}$, and in fact can be described by function $P(p)$! Thus, rewriting the above equation we have

$$P(p) \propto \left[2p(1-p) F_{N-1}^{(N-1)/2}[P(p)]\right]^{-1}. \qquad (8.51)$$

Here we drop the suffix k for a clearer, more general notation. We also explicitly write down the dependence of F on strategy distribution $P(p)$ as a functional. Equation (8.51) clearly exhibits an interactive nature, and we indeed calculate it iteratively. The steps are the following:

1. Assume an initial form for $P(p)$, say a uniform distribution $P(p) = 1$.
2. Obtain $F_N^n[P(p)]$—this step is easier said than done because calculating $F_N^n[P(p)]$ is extremely complicated due to its numerous possible distributions. Our way around it is inspired by eqn (8.30), where we find that the average distribution $\langle F_N^n[P(p)]\rangle$ is simply binomial. Here, for particular values of $F_N^n[P(p)]$ we also

approximate it with a Gaussian distribution using the central limit theorem, by evaluating mean $\bar{p}$ and the standard deviation from the assumed $P(p)$.

3. Use eqn (8.51), together with eqns (8.48) and (8.49), if necessary, to obtain the new $P(p)$ and normalize it accordingly.
4. Check for convergence of $P(p)$ and, if necessary, repeat the steps until convergence is obtained.

By following the above steps, we are actually doing a much simplified simulation of the evolutionary minority game over time (i.e. iterations). Instead of simulating the movements of every agent, we have found mathematically a proper approximation through evolution of the system-level strategy distribution. Results from our theory are in good agreement with numerical data. A further test of the theory is obtained by comparing results for $\tau(p)$ as a function of p with numerical data for different N. $\tau(p)$ provides a better test than $P(p)$ for the validity of any theory, since many forms of $\tau(p)$ can give rise to similar forms for $P(p)$. Simulations suggest that the correct $\tau(p)$ in the steady state, which follows from eqn (8.42), has the form

$$\tau(p) \sim \frac{1}{2} - \mathcal{A}(N) p (1-p), \qquad (8.52)$$

where $\mathcal{A}(N)$ is an N-dependent constant which decreases with N as $1/\sqrt{N}$. Such a scaling with N makes sense from random walk arguments.

8.5 Rewards for winning and the 'softness' of news

The discussion so far suggests that for individual entities in any human–technology–AI system that are trying to grab some limited resources for themselves—and hence play the evolutionary minority game—they must become radicalized and self-segregated to maximize the likelihood of winning. There are indeed many real-world examples where a population is divided into two opposing parties to fight for one resource, like some of the conflicts happening in today's world. On the other hand, there do also exist scenarios where a population becomes united facing a minority-like challenge, e.g. when facing a severe existential threat. Our theory cannot yet explain the latter, but a generalization by Hod and Nakar (2002) enables us to gain insight. Here we provide a brief summary.

Our original theory follows a simple points award scheme: 1 point for winners and -1 point for losers. This may be an oversimplification in many cases though. We instead generalize it to gaining ρ_w points if winning, and $-\rho_l$ points when losing ($\rho_{w,l} > 0$). We then define the *prize-to-fine* ratio as

$$R = \frac{\rho_w}{\rho_l}. \qquad (8.53)$$

It measures the relative reward of a game: larger R values correspond to situations where the gains from winning the game outweigh potential losses and hence measure the game as a more rewarding one. Smaller R values correspond to the reverse, hence measuring the game as a less rewarding one. Hod and Nakar found that different R values give rise to 3 distinct phases, separated by 2 critical values $R_c^{(1)}$ and $R_c^{(2)}$:

1. For large prize-to-fine ratio $R > R_c^{(1)}$, the game is rewarding and the corresponding strategy distribution is extreme and self-segregating. Our original $R = 1$ model falls in this phase. A large enough R prevents adaptation from being too fast and drives the formation of similarly sized groups around the extremes $p \sim 1$ and $p \sim 0$ (crowd and anticrowd). The closer individuals are to $p \sim 1$ and $p \sim 0$ respectively, the more likely they are to take opposite positions from each other.

2. For intermediate values $R_c^{(2)} < R < R_c^{(1)}$, $P(p)$ takes up an M-shaped distribution. It still exhibits self-segregation, but not in the extremes. As R decreases, the 2 peaks of the distribution move to the centre, while the trough in the middle becomes less prominent. This indicates a decrease in self-segregation and an increase in clustering.

3. For small $R < R_c^{(2)}$, the trough in the M-shaped curve disappears and we instead have an inverted U-shaped distribution. Now the game yields poor rewards, and the population evolves to a clustered, aligned choice of strategies.

Figure 8.6 shows strategy distributions following regimes 1 (A) and 3 (C) respectively. Also shown are 2 figures (B, D) from an empirical study by Facebook (Bakshy et al., 2015). This empirical study aims at finding the relationship between news stories

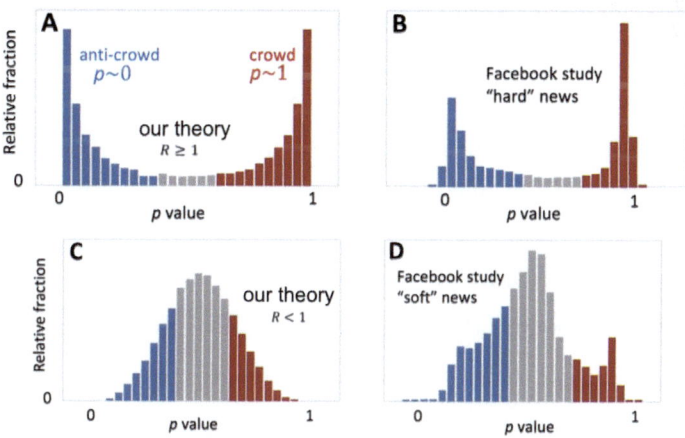

Fig. 8.6: Our theory with different prize-to-fine ratios, compared with empirical data of individuals' strategy distributions when faced with different types of news inputs. Source: data from Johnson et al. (2019b).

circulating on the Facebook platform and the ideology of internet users consuming these online news stories. Specifically, they classify the news into 'hard' and 'soft' stories using machine-learning methods: *hard* stories are more political, while *soft* stories are typically those on, for example, travel, lifestyle, sports, etc. As shown in Fig. 8.6, the distributions of ideology alignments adopt similar trends to the results of our evolutionary minority game. Hard news gives rise to a more bipolar distribution akin to the high prize-to-fine ratio regime, whereas soft news corresponds to a more clustered distribution akin when R is small.

This shows that our theory provides a parsimonious explanation for the findings of Bakshy *et al.*, since 'hard' news (Fig. 8.6(B)) is directly relevant to the socioeconomic climate at any given moment, and hence will be considered as relevant to an individual in deciding whether to pursue L at that moment: hence R would be larger which is consistent with $R > R_c^{(1)}$. 'Soft' news (Fig 8.6(D)) is not directly relevant to the socioeconomic climate at any given moment, and hence will play less of a role in whether an individual decides to pursue L at that moment: hence R would be smaller, consistent with $R < R_c^{(1)}$.

In stark contrast to the key role of R, the emergence of the polarized U-shape is not sensitive to the other parameters in the model. We mentioned at the beginning of the chapter how the system is insensitive to external information. But with a closer look at the simulations, we reaffirm that it is not sensitive to individuals' threshold value d, or the size of the change of p upon adaptation (i.e. the window D), or the absolute values of L or N at fixed $L/N = 0.5$, or the nature of the information I. It is also not sensitive to:

1. whether the information I is fabricated to provide artificially high levels of 'good' news (e.g. all 1s) or 'bad' news (e.g. all 0s), or whether it is true or false, or whether it is released in a certain sequence;

2. whether we include population heterogeneity from the outset by randomly assigning p values, or set them all equal to 0.5 to mimic maximum initial uncertainty of all individuals;

3. the origin or precise nature of I;

4. the amount of clock-time that passes between the new arrival of information I;

5. the absolute value of the reward and penalty (only the ratio R is relevant);

6. the way in which individuals adapt and hence choose a new p value, as long as they do so independently;

7. whether different sectors of the society receive different I and hence exist in different news bubbles. As long as each bubble remains self-contained and has the

same R, each U or inverted U will just add together and hence preserve the same shape;

8. other 'bubble' mechanisms, such as allowing individuals with similar p values to mimic each other. This applies even if the population gets broken up into only three bubbles.

In short, while any of these factors may impact the way in which $P(p;t)$ evolves and the time it takes to reach a steady state, they do not themselves dictate whether the final $P(p;t)$ is U-shaped or not. Since a U and inverted U are symmetric around $p = 0.5$, even a switch in definition of what is extreme 'left' and what is extreme 'right' would not change the resulting $P(p;t)$. We also note that generalizing the characterization of people's strategies and hence ideologies from a single p value to a vector of attributes of arbitrary length, and/or allowing I to be a vector, does not change the emergence of crowds and anticrowds at the extremes.

8.6 Impact of future social media algorithms

We end this chapter by looking at the impact of future social media algorithms which are likely to actively create new bonds (links) between human–technology–AI entities in Fig. 1.2, in order to build new types of hybrid online–offline communities as part of some higher Metaverse. Specifically, we simulate the likely impact that such social

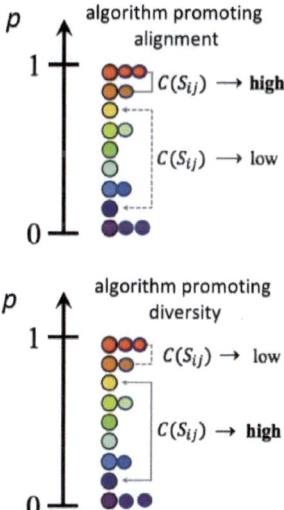

Fig. 8.7: An illustration of next-generation bonding algorithms. Source: data from Johnson et al. (2019b).

228 *Adding adaptation: emergence of anticrowds*

bonding algorithms (Fig. 8.7) will have, by building on the theory in this chapter together with the fusion theory from Chap. 3. We assume that they start from the present-day $P(p;t)$ profiles shown in Fig. 8.6(B,D). Since any machine-based mechanism can create links quickly once implemented, we assume a quasi-static $P(p;t) \equiv P(p)$ while these algorithms are operating. As shown in Fig. 8.7, the alignment-algorithm links together individuals based on the similarity of their p, and hence promotes the growth of clusters with high internal alignment, whereas the diversity-algorithm links together those based on dissimilarity of p, hence promotes cluster growth with high internal diversity. This calls for our heterogeneous approach of fusion from Chap. 3 to describe these opposite cluster growth schemes.

The next-generation algorithms are implemented exactly with that approach. We recall the definition of *similarity* $S_{ij} = 1 - |p_i - p_j|$ between individuals i and j. Here $p_{i,j}$ follow the strategy distribution $P(p)$, and the corresponding heterogeneity factor F is simply given by averaging S_{ij} (alignment, or homophily) or $(1 - S_{ij})$ (diversity,

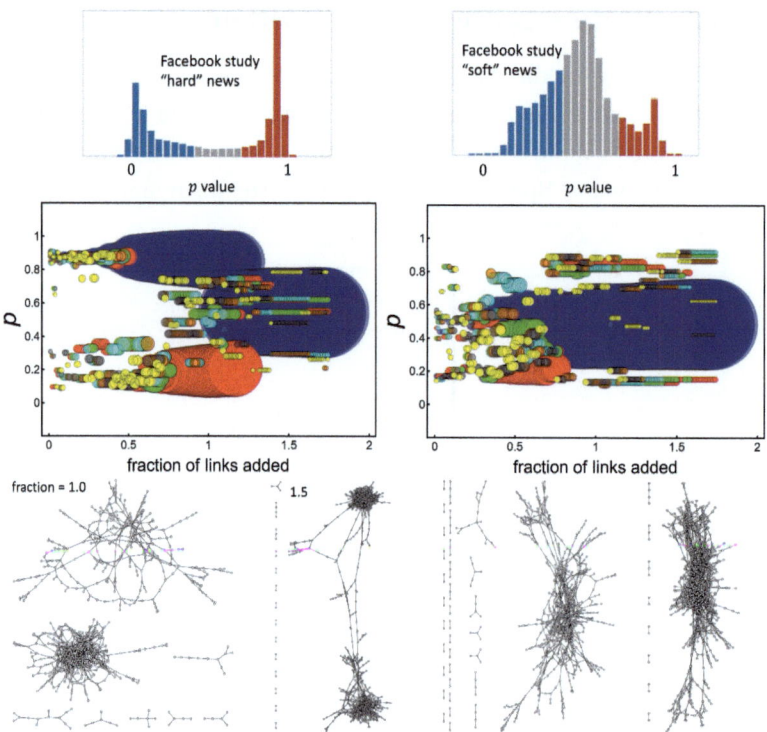

Fig. 8.8: Impact of adding links using next-generation algorithm promoting alignment (homophily) on the heterogeneity distributions from 'hard' (left-hand side) and 'soft' (right-hand side) news. Source: data from Johnson *et al.* (2019*b*).

or heterophily), over $P(p)$, for either 'soft' or 'hard' distributions. This is a simple mean-field example of how we study the complete cycle of online extremism from its appearance to its growth.

Results of the above simulations are shown in Figs 8.8 and 8.9. In these graphs, the upper panels (coloured bubble diagrams) use circles to represent each of the seven largest clusters at a specific time instant (shown in the diagram as a specific fraction of links added, which is proportional to time evolution). The centre of a circle sits at the average p values for a cluster, and its radius is proportional to the number of individuals in that cluster. Different colours are used to denote different clusters. Meanwhile, the lower panels (network diagrams) show snapshots of the actual network at two values of the added-link fraction.

1. For the homophily case, from Fig. 8.8 it is clear that, for both 'hard' and 'soft' scenarios, small isolated clusters appear with more extreme average p values away from $p = 0.5$ even when there is a single large cluster formed around $p = 0.5$ and even though the fraction of links added is greater than 1.0 so that each individual is included in at least one link on average, i.e. the beginning of the gel phase for simple fusion dynamics. We also find from the network for the 'hard' distribution that even the single large cluster that exists at added-link fraction 1.5 is actually internally highly segregated. It requires only one weak link to be broken in order to fragment the cluster into two separate pieces.

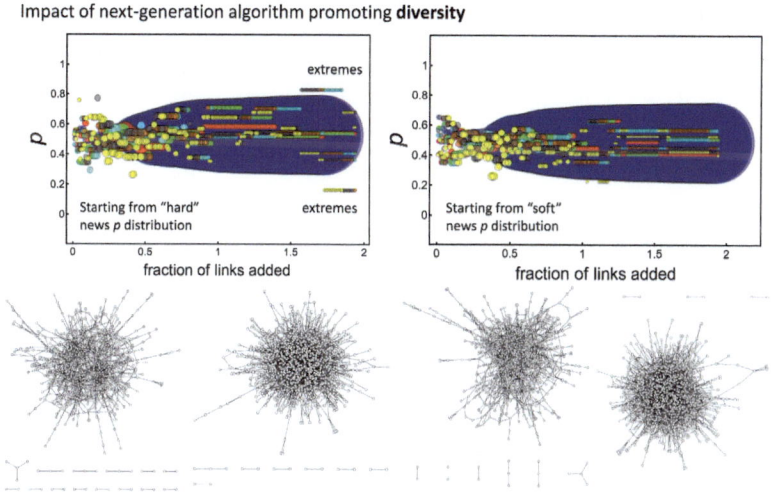

Fig. 8.9: Impact of adding links using next-generation algorithm promoting diversity (heterophily) on the heterogeneity distributions from 'hard' (left-hand side) and 'soft' (right-hand side) news. Source: data from Johnson et al. (2019b).

2. For heterophily, the main cluster forms quickly around $p = 0.5$, but it leaves a number of small isolated clusters with more extreme average p values that again survive until a surprisingly large fraction of links is added (i.e. $\gg 1.0$). However, this time the network of the dominating cluster looks much more homogeneous due to the diversity of individuals that join the cluster. This makes it hard to break into smaller extreme and conflicting clusters.

To sum up, although these algorithms can eventually connect together the majority of the population, our analysis shows that the process will likely generate *new* pockets of isolated extremes, and this is especially true for the algorithms promoting alignment.

Exercises

8.1 Show that in an evolutionary minority game, for a 3-entity system with possible strategies $\{0, \frac{1}{2}, 1\}$, the particle distribution at late times $P(p)$ is 1.1 : 1 : 1.1 if assuming that the likelihood of each class is weighted by its corresponding points.

8.2 Find the second-order strategy distribution $P^{(2)}(p)$ starting from eqn (8.20). (**Hint:** don't forget the extreme individuals with strategies $p = 0, 1$!)

9
Controlling human–technology–AI systems

We have now established theoretical descriptions of interacting entities that can mimic the observed empirical behaviour (patterns) of humans, technology, and/or AI presented in Chap. 2 and elsewhere in the book. This chapter focuses on the key issue for policymakers and engineers alike, of how to control current and future human–technology–AI systems. Some of the mathematics used here has been mentioned already in the respective sections, but is presented here in order to provide an integrated, self-contained response to the control question. Specifically, we look at the following control options for human–technology–AI systems based on the contents of the book:

1. Control based on the generating functions and their generative PDEs, or their linear reductions.
2. Control based on steering individuals' strategies.
3. Control based on engineering the human–technology–AI system's network properties. Specifically, we consider controlling malign activity across the multi-platform ecosystem, by adding an effective cost for creating links (hyperlinks) to in-built communities on particular platforms.

9.1 Control via the generative equation

We start with the generative equations, which were one of the first analytical tools we discussed quantitatively in this book. We know from Chap. 3 that gelation (i.e. a cohesive unit, shock wave, or giant connected component) emerges at time $t_{\text{sw}} = N/(2F)$. We have also seen the expression for t_{sw} when N changes linearly with time, in eqn (3.131). This gives us 2 straightforward handles to control: tuning population size N and/or the heterogeneity factor F. In this section, we briefly discuss these control schemes.

9.1.1 Mitigation scheme 1: control through population change

Equation (3.131) is based on the assumption that the system with fusion is not closed—that is, there are entities (e.g. users and hence potential recruits) flowing in or out of the system over time, and that the flux can be controlled. The mathematics of a

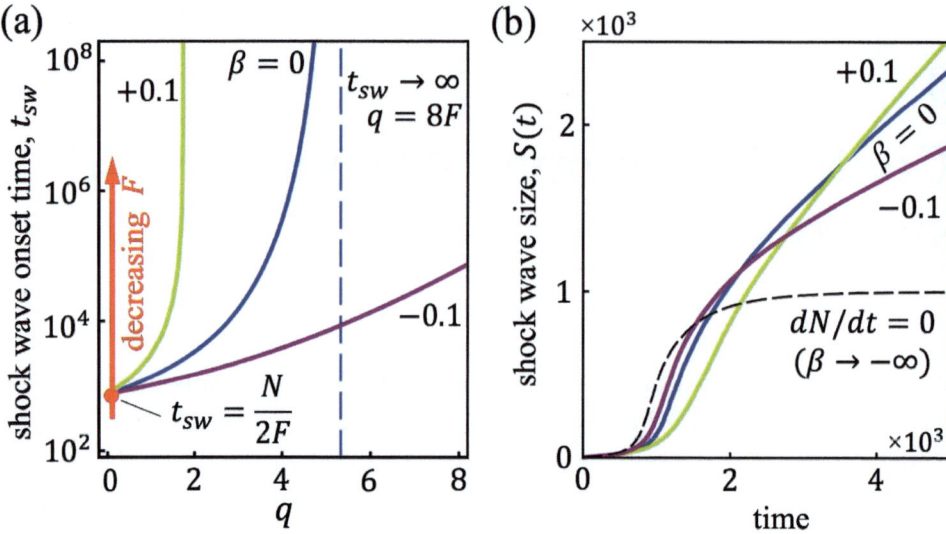

Fig. 9.1: Mitigation of gelation by changing newcomer (i.e. new recruits) flux $\dot{N}(t) = qt^\beta$, with uniform user distribution and initial population size $N(0) = 10^3$. In (b) we set $q = 0.5$. Source: data from Manrique, Huo, El Oud, Zheng, Illari, and Johnson (2023).

simple linear flux $\dot{N}(t) = q$ has been discussed in depth in Section 3.7.1. To briefly summarize, as the influx of monomers (e.g. new recruits) grows in rate, the system becomes increasingly diluted with more heterogeneous individuals, and hence the ability of a gel (i.e. cohesive unit, shock wave, giant connected component) to organize macroscopically slows down. The critical rate stands at $q = 8F$, above which there is no gelation. This is reflected in the blue curve in Fig. 9.1(a).

Here in this section, aside from this familiar example, we show results for a more general, non-linear influx rate

$$\dot{N}(t) = qt^\beta, \qquad (9.1)$$

where β is a parameter. Analytics can no longer handle this form of $\dot{N}(t)$, but as shown in Fig. 9.1(a), increasing β, with $\beta > 0$, makes $t_{sw} \to \infty$ at smaller q. In other words, a higher order of $\dot{N}(t)$ time dependence corresponds to a smaller critical rate parameter q for gelation. This makes sense: increasing β accelerates dilution as time progress. By contrast, $\beta < 0$ appears to remove this transition altogether, as shown in the purple curve in Fig. 9.1(a).

Figure 9.1(b), on the other hand, shows the corresponding growth $G(t)$ for the entire gelation period. For $\beta > 0$, $G(t)$ has a comparatively delayed gelation onset and initially rises more slowly than for $\beta \leq 0$, but it eventually overtakes. As β becomes more negative, $S(t)$ rises quicker but flattens faster.

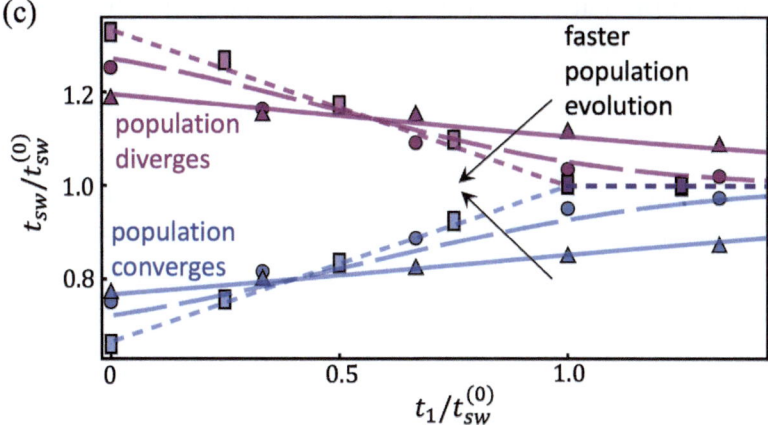

Fig. 9.2: Mitigations by changing user population distribution $q(\vec{y}, \tau)$ under 2 different schemes. Both starting with uniform user distribution, but evolving to converging or diverging population, respectively. t_1 is the median time for the change. Analytic results (curves) and microscopic simulations (symbols). Source: data from Manrique, Huo, El Oud, Zheng, Illari, and Johnson (2023).

9.1.2 Mitigation scheme 2: controlling heterogeneity distribution

Another explicit control parameter is the heterogeneity factor $F(t)$. This can be tuned, albeit indirectly since it is a mean-field parameter, by changing the characteristics distribution $q(\vec{y}_i, \tau)$ across the population. This is like changing the 'collective chemistry' of a population, e.g. a platform's user base. Figure 9.2 shows two opposite cases starting from uniform $q(\vec{y}, t)$: *converge* to a single delta function

$$\lim_{\tau \to \infty} q(\vec{y}, \tau) = \delta(\vec{y} - \vec{y}_0) \qquad (9.2)$$

for a constant vector $\vec{y}_0$, i.e. users become all identical; or *diverge* to two maximally separated, opposite delta functions

$$\lim_{t \to \infty} q(\vec{y}, t) = \frac{1}{2}\left[\delta(\vec{y} - \vec{0}) + \delta(\vec{y} - \vec{1})\right], \qquad (9.3)$$

i.e. users polarized. These 2 schemes are inspired directly by the results in Chap. 8, and they correspond, respectively, to $F(t \to \infty)$ values of 1 and 1/2. If we model the change from initial $F(0) = 2/3$ into the respective final values using the sigmoid function, we can write down 2 contrasting time-dependent heterogeneous factors

$$F_1(\tau) = \frac{2}{3} + \frac{1}{3\left[1 + e^{-\gamma(\tau - \tau_1)}\right]}, \quad F_2(\tau) = \frac{2}{3} - \frac{1}{6\left[1 + e^{-\gamma(\tau - \tau_1)}\right]}, \qquad (9.4)$$

where γ quantifies the rate of change and τ_1 is the median time for the change.

The effect of this mitigation scheme can be seen using the physics from Chap. 5, which allows time-varying $F(\tau)$. From there we know that the onset time is given by

$$\Phi(\tau_c) \equiv \int_0^{\tau_c} F(\tau)\,\mathrm{d}\tau = 1. \tag{9.5}$$

For F_1 and F_2 eqn (9.5) is actually in closed form and can be simplified into a cubic equation in terms of $e^{\gamma\tau}$—which we shall do in the exercises. If the switch of heterogeneity is very quick ($\gamma \to \infty$), $F_{1,2}(\tau)$ are then approximately step functions, and the corresponding $\tau_c^{(1,2)}$ are linear in terms of the median time τ_1. This is not hard to see in terms of $\Phi(\tau)$, which is continuous in τ but consists of 2 linear branches bent at τ_1. For F_1 the second branch is steeper, hence the system has an earlier onset, while for F_2 it is the opposite. When $\tau_1 > 1/F(0)$, the bend no longer affects the onset time since the system is already post-onset at τ_1. This confirms that $G(\tau)$ grows slower with a later onset as $F(\tau)$ decreases, i.e. making the population more diverse and heterogeneous will delay the anti-X shock wave onset and flatten its growth.

As a summary, controlling the heterogeneity distribution of a system fundamentally changes its pattern of interactions. This, which is reflected in the macroscopic generative equations, is equivalent to controlling the parameter $F(\tau)$. It can in turn result in all kinds of fusion–fission behaviours as we have seen in Chap. 5, and it is not limited to a single species as in our simple example. *Controlling the collective chemistry is therefore key to moderating a complex system of mixed entities such as humans, machines, software, and/or AI.* Different entities within such a system interact with different degrees of entanglement, forming D coarse-grained species. The control of this system then comes down to controlling the entries of the $D \times D$ symmetric matrix $\mathsf{F}(\tau)$.

9.2 Control by steering population strategies

This section is all about 'steering' entities that have internal strategies, along a given trajectory in some operating space. As discussed in Chap. 8, changing the heterogeneity distribution, especially in systems with crowds and anticrowds, occurs by individuals shifting strategies based on prior wins and losses. Here we present a specific application of the crowd–anticrowd theory with a more detailed picture of the strategy space than in Chap. 8, in order to show how the anticrowd can be 'steered' in some other direction in the 'character' space in which potentially dangerous anti-X ideas form.

Our 'steering' model actually has a seemingly irrelevant, living systems inspiration. Our description of this draws from Manrique, Klein, Li, Xu, Hui, and Johnson (2019). We know that in nature, organisms are able to exhibit a wide range of *taxes*, i.e. movements induced by stimuli. Unlike many higher-order movements of organisms that are centralized, such as high-level human activities that rely on a central processor (brain), a number of basic taxis behaviours are shown to be decentralized, such as klinotaxis by the *Drosophila* larva. A *Drosophila* larva explores space via aggregate actions of its decentralized constituent entities, i.e. abdominal and thoracic network segments. Though on the surface this seems very different to a human–technology–AI system, they are both examples of a system comprising adaptive, heterogeneous components, hence it should be no surprise that the emergent mechanism for the larva's steering and hence movement is analogous to our crowd–anticrowd model in Chap. 8. In both cases the decentralized interacting entities that comprise them each make decisions induced by an external stimulus. Of course, we do not seek for a precise model here that replicates the biological details of a *Drosophila* larva—instead, we want to abstract the decentralized nature of the system and apply our bottom-up, aggregate perspective to a similar scenario. Since the entities are mechanistic and hence deterministic, we use a slightly modified crowd–anticrowd model based on deterministic strategies, as mentioned briefly in Chap. 8, in order to describe this decentralized phenomenon of klinotaxis. This gives us a visible demonstration of control with our theory, since we can now see in physical space how the system can be steered closer to its goal. This physical space can also be seen as representing a more abstract control space in which the system's goal is represented by a particular location.

9.2.1 The taxis model

We start by briefly describing our model, illustrated in Fig. 9.3. The notation is largely inherited from Chap. 8, together with kinematic details of this specific system. We assume that there are N decentralized entities that together form the system responsible for rotation and hence steering—like some futuristic vehicle in which there are N steering wheels and where the driver behind each one is independent and only fed some limited information about the overall system. At each time step, every entity determines independently if it rotates anticlockwise $(+1)$ or clockwise (-1) by a small angle δ, while the entire organism advances a step of 1 unit in space. These micro-rotations add up together and all entities collectively determine how much the organism rotates. Denoting the number of entities choosing ± 1 at time t to be $n_{\pm 1}(t)$, respectively, then the resulting rotation after the time step is $[n_{+1}(t) - n_{-1}(t)]\delta$. The choice of δ reflects the rotation range of the system: for $\delta = \pi/N$ for example, the system can rotate as much as an angle π, i.e. it can change moving direction completely. For $\delta = \pi/(2N)$, the system can rotate up to a right angle each time step.

Each rotation results in a binary outcome of whether the organism becomes more or less aligned with the target (stimulus). If it maintains a relatively high alignment, it will take a short time for the organism to reach its goal (i.e. the target). For individual entities to make decisions, each entity has a memory μ of m previous outcomes—1 if

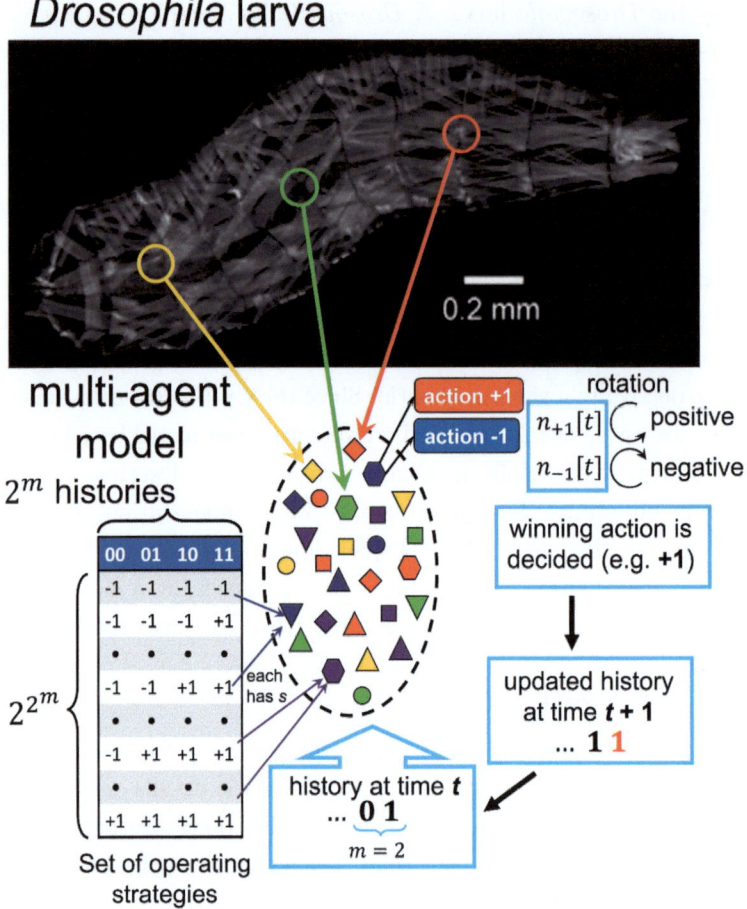

Fig. 9.3: Illustration of our taxis model. Source: data from Manrique, Klein, Li, Xu, Hui, and Johnson (2019).

more aligned and 0 if less, after a prior rotation event. The collective decision process follows the crowd–anticrowd model in Section 8.1.

Figure 9.4 compares the simulated motion trajectories using our model with real *Drosophila* larva trajectories from experiments. We see that the model reflects well the general directionality of the system as it guides itself towards its objective, while also reproducing the reversing loops and zig-zags along the way. If the resemblance of trajectories is not convincing enough, Fig. 9.5 provides a comparison on how much the model and the *Drosophila* larva turn, respectively, depending on their bearing angles. It shows a striking similarity: our theoretical model follows a sinusoidal relationship

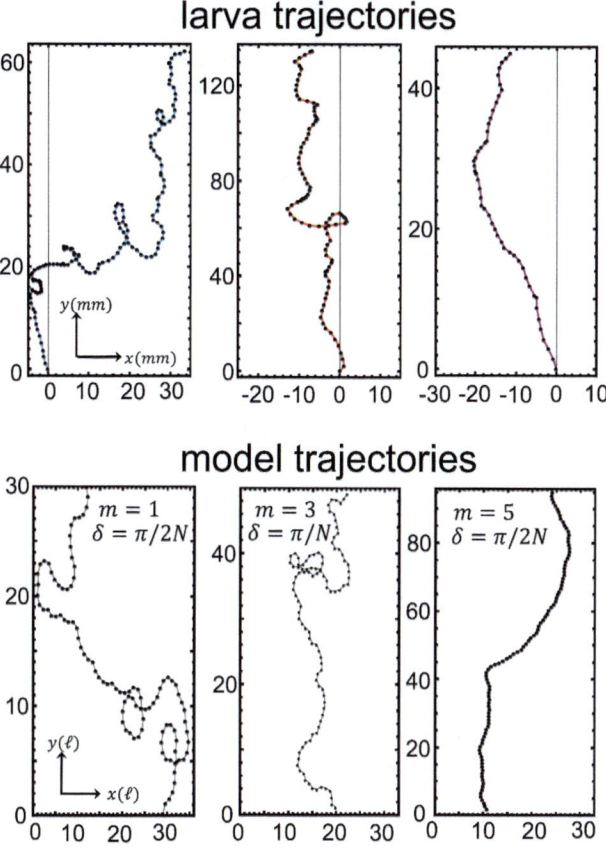

Fig. 9.4: Example trajectories of actual larva movements compared with our model results. Indicated in the plots are the memory m and angle δ values taken for each trajectory. Source: data from Manrique et al. (2019).

between ϕ and β, as much as a real *Drosophila* larva.[1] A comprehensive comparison can be found in an earlier paper of ours (Manrique et al., 2019), on which this section is based.

Figure 9.4 shows an improvement in the efficiency of pathfinding as the memory m increases. This is quite remarkable: as m increases, the system (like a 'boat') seems to be more efficiently steering itself towards its objective and freeing itself from wasteful circling loops, and this seems to be arising purely by controlling the m value of the internal, decentralized entities within the system. But this prompts the question

---

[1]The sinusoidal relationship for the larva is no coincidence. It is due to a navigation approach called *proportional navigation* which is used in sailing and missile navigation, etc. See (Martinez, 2014).

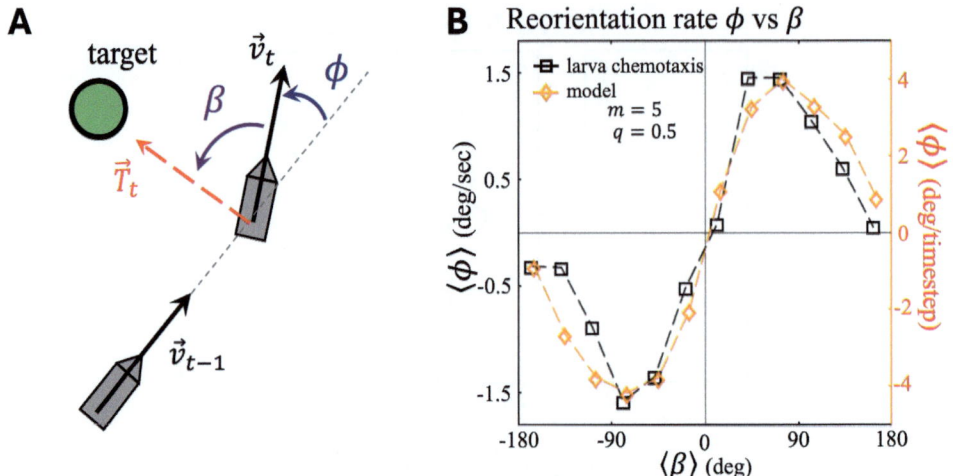

Fig. 9.5: A: Illustration of rotation showing notation for the relevant angles ϕ and β. B: Dependence of reorientation rate (i.e. rotation angle per time step) ϕ on the bearing angle β for experimental larva chemotaxis trajectories (Kane et al., 2013) and our model. Source: data from Manrique et al. (2019).

that we now address mathematically. Is it strictly true that pathfinding improves monotonically with increasing m? If this is indeed the case, controlling the system will be straightforward. All we would need is more memory capacity for each and every entity in the system. This is a little like the brute-force argument that a Large Language Model (LLM) is always more powerful the larger, deeper, and more complex its construction is. But is that true? This is what we now explore.

Figure 9.6 illustrates how increasing m indefinitely does not improve the steering efficiency. They show portions of model trajectories for the parameters $s = 2$, $N = 101$, with different values of the memory (or 'capability') m, rotation range δ, and information noise q. This noise term q represents the probability that the outcomes of rotations (0 or 1) are passed down to the component entities incorrectly. It mimics disturbances in signal interactions and will always be present in biological systems and real-world human–technology–AI systems operating in real-world environments (Manrique et al., 2019). From these plots it appears that irrespective of the noise term q, there is an optimal memory capacity, which in these cases is around $m = 5$. Above that, the task of finding the target (i.e. goal) becomes harder again, as evidenced by the observed zig-zagging or curvy trajectories for $m > 5$. The key to this non-monotonic dependence on m stems from the crowd–anticrowd interplay, as explained below.

1. For small m the zig-zag motion is clearly observed. Small m means that the strategy pool (size 2^{2^m}) is small and hence all the strategies are likely being used.

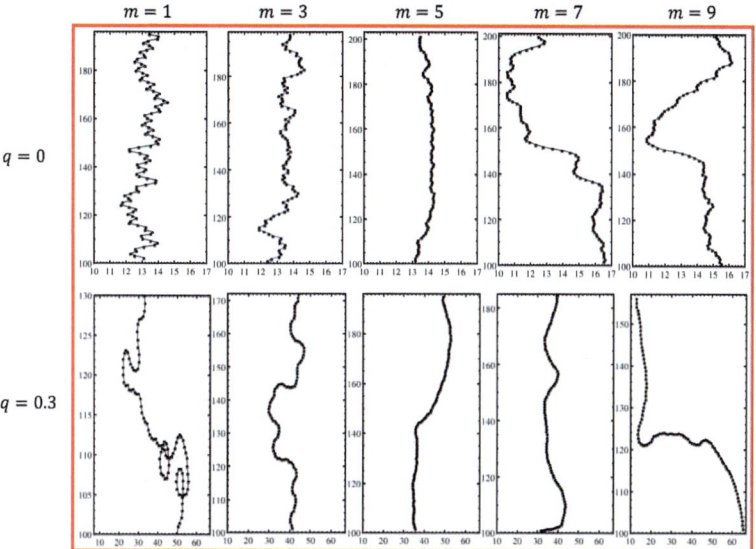

Fig. 9.6: Portion of trajectories for different values of m. The initial location is at $(0,0)$ and target at $(0,10^4)$, both outside the plots. The initial direction is $\theta = \pi/2$. q is the information noise. Source: data from Manrique et al. (2019).

As a consequence, the same strategy is selected by many agents hence creating a large crowd that all take the same action. When this strategy then loses, all the agents switch to another strategy which produces another large crowd. Only a small portion of agents are stuck with the opposite strategies and hence form a small anticrowd. This switching continues, meaning that the system advances in an inefficient loopy zig-zag. Adding noise (i.e. increasing q) breaks up the zig-zags, but allows more loops to form and hence maintains the inefficient state.

2. As m increases, the size of the strategy space increases (size 2^{2^m}) and hence there are significant subpopulations of agents that can use strategies that are opposite to each other. For some intermediate m (~ 5), the crowd and anticrowd have similar sizes and hence cancel out in terms of their rotation actions. The system therefore moves in a rather direct way towards the target with small fluctuations. In this optimal situation, the end point after a finite number of steps is nearest to the target. This is similar to the situation in Chap. 8 where the symmetric heterogeneity distribution became achievable.

3. For large $m > 5$, the strategy pool becomes huge and only a small portion of all available strategies are held and hence used by the agent population. Since the momentarily best strategies are not necessarily in play, the agents end up using strategies that act like a coin-toss in terms of being correct, which then leads to a

rather random output at the system level in terms of the overall motion. This also leads to the end point being farther away from the target.

9.2.2 A theoretical approach

We now quantify the impact of memory capacity m on this pathfinding. Since the net steering is the key mechanism at each timestep, our focus is not on the angles or the trajectories, but rather the aggregate decision of $+1$ or -1. For this we employ the multi-entity model from Chap. 8 using deterministic strategies as shown in Fig. 9.3. For a given m, each component entity (agent) picks its action using the momentarily best strategy from its set of s strategies that it was initially assigned at random from the strategy pool. This strategy pool is the set of all 2^{2^m} possible strategies for a given m, and is called the *Full Strategy Space* (FSS). Given their binary form (Fig. 9.3) we can also represent these strategies with vectors σ, since in a 2^m-dimensional *memory space* there are a total of 2^{2^m} vectors composed of ± 1 digits that represent all strategies. Let's take the simplest example of $m = 1$: the only 2 permitted pieces of memory are 0 and 1, i.e. the previous outcome. This gives rise to a total of 4 strategies: a strategy that always picks $+1$ is represented in vector form as $[+1, +1]$. The strategy $[-1, -1]$ always has the opposite action to this. Then there are two intermediate strategies $[+1, -1]$ and $[-1, +1]$. These 4 vectors constitute the FSS of $m = 1$. Because of its key role in quantifying the crowd and anticrowd effects in this model with deterministic strategies, we must first establish the properties of this strategy space.

Strategy space. We can write a strategy vector as a binary number (string) simply by representing $+1$ components with digit 0 and -1 components with digit 1. [2] The above $m = 1$ FSS can then be represented as $00, 11, 01, 10$. The binary representation then yields a decimal representation, where each strategy can be represented by the decimal value of the above binary string. Again following the above example, the 4 strategies are now $0, 3, 1, 2$. We can use the decimal representation to label the vectorial representation as σ_i, where $i = 0, \ldots, 2^{2^m} - 1$ for better organization. Table 9.1 summarizes all the strategies and their representations for $m = 1$.

Table 9.1 Representations of the Full Strategy Space (FSS) for $m = 1$.

Label	σ_0	σ_1	σ_2	σ_3
Decimal	0	1	2	3
Binary	00	01	10	11
Vector	$[+1, +1]$	$[+1, -1]$	$[-1, +1]$	$[-1, -1]$

[2] It might seem more natural to flip the digits: 1 for $+1$ and 0 for -1. But this particular representation will later turn out to be convenient.

The problem with the FSS is that as m increases, it grows exponentially. It would be useful to work instead with a smaller 'skeleton' subset that just retains the core features of the FSS. This *reduced strategy space* (RSS) needs to be representative in that strategies within the RSS cannot be too similar, since similar strategies result in biases in agents' decision-making if they feature disproportionately in the RSS. To help build the RSS, we note that the similarity, or alignment, of strategies i, j is measured naturally by their dot product $\boldsymbol{\sigma}_i \cdot \boldsymbol{\sigma}_j$. i and j have more identical than opposite digits if $\boldsymbol{\sigma}_i \cdot \boldsymbol{\sigma}_j > 0$ and vice versa. We also note that strategies i, j are *correlated* if $\boldsymbol{\sigma}_i \cdot \boldsymbol{\sigma}_j > 0$, *uncorrelated* if $\boldsymbol{\sigma}_i \cdot \boldsymbol{\sigma}_j = 0$ and *anticorrelated* if $\boldsymbol{\sigma}_i \cdot \boldsymbol{\sigma}_j < 0$. The strategy space is closed in that it contains all possible strategies. In fact, it is a group and closedness is just one of its properties. The reader can verify with the simple $m = 1$ model that the FSS forms a group under the XOR operation in terms of the binary representation, or equivalently, direct product for the vector representation. [3] Essentially, bit-wise operations recover all elements in the FSS, and $\boldsymbol{\sigma}_0$ is the identity. Precisely due to the bit-wise operations, a special property for this group is that each element is its own inverse. The reader may have realized already that the FSS for $m = 1$ is precisely the *Klein four-group* $K_4 \equiv (\mathbb{Z}_2)^2$. Generalized to arbitrary m, the FSS is a $(\mathbb{Z}_2)^{2^m}$ group under bit-wise XOR operations. Hence, the RSS that we hope to find should be a subgroup of the FSS in which there are no correlated elements, or cosets of the subgroup.

The RSS can be found using mathematical induction. We begin by explicitly studying the simplest cases $m = 1$ and $m = 2$.

1. For $m = 1$, the FSS itself is the RSS since all elements are uncorrelated or anticorrelated. It can be further partitioned into 2 purely uncorrelated, disjoint sets

$$U = \{00, 01\} \quad \bar{U} = \{10, 11\}. \tag{9.6}$$

    Note that there is more than one way of partition. $\bar{U}$ turns out to be the negation (i.e. the anticorrelated strategies) of U. It is technically not a group, unlike U, but a coset of U. There is only one RSS in this case, the FSS itself, given by

$$\mathcal{R} = U \cup \bar{U}. \tag{9.7}$$

    It contains $4 = 2^{m+1}$ strategies.

2. For $m = 2$, there are in total 16 different strategies in the FSS. The reader is asked to find one possible way to partition the FSS into an RSS and one coset. It can be found that the RSS contains $8 = 2^{m+1}$ strategies.

---

[3] Now we can see the convenience of this particular vector representation—it unifies the XOR and direct product operations so that we do not need to take any complement operations.

Given this, we assume that for general m values an RSS has 2^{m+1} strategies. The one $\mathcal{R}$ that includes the identity 0 is a subgroup of the FSS, and it can be decomposed into a group U whose elements are strictly uncorrelated, and its negation $\bar{U}$, each of them containing 2^m elements. $\mathcal{R}$ partitions the FSS into disjoint sets. The trivial case $m = 1$ serves as the base case of the induction. To prove the above assumptions, we assume that for $m = m_0$, $m_0 \in \mathbb{Z}_+$, we can indeed find the subgroup U_0 as described with size 2^{m_0}. We first prove that for $m = m_0 + 1$, the uncorrelated subgroup U does indeed exist. First we note that the length of the strategy string is 2^m. We also note that within U_0, any pair of strategies must have half of the digits identical and the other half opposite (i.e. the digits in the same position, for example 1001 and 1010). Hence we can construct the group U as follows.

We first permute all the elements in U_0 except the identity 0 arbitrarily into a sequence of strings a_i, where $i = 1, 2, \ldots, 2^{m_0} - 1$. A simple permutation doesn't change anything, and we have

$$U_0 = \{0 \equiv a_0, a_1, \ldots, a_{(2^{m_0}-1)}\}. \tag{9.8}$$

Now we append a copy of string a_i behind itself for all i, and denote the combined bit string as

$$c_i = a_i a_i, \quad c_0 = 0. \tag{9.9}$$

From the assumption, we know that $\forall\ i \neq j$, a_i, a_j are uncorrelated. In other words, there must be 2^{m_0-1} digits in each of a_i, a_j that are identical and the other 2^{m_0-1} opposite. This means that c_i and c_j must be strictly uncorrelated.

We have so far constructed 2^{m_0} elements for subgroup U, but this only accounts for half of the required elements. The other half is even easier to construct: simply taking the negation of one of the a_i strings will do the trick, and we write

$$d_i = \bar{a}_i a_i \tag{9.10}$$

(or $a_i \bar{a}_i$). This guarantees that both d_i, d_j, and d_i, c_j, are uncorrelated. This completes the subgroup

$$U = \{c_0, \ldots, c_{(2^{m_0}-1)}, d_0, \ldots, d_{(2^{m_0}-1)}\} \tag{9.11}$$

and its size is indeed 2^{m_0+1}. We still need to prove that U preserves group properties. The identity (0) and commutativity are trivial. So is the inverse, which is still the element itself, preserving this distinct group property. We do need to inspect closedness more closely. This comes down to checking 3 quantities ($i \neq j$ for below):

1. c_i and c_j. For $i \neq j$, group operation yields

$$c_i \oplus c_j = (a_i \oplus a_j)(a_i \oplus a_j) = a_k a_k \equiv c_k \in U, \tag{9.12}$$

for some k.

2. d_i and d_j. Following the same proof as above, it is clear that the XOR operations ($\oplus$) between these quantities are also closed.
3. c_i and d_j.

$$c_i \oplus d_j = (a_i \oplus \bar{a}_j)(a_i \oplus a_j) = (a_i \bar{\oplus} a_j)a_k = \bar{a}_k a_k \equiv d_k \qquad (9.13)$$

for some k. Note that in the second equality we applied the Boolean property of XOR.

4. c_i and d_i.

$$c_i \oplus d_i = (a_i \bar{\oplus} a_i)(a_i \oplus a_i) = \bar{a}_0 a_0 \equiv d_0. \qquad (9.14)$$

The induction is hence completed and we have successfully proven the existence of U. Together with $\bar{U}$, the RSS

$$\mathcal{R} \equiv U \cup \bar{U} \qquad (9.15)$$

does indeed contain 2^{m+1} elements.[4] $\mathcal{R}$ partitions the FSS into $2^{2^{m+1}-m-1}$ disjoint sets, each of which can be chosen as the RSS. We have thus successfully reduced the dimension of this system from 2^m to $(m+1)$: a huge simplification for our multi-agent system, without introducing bias.

Strategy choice and the emergence of fluctuations. So far we have only set the scene where actions emerge, i.e. the strategy space. The core of our multi-agent theory is to study the collective behaviour of how agents choose within the strategy space. Our model assumes that an agent can have S different strategies R from the RSS, with $1 \leqslant S \leqslant 2^{m+1}$ generally. We represent the strategy choice using a rank-S *strategy allocation matrix* Ψ, such that its entry $\Psi_{i_1...i_S}$ denotes the number of agents that choose a specific combination of strategies $\{i_1, \ldots, i_S\}$. The size of Ψ is, naturally, $2^{m+1} \times \cdots \times 2^{m+1}$. Hence,

$$\sum_{i_1=1}^{2^{m+1}} \cdots \sum_{i_S=1}^{2^{m+1}} \Psi_{i_1...i_S} = N, \qquad (9.16)$$

the total population. Note that Ψ is totally symmetric since there doesn't exist an order in strategy choice.

This matrix Ψ may seem ordinary: what can it possibly tell us apart from logging agents' strategy choice? In fact, it behaves very differently for different m values. When m is small, there are few possible combinations available, i.e. the dimension of the matrix Ψ is small. Consequently, each entry (bin) in Ψ typically has a high number—hence the fluctuations in each bin are smaller than the mean entry in that bin.

---
[4] One can prove that $\mathcal{R}$ is indeed a subgroup of the FSS following the same method as for U.

This means that we can regard it as rather a 'flat' matrix. As almost every strategy has its supporters, by definition there always exist opposite agents that pick the completely anticorrelated strategies. On the other hand, when m is large, there are numerous available strategies, and this in turn makes Ψ sparse (i.e. composed of mostly 0 entries) with large fluctuations in each bin compared to its mean. This means that for larger m, we will see increasing fluctuations in the collective decision.

The following quantitative discussions illustrate exactly the above arguments. To start with we need a measurement for the level of fluctuations. At the system level, the fluctuations that we can observe are those of the general outcome, i.e. how many agents choose $+1$ and how many -1. We can then describe, at each time step, the net points assigned

$$D(t) = n_{+1}(t) - n_{-1}(t) \equiv \sum_{R=1}^{2P} a_R^{\mu(t)} n_R^{\underline{S}(t);\Psi}, \tag{9.17}$$

where $P = 2^m$. We suddenly have a explosion of new symbols and here we briefly explain their meanings: $n_{\pm 1}(t)$ represents the number of agents that choose ± 1 at time t respectively; $a_R^{\mu(t)} = \pm 1$ is the response of strategy R to the history bit-string μ at time t, i.e. the function value $\sigma(\mu(t))$; $n_R^{\underline{S}(t)}$ is the number of agents using strategy R at time t, following the strategy allocation matrix Ψ. The superscript is a reminder that this number of agents will depend on the strategy score $\underline{S}(t)$ and the strategy allocation Ψ at time t. Since the number of strategies scales exponentially with m, it quickly gets huge and we can hence work with a continuum version of eqn (9.17):

$$D(t) \equiv \int_{R=1}^{R_{\max}} dR\, a_R^{\mu(t)} n_R^{\underline{S}(t);\Psi}. \tag{9.18}$$

In conventional many-body physics, we are typically interested in the following statistical properties of macroscopic measurable quantities such as $D(t)$: (i) the moments of the probability distribution function (PDF) of $D(t)$ (e.g. mean, variance, kurtosis) and (ii) the correlation functions that are products of $D(t)$ at different times $t_1 = t$, $t_2 = t + \tau$, $t_3 = t + \tau'$ etc. (e.g. autocorrelation). Numerical multi agent simulations typically average over time t and then over configurations $\{\Psi\}$. A general expression to generate all such functions is therefore

$$D_p^{(\tau,\tau',\tau'',\ldots)} \equiv \langle\langle D(t_1)D(t_2)\ldots D(t_p)\rangle_t\rangle_\Psi$$

$$= \left\langle\left\langle \int\ldots\int_1^{R_{\max}} dR_1\ldots dR_p\, a_{R_1}^{\mu(t_1)}\ldots a_{R_p}^{\mu(t_p)} n_{R_1}^{\underline{S}(t_1);\Psi}\ldots n_{R_p}^{\underline{S}(t_p);\Psi}\right\rangle_t\right\rangle_\Psi$$

$$\equiv \int\ldots\int_1^{R_{\max}} dR_1\ldots dR_p \left\langle\left\langle V^{(p)}(R_1,\ldots,R_p;t_1,\ldots,t_p) n_{R_1}^{\underline{S}(t_1);\Psi}\ldots n_{R_p}^{\underline{S}(t_p);\Psi}\right\rangle_t\right\rangle_\Psi,$$

(9.19)

where
$$V^{(p)}(R_1, R_2 \ldots, R_p; t_1, t_2 \ldots, t_p) \equiv a_{R_1}^{\mu(t_1)} a_{R_2}^{\mu(t_2)} \ldots a_{R_p}^{\mu(t_p)} \qquad (9.20)$$

resembles a time-dependent, non-translationally invariant, p-body interaction potential in $\boldsymbol{R} \equiv (R_1, R_2 \ldots, R_p)$-space (i.e. in $\mathcal{R}^p$), between p charge-densities $\left\{n_{R_i}^{\underline{S}(t_i);\Psi}\right\}$ of like-minded agents. The potential function does not depend on Ψ since it is determined straightforwardly by the exact strategies in use. Hence, we can simplify the averaging into

$$D_p^{(\tau,\tau',\tau'',\ldots)} = \int \cdots \int_1^{R_{\max}} d^p \boldsymbol{R} \left\langle V^{(p)}(\boldsymbol{R}; t_1, \ldots, t_p) \left\langle n_{R_1}^{\underline{S}(t_1);\Psi} \ldots n_{R_p}^{\underline{S}(t_p);\Psi} \right\rangle_{\Psi} \right\rangle_t. \qquad (9.21)$$

We emphasize again that the integration is simply another way of writing the sum when m is large: the theory works in exactly the same way exchanging the integrals with summations. Since $\left\{n_{R_i}^{\underline{S}(t_i);\Psi}\right\}$ are determined by the game rules, eqn (9.19) can actually be applied to *any* multi-agent game, although in the following discussions we remain focused on the crowd–anticrowd model of a minority game-like setup. In this section we focus on the fluctuations of $D(t)$, so temporal correlation is beyond the scope of this book and we set $\{t_i\} \equiv t$, i.e. $\{\tau\} = 0$. Fluctuation is measured by the variance

$$\sigma^2 = D_2 - D_1^2, \qquad (9.22)$$

where D_1 is the mean given by

$$D_1 = \int_{R=1}^{R_{\max}} dR \left\langle V^{(1)}(R; t) \left\langle n_R^{\underline{S}(t);\Psi} \right\rangle_\Psi \right\rangle_t. \qquad (9.23)$$

If the game's output is unbiased, the average $D_1 = 0$. This condition is not necessary for calculating the variance—one can simply employ eqn (9.22)—however, we will take $D_1 = 0$ for convenience. The variance is then reduced to

$$D_2 = \int_1^{R_{\max}} \int_1^{R_{\max}} dR\, dR' \left\langle V^{(2)}(R, R'; t) \left\langle n_R^{\underline{S}(t);\Psi} n_{R'}^{\underline{S}(t);\Psi} \right\rangle_\Psi \right\rangle_t. \qquad (9.24)$$

There are 2 averages left to resolve. We start with time, in which both the effective charge-densities and potential fluctuate. A significant reduction would be to rewrite the charge-densities as a mean value plus a fluctuating term

$$n_R^{\underline{S}(t);\Psi} n_{R'}^{\underline{S}(t);\Psi} = n_R^\Psi n_{R'}^\Psi + \varepsilon_{RR'}^{\underline{S}(t);\Psi}(t), \qquad (9.25)$$

if the fluctuation is small and $n_R^{\underline{S}(t);\Psi}$ is relatively constant. There is no guarantee that this is the case, but here is a crucial observation: the ordering of the RSS $\mathcal{R}$

does not matter since we simply sum up all strategies. In fact, we can reorder the group arbitrarily as we want. We don't even need to assign the same strategy R with a constant label throughout the time. This is exactly what we shall do, since if we rank strategies based on their success over time, or based on their popularity over time, the fluctuations are considerably smaller. Let's explain this in detail: for each time step we assign a new label K to the strategies based on their performance—the best strategy is thus used by n_1 agents, the second best n_2, ..., n_K, etc. Note that this label is abstract in that it doesn't consistently represent a single strategy: the strategy K at t_0 and t_1 could correspond to physical strategies R_0 and R_1 respectively, but it doesn't matter since we sum over all strategies and average over time. The key is that now n_K fluctuates much less, and our approximations become more accurate. We can do the same thing also with a popularity ranking Q, with now strategy $Q = 1$ being the most popular. The additional property that comes with this ranking is that we know for sure that

$$n_{Q=1}^{S(t);\Psi} \geqslant n_{Q=2}^{S(t);\Psi} \geqslant n_{Q=3}^{S(t);\Psi} \geqslant \ldots, \tag{9.26}$$

thereby constraining the magnitude of fluctuations in the charge-density $n_Q^{S(t);\Psi}$.

There is one particular argument to make about the rankings K and Q: when m is very small and the RSS is packed, the two orders are highly correlated and hence very similar, since the limited choice of strategies results in a consensus among the agents on the relative performance of every strategy. As m grows large, however, this unanimous consensus diminishes as the increasing number of available strategies increase the fluctuations of the best performing strategies (K ranking). n_Q, on the other hand, remains relatively stable. When m becomes very big though, none of this matters since the strategy space is too big for the population. That is, the matrix Ψ becomes very sparse when m is large—most of its entries are 0 and a few fluctuate between, most likely, 0 and 1.

The above discussions will soon be applied to finding explicit expressions for D_2, but for now let's complete its simplification first. With temporal fluctuations controlled, we assume that they average to 0, and hence

$$\begin{aligned} D_2 &= \int_{Q \in \mathcal{R}} \int_{Q' \in \mathcal{R}} \mathrm{d}Q \, \mathrm{d}Q' \left\langle V^{(2)}(Q, Q'; t) \left\langle n_Q^\Psi n_{Q'}^\Psi + \varepsilon_{QQ'}^{S(t);\Psi}(t) \right\rangle_\Psi \right\rangle_t \\ &= \int_{Q \in \mathcal{R}} \int_{Q' \in \mathcal{R}} \mathrm{d}Q \, \mathrm{d}Q' \left\langle V^{(2)}(Q, Q'; t) \right\rangle_t \left\langle n_Q^\Psi n_{Q'}^\Psi \right\rangle_\Psi. \end{aligned} \tag{9.27}$$

Under the RSS the double integration over the whole space $\mathcal{R}$ is actually redundant due to the minimum correlations between strategies. To show that, recall that in eqn (9.15) we partition $\mathcal{R}$ into 2 completely uncorrelated subsets U and $\bar{U}$, where every element u of U has 1 and only 1 anticorrelated counterpart $\bar{u}$ in U. By definition, the integrand

$$\left\langle V^{(2)}(Q,Q';t)\right\rangle_t \langle n_Q^\Psi n_{Q'}^\Psi \rangle_\Psi = \begin{cases} \left\langle (n_Q^\Psi)^2 \right\rangle_\Psi & Q' = Q \\ -\left\langle n_Q^\Psi n_{\bar{Q}}^\Psi \right\rangle_\Psi & Q' = \bar{Q} \\ 0 & \text{otherwise} \end{cases} \quad (9.28)$$

corresponding to the fully correlated, fully anticorrelated, and uncorrelated cases, respectively. Therefore, D_2 can be reduced to a single integral

$$\begin{aligned} D_2 &= \int_{Q \in \mathcal{R}} dQ \left\langle n_Q^\Psi n_Q^\Psi - n_Q^\Psi n_{\bar{Q}}^\Psi \right\rangle_\Psi \\ &= \int_{Q \in U} dQ \left\langle n_Q^\Psi n_Q^\Psi - n_Q^\Psi n_{\bar{Q}}^\Psi + n_{\bar{Q}}^\Psi n_{\bar{Q}}^\Psi - n_{\bar{Q}}^\Psi n_Q^\Psi \right\rangle_\Psi \\ &= \int_{Q \in U} dQ \left\langle \left(n_Q^\Psi - n_{\bar{Q}}^\Psi \right)^2 \right\rangle_\Psi. \end{aligned} \quad (9.29)$$

Equation (9.29) has a simple interpretation. Since n_Q and $n_{\bar{Q}}$ have opposite sign, they act like two charge-densities of opposite charge which tend to cancel each other out: n_Q represents a **crowd** of like-minded individuals (entities), while $n_{\bar{Q}}$ corresponds to a like-minded **anticrowd** who always do exactly the *opposite* of the crowd regardless of the specific $\mu(t)$. We have effectively renormalized the charge-densities $n_R^{S(t);\Psi}$ and $n_{R'}^{S(t);\Psi}$ and their time- and position-dependent two-body interaction $V^{(2)}(R,R';t) \equiv a_R^{\mu(t)} a_{R'}^{\mu(t)}$, to give two identical *crowd–anticrowd quasiparticles* of charge-density $(n_R - n_{\bar{R}})$ (dropping the suffixes to simplify the notation) which interact via a *time-independent* and *position-independent* interaction term $V_{\text{eff}}^{(2)} \equiv 1$. The different types of crowd–anticrowd quasiparticle in eqn (9.29) do not interact with each other, i.e. $(n_R - n_{\bar{R}})$ does not interact with $(n_{R'} - n_{\bar{R'}})$ if $R \neq R'$. Interestingly, this situation could *not* arise in a conventional physical system containing collections of just two types of charge (i.e. positive and negative).

Evaluating the fluctuations. Finally we evaluate the average over the ensemble of strategy allocation matrices Ψ. As discussed, the number of strategies explodes as m increases and Ψ tends to become increasingly disordered, i.e. increasingly non-uniform, less 'flat'. We can mathematically describe the two extremes: small m such that $P \equiv 2^m \ll N$, and large m such that $P \gg N$. The former is the 'high-density' regime where the charge-densities $\{n_K\}$ (or $\{n_Q\}$ depending on our perspective) tend to be large, non-zero values which monotonically decrease with increasing K (Q). Hence, the set $\{n_K\}$ acts like a smooth function $n(K) \equiv \{n_K\}$ (or in terms of Q). By contrast, the latter is the 'low-density' regime where Ψ becomes sparse with each element $\Psi_{i_1,\ldots,i_S}$

reduced to 0 or 1. In this case $\{n_R\}$ should therefore be written as 1s or 0s in order to retain the discrete nature of the agents. Below we study these 2 regimes respectively.

I. The 'high-density' regime. For $P \ll N$, Ψ fluctuates relatively little, especially under the popularity ranking $Q \leqslant 2P$. Recall that n_Q^Ψ records the number of agents choosing the Qth popular strategy. Since in a game like this agents naturally tend to apply its best-performing, or most popular, strategy out of its S available ones, a relatively 'flat' $\Psi_{Q_1...Q_S}$ matrix does indeed result in $Q_1 \geqslant \cdots \geqslant Q_S$. We thus take the ensemble average of Ψ to be flat and each entry of $\Psi_{Q_1...Q_S}$ has an average value of

$$\langle \Psi_{Q_1...Q_S} \rangle = \frac{N}{(2P)^S}. \tag{9.30}$$

It follows that we can obtain n_Q^Ψ by summing up the entries $\Psi_{Q_1...Q_S}$ for $\{Q_1 = Q, Q_{2,...,S} \geqslant Q\}$, or $\{Q_2 = Q, Q_{1,3,...,S} \geqslant Q\}$, ..., $\{Q_S = Q, Q_{1,2,...,S-1} \geqslant Q\}$. For the S-dimensional matrix, this is an $(S-1)$-dimensional surface. An $S = 2$ example is shown in Fig. 9.7. The resultant value of n_Q^Ψ is thus

$$n_Q^\Psi = \langle \Psi_{Q_1...Q_S} \rangle \left[(2P - Q + 1)^S - (2P - Q)^S \right]. \tag{9.31}$$

How about $n_{\bar{Q}}^\Psi$? This is quite tricky since we know that $\bar{Q}$ must be related to Q, but we don't know how this gets reflected in the matrix Ψ. For some particular strategy rankings, the anticorrelated strategy $\bar{Q}$ may be located at some fixed entries, but it is generally not the case for the popularity ranking. After all, the opposite strategy of the most popular one can be the second most popular, just as it can be the least popular. Unfortunately as a result, we can only gauge the range of $n_{\bar{Q}}^\Psi$ without finding a more precise expression. We shall soon see that this is enough to capture the empirical findings very accurately. We consider the following two extremes of the collective behaviours:

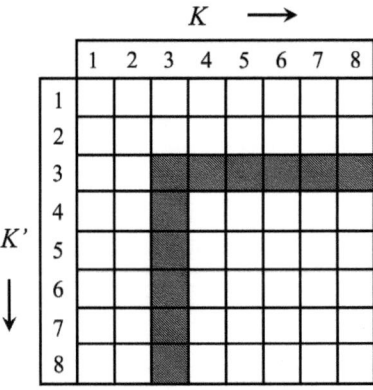

Fig. 9.7: Schematic representation of the matrix Ψ for $m = 2$ and $s = 2$ in the RSS. $N \gg 8$.

1. The population is extremely ordered in the sense that the popularity ranking is exactly the same as the performance ranking. This represents a system where agents follow closely the outcomes of strategies, which is likely to arise when strategy choices are very limited. In this case the Qth most popular strategy is the Qth best-performing, and since $\bar{Q}$ is its exact opposite, it must be the Qth worst-performing strategy, and hence satisfies $\bar{Q} = 2P + 1 - Q$. Therefore,

$$n_{\bar{Q}}^{\Psi} = \langle \Psi_{Q_1...Q_S} \rangle \left[Q^S - (Q-1)^S \right]. \qquad (9.32)$$

Together with n_Q^{Ψ}, we obtain

$$D_2 = \frac{N^2}{(2P)^{2S}} \int_{Q \in U} dQ \left[(2P+1-Q)^S - (2P-Q)^S - Q^S + (Q-1)^S \right]^2. \qquad (9.33)$$

For the simple case of $S = 2$, this can be easily calculated:

$$D_2 = \frac{N^2}{(2P)^4} \sum_{Q=1}^{P} (4P - 4Q + 2)^2$$

$$= \frac{N^2}{4P^4} \sum_{Q=1}^{P} (2P + 1 - 2Q)^2 = \frac{N^2}{4P^4} \left[\sum_{x=1}^{2P} x^2 - \sum_{x=1}^{P} (2x)^2 \right]$$

$$= \frac{N^2}{4P^4} \left[\frac{2P(2P+1)(4P+1)}{6} - 4 \frac{P(P+1)(2P+1)}{6} \right]$$

$$= \frac{N^2}{12P^3} \left(4P^2 - 1 \right) = \frac{N^2}{3 \times 2^m} \left(1 - 2^{-2m-2} \right). \qquad (9.34)$$

Note that we switch the integral into a summation since m is small and the system is not yet in the continuum limit. The charge-density is therefore significantly biased towards the high-performance strategies. This leads to higher-than-normal fluctuations and hence yields an overestimate:

$$\sigma^{\text{upperbound}} = \frac{N}{\sqrt{3 \times 2^{m/2}}} \left(1 - 2^{-2m-2} \right)^{\frac{1}{2}}. \qquad (9.35)$$

As a matter of fact, order is not the only source of overestimation. In this scenario it is also caused by tie-breaking of strategies, which is beyond the focus of this book. We refer to our earlier work (Johnson and Hui, 2003) on this matter.

2. The contrasting case is when the system has extreme disorder, and hence $\bar{Q}$ can be anywhere in the strategy allocation matrix. This means that each Ψ entry is equally likely to hold and play strategy $\bar{Q}$. Hence $n_{\bar{Q}}^{\Psi}$ is simply taken to be the mean:

$$n_{\bar{Q}}^{\Psi} = \frac{N}{2P}, \qquad (9.36)$$

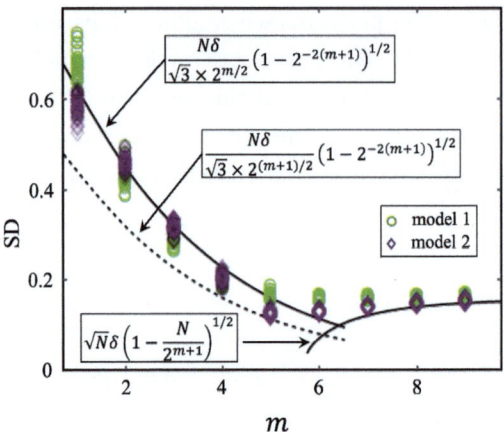

Fig. 9.8: Crowd–anticrowd theory against numerical simulations for $N = 101$ agents, $S = 2$ and $\delta = \pi/2N$. The theoretical predictions match the simulation results very well. For the detailed meanings of model 1 and model 2 and the details of the connection to the angles in the steering problem, see the adapted version of this figure in Manrique, Klein, Li, Xu, Hui, and Johnson (2019).

and it follows that

$$D_2 = \frac{N^2}{(2P)^{2S}} \int_{Q \in U} dQ \left[(2P + 1 - Q)^S - (2P - Q)^S - (2P)^{S-1} \right]^2. \quad (9.37)$$

For $S = 2$,

$$D_2 = \frac{N^2}{(2P)^4} \sum_{Q=1}^{P} (2P - 2Q + 1)^2, \quad (9.38)$$

which is half the value of eqn (9.34). This provides the lower bound of our estimate:

$$\sigma^{\text{lowerbound}} = \frac{N}{\sqrt{3} \times 2^{(m+1)/2}} (1 - 2^{-2(m+1)})^{\frac{1}{2}} = \frac{1}{\sqrt{2}} \sigma^{\text{upperbound}}. \quad (9.39)$$

II. The 'low-density' regime. After finding both bounds for the small-m regime, let's now focus on large m values. As have already been mentioned, Ψ entries are simply 0s and 1s. In this limit, there will tend to be $O(N)$ different crowds, with each crowd having $O(1)$ agent. Hence the popularity ordering is highly degenerate since $n_Q = 0, 1$ for all Q, i.e. highly disordered. From eqn (9.29), we are to sum over all strategies $Q \in U$. However, since $N \ll P$, most of the strategies have $n_Q = 0$. Hence this reduces to summing over only the $n_Q = 1$ terms, totalling N. We therefore have

$$D_2 = \sum_{Q=1}^{N} \left\langle \left(1 - n_{\overline{Q}}\right)^2 \right\rangle. \quad (9.40)$$

We know from the small-m scenario that the average number of $\langle n_{\bar{Q}} \rangle$ is $N/2P$. But now this value is much less than 1, meaning that $\bar{Q}$ is distributed very sparsely all over the matrix Ψ. Since $n_Q = 0, 1$ for all Q, including of course $\bar{Q}$, $N/2P$ becomes a probability distribution where only this fraction of $n_{\bar{Q}}$ is 1. This yields

$$D_2 = \sum_{Q=1}^{N} [1 - (n_{Q'} = 1)]^2 \frac{N}{2P} + [1 - (n_{Q'} = 0)]^2 \frac{2P - N}{2P} = \frac{N(2P - N)}{2P}$$

$$\Rightarrow \sigma^{m \gg 1} = \sqrt{N} \left(1 - \frac{N}{2^{m+1}}\right)^{\frac{1}{2}}. \quad (9.41)$$

The above two regimes agree with our qualitative discussions of the *Drosophila* model at the very beginning of this section. In terms of the quantitative results, Fig. 9.8 shows how the theory of fluctuations compares with numerical simulations of the step-rotation for different values of m. It fits very well indeed. But most importantly, we see that there is a 'sweet spot' in the m value that corresponds to minimal fluctuations, and hence leads to the target most quickly and hence most efficiently. This does not keep improving as agents possess better memory and hence higher m: instead when m gets too big, it interferes with the decision making. We note that the sweet spot here is more of a transitionary period than a critical point of any phase transition, since there isn't a clearly defined switch of regimes. Nevertheless, by controlling the memory capacity m away from being too small or too large, we are able to better steer a multi-agent, collective system in the desired direction and hence it is more likely reach its goal.

A very similar argument and derivation can also be used to steer the system *away* from certain dangerous regimes in an abstract operating space. Hence our theory opens the door to new forms of quantitative risk assessment for future human–technology–AI systems. We leave this exciting risk and hence insurance application for future work.

9.3 Control via link costs

The following draws heavily from our prior publication (Jarrett *et al.*, 2006) and the physics papers referenced within that. Control in this context is focused around networks, which underlie the online empirical data (e.g. Chap. 6) and also arose when discussing the fusion–fission model in Chaps 3–5. Our focus here is very much on the online world (Chap. 6) where our prior work shows (Johnson *et al.*, 2019a) that in-built communities engaging in malign activity tend to link to each other across platforms.

In addition to broadening their audience, links *between* platforms can provide shortcuts to better spread their content in a way that bypasses single-platform moderators. Hence we focus on this specific scenario of linked in-built communities (nodes) on one platform, seeking new links to an in-built community on another platform (node) in order to reduce the effective path-length between them. We abstract it into a *ring-and-star* model. The ring can be thought of as the linked communities (nodes) on platform 1, whose chain-like structure provides a pathway for malign content, etc., while the central node (star centre) is an in-built community on platform 2 which can provide a shortcut that would bypass platform 1's moderators. Hence this simple network model consists of a hub P at the centre of a chain of n nodes which form a ring-shaped network. We consider the ring as closed since this could be a real-world feature, as well as acting as a useful theoretical approximation. This model (Dorogovtsev and Mendes, 2000) stems from a simplification of the well-known small-world network.[5] In addition to online behaviour, this topology has a wide range of possible applications in human–technology–AI systems—from more abstract communications networks formed by decentralized linked entities, to real transport networks akin to cities with ring roads (e.g. London, Washington, D.C.). It can even be extended to describe multiple rings (e.g. Moscow, Beijing).

The model works as follows. Every pair of adjacent nodes on the ring are linked by a unit 'distance' which could be a measure of latency in some abstract network, or simply a physical separation. There is also a hub platform P. The nodes on the ring can create connections with P, and hence form shortcuts (e.g. motorways M4 and M11 that connect M25 to central London). The resulting shortcuts between ring nodes are also of unit length, i.e. the hub P is half a unit distance away from the ring nodes. The links between hub P and nodes on the ring are each formed with a probability p. An illustration is shown in Fig. 9.9. The practical questions that we will answer here are: (1) how does the presence of P impact the shortest path between two nodes along the chain (i.e. ring) and hence facilitate the spread of malign content and bad-actor activity online, and (2) how can we control this effectively?

To answer the first question, we start by studying the shortest path l between 2 nodes A, B on the ring. The value of l depends on probability p, and should hence be probabilistic. Of immediate interest is then its probability distribution, $\mathcal{P}(l, p)$, i.e. starting from a node A, the probability distribution of distance l to the other nodes on the ring, given the connection probability p to the hub P. Due to the homogeneity of the topology (circular symmetry) we can choose node A *wlog*, but the definition of $\mathcal{P}(l, p)$ requires us to pick and start from this node a priori. Clearly l has an upper bound, the distance on the ring between A and B, denoted as m (i.e. there are $m - 1$ nodes between A and B). The range of l is therefore $1 \leqslant l \leqslant m$. There is some additional complexity regarding the value of m: a network like this is generally without direction.

---

[5] See (Watts and Strogatz, 1998) for the detailed model. Notably, the small-world network transforms to a random network when the randomness of the model is fully turned on.

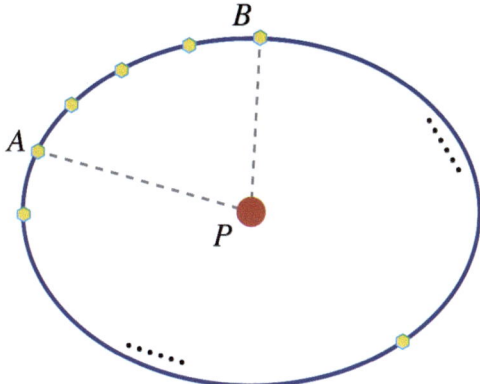

Fig. 9.9: Illustration of the ring-and-star model. Shortcuts with the hub P are shown with dashed lines.

This means that $m \leqslant n/2$, and it incurs extra burden when calculating $\mathcal{P}(l,p)$ since we are required to consider routes both from the left and the right of nodes A and B. In contrast, considering a directed graph here should considerably ease the calculation (by half at least), and we should be fairly certain that it won't yield something utterly alien—since we are simply muting a direction, there should be a factor of 2 somewhere when translating one to the other. We will see later that this is indeed the case. But for now, let's look into the directional case, where *wlog* we assume that A passes on information to B but there is no reverse communication. We note that directional circular networks do of course exist in the real world, such as the mono-directional bus route 10 in Segovia, Spain surrounding its old town.

We proceed by first finding the probability $\mathcal{P}(l|m)$ that the shortest path between any two nodes on the ring is l, given that they are separated around the ring by length m. Note that it can be found explicitly for undirected models too; however, this is a bit more tedious. Summing over all m for a given l and normalizing yields the probability

$$\mathcal{P}(l,p) = \frac{1}{n-1} \sum_{m=1}^{n-1} \mathcal{P}(l|m) = \frac{1}{n-1} \sum_{m=l}^{n-1} \mathcal{P}(l|m), \qquad (9.42)$$

where the summation can be truncated to get rid of $m < l$, since the shortcut cannot be longer than the original distance. The normalizing factor is simply the number of possible m values.

Figure 9.10 illustrates how we can calculate $\mathcal{P}(l|m)$ for both directed and undirected models. For convenience we can transform the ring of nodes into straight lines without altering the results. For the directed ring, node A can be put at one end of the straight line (which is left *wlog* throughout the discussion, with the directed edges pointing

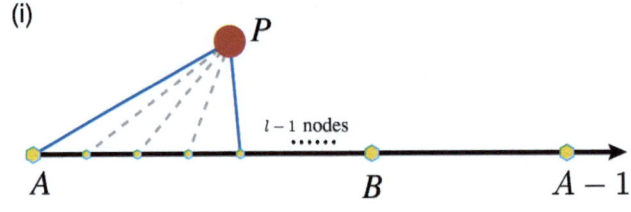

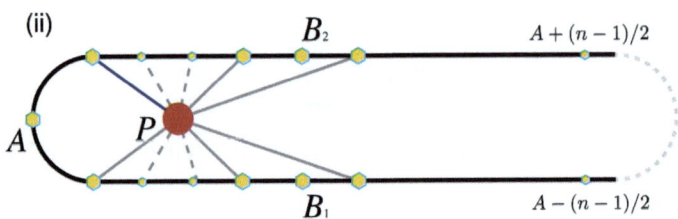

Fig. 9.10: We can transform the problem to straight-line models. This shows how we calculate $\mathcal{P}(l|m)$ for (i) directed and (ii) undirected models, respectively. The dashed lines are possible shortcuts that do not affect the distance between A and B. The solid grey lines in (ii) are the necessary shortcuts to satisfy the required distance, but all of them need not coexist—one from the left of P plus one from the right suffix.

towards the right) since the distance from A to all the other nodes increases following the direction of the edges, while for the undirected model A has to be placed at the centre.

9.3.1 Distribution of distance for the directed model

Here we present the detailed derivations of the directed model. The undirected model follows the same method. The first thing we notice is that $\mathcal{P}(l = m, m)$ may have a different expression than for other l values since when $l = m$ we do not need, and cannot have, shortcuts through P. How shortcuts work is shown in Fig. 9.10: since all nodes that are connected to P are taken to be adjacent with distance 1, the outer-most pair of P connections determine the shortcut. This is to say that, among all the nodes that are connected to P, we find the left-most and the right-most points on the line, and their distance is reduced to 1. This in turn means that, in order to construct a reduced distance $l < m$ between nodes A and B, we need to partition the $(m + 1)$ nodes (including A, B) into l parts:

1. One connected component $\mathcal{C}$ of $(m - l + 2)$ nodes, whose left-most and right-most nodes are connected to hub P. These are the shortcut nodes, and are shown in Fig. 9.10 by the solid lines connected to P.

2. The rest $(l-1)$ nodes, which are **forbidden** to connect to P. They do not have to be mutually connected, hence there are $(l-1)$ different components.

Within the component $\mathcal{C}$, the only hard requirement is the connection of its left-most and right-most nodes to P. Connections of its other nodes do not affect the distance l. This means that, for each partition, its probability is $p^2(1-p)^{l-1}$. And since there are in total l different partitions (i.e. l different ways of shortcuts, or l different ways to place component $\mathcal{C}$ within the $(m+1)$ nodes), the total probability is

$$\mathcal{P}(l|m) = lp^2(1-p)^{l-1}, \quad 1 \leq l < m. \tag{9.43}$$

Having sorted out the $l < m$ case, we now focus on the scenario $l = m$. Here we do not *need* shortcuts, but it doesn't mean that we cannot allow for their existence, provided that the existing shortcuts do not shorten the distance between A and B. Obviously to the right of B there can be as many nodes connected to P as it likes without affecting the probability due to the binomial nature of each connection (see exercises). Between A and B, however, there can be at most 2 nodes connected to P and they must be adjacent. This results in 3 regimes:

1. No nodes between A and B are connected to P. Its associated probability is $(1-p)^{l+1}$.
2. Only one node between A and B is connected to P, thus failing to create a shortcut. Its probability is $(l+1)p(1-p)^l$.
3. Two adjacent nodes between A and B are connected to P, with probability $lp^2(1-p)^{l-1}$.

Adding them up, we have

$$\begin{aligned}\mathcal{P}(l=m,m) &= (1-p)^{l-1}\left[(1-p)^2 + (l+1)p(1-p) + lp^2\right] \\ &= (1-p)^{l-1}[1 + (l-1)p]. \end{aligned} \tag{9.44}$$

Hence we can write down $\mathcal{P}(l|m)$ as

$$\mathcal{P}(l|m) = \begin{cases} lp^2(1-p)^{l-1} & 1 \leq l < m \\ (1-p)^{l-1}[1+(l-1)p] & l = m \end{cases}. \tag{9.45}$$

Note that $\mathcal{P}(l|m)$ does not depend on m at all, precisely due to the fact that nodes in component $\mathcal{C}$ do not matter. It then follows that

$$\begin{aligned}\mathcal{P}(l,p) &= \frac{1}{n-1}\sum_{m=l}^{n-1}\mathcal{P}(l|m) \\ &= \frac{1}{n-1}\left[(n-l-1)lp^2(1-p)^{l-1} + (1-p)^{l-1}[1+(l-1)p]\right]\end{aligned}$$

$$= \frac{1}{n-1}(1-p)^{l-1}\left[1+(l-1)p+l(n-l-1)p^2\right]$$
$$= \frac{1}{n-1}(1-p)^{l-1}\left[(1-p)+\left(p+(n-1)p^2\right)l-p^2l^2\right]. \quad (9.46)$$

With this we can calculate exactly the expected path length with shortcuts across the network

$$\bar{l}_n = \sum_{l=1}^{n-1} l\mathcal{P}(l,p), \quad (9.47)$$

where the suffix n denotes the total number of nodes n. The derivations are tedious but analytic nonetheless. We have, in the final step of eqn (9.46), organized the expression into 3 terms with different powers of l. They give 3 sequences which can all be evaluated exactly, resulting in

$$\bar{l}_n = \frac{1}{n-1}\left[\frac{2n+2}{p}-\frac{3}{p^2}-n+\frac{(1-p)^n}{p}\left(n-2+\frac{3}{p}\right)\right]. \quad (9.48)$$

9.3.2 Distance distribution for the undirected model

For the undirected model, exact calculations can still be carried out, but with a considerable increase in complexity. Let's take a look at the complexity without solving the system, since the gain of a complete calculation is small compared with the effort. As shown in Fig. 9.10, the loss of direction is responsible for the increased computational labour. In the undirected model there are 2 candidates of node B (denoted as $B_{1,2}$ in Fig. 9.10) for each node A, with the same distance m. This is why we represent it using a linear model with A positioned at the *centre* instead of the end. Consequently, the range of m for the undirected model is half that of the directed one, i.e.

$$1 \leqslant m \leqslant \frac{n}{2}. \quad (9.49)$$

Symmetries like this always require an additional discussion on the oddness of n, although for very large $n \gg 1$ this hardly matters. Just for completeness, the equality $m = n/2$ can be reached if and only if n is even. For odd n, it rounds down to $(n-1)/2$.

If we take a crude and qualitative look at the undirected model, it seems that its probability distribution should be related to that of the simpler directed model. To see this more clearly, we put the model in the many-body limit $n \gg 1$ for which we have a quasi-continuous model. In other words, there are now so many nodes on the ring that their distances relative to the system size, m/n and l/n, are almost continuous. This is of course a very common working condition in statistical physics. We continue by defining continuous variables

$$\rho \equiv pn, \quad z \equiv \frac{l}{n}. \quad (9.50)$$

Here ρ represents the average number of nodes connected to the hub P (akin to the definition of the mean degree c in a random network, recalling Chap. 5), and z

represents the relative distance between 2 nodes on a scale of 0 to 1. To round up our new definitions, we further define the average number of nodes with a distance l from node A

$$Q(z,\rho) \equiv (n-1)\mathcal{P}(l,p) \approx n\mathcal{P}(l,p). \tag{9.51}$$

As we have stated above, in the undirected model the maximum distance is $z_{\max}^{\text{undir}} = 1/2 = z_{\max}^{\text{dir}}/2$. If we take one side, say the right side of node A and for the undirected model only, now A becomes the end point also—and it appears *as if* we can renormalize this half of the model into a directed model, with a renormalization factor of 2 in the distance z. Note that this is not a strict renormalization group operation—we shall see later that it only works in leading order and for small p. Together with the other half of the undirected model, we can obtain the following approximate scaling rule:

$$Q_{\text{undir}}(z,\rho) = 2Q_{\text{dir}}(2z,\rho), \quad z \in (0, 1/2]. \tag{9.52}$$

This means that we can reduce the calculations of an undirected model into a directed one. We stress that this relation becomes accurate in the limit $n \to \infty$ and $p \to 0$, in which the population becomes huge and shortcuts are sparse. The reason is again due to the loss of direction. In the undirected model, a shortcut can be formed from the left branch to the right to meet the requirement of distance; they can also connect a node between A and B with a node beyond while also shortening the distance (see Fig. 9.10 (ii)). These two new phenomena break the equivalence between the two models, and make our scaling rule a mere approximation. Moreover, they make the calculations a statistical nightmare—to this end, even the original paper that devises this model (Dorogovtsev and Mendes, 2000) unavoidably misses some of these contributions towards $\mathcal{P}(l|m)$ and $\mathcal{P}(l,p)$, and hence their stated results are not exactly accurate. Nevertheless the beauty of this theory, and of statistical physics in general, is that these misses do not affect the macroscopic behaviours: the results such as the scaling law derived in the paper, notwithstanding the mistakes, are absolutely correct. This becomes an unintentional proof of the robustness of the theory and applicability of the assumptions. In the exercises, the reader is asked to calculate some of the simplest specific $\mathcal{P}(l|m)$ values to get an idea of counting all the different routes that all lead to the shortest distance l.

9.3.3 Control with costs on shortcuts

Now on to the control of this system. Under the context of harmful information online, the model can describe a connected, 'small-world'-like network of platforms that can pass around harmful information with a hub which exacerbates the spread by providing fast-track connections. In this scenario $\bar{l}_n$ is the average path length of the spread of harmful information. In terms of control, we naturally want to increase $\bar{l}_n$, making propagation of harmful information lengthier and thus harder. Equation (9.48) provides us with the relationship between $\bar{l}_n$ and the likelihood p of the nodes being connected to the hub, hence it provides us with a means of controlling $\bar{l}_n$ by moderating the hub P to control p. Therefore, a thorough understanding of the function $\bar{l}_n(p)$ is necessary. The first things to check are the values of $\bar{l}_n(p)$ for the limits of very big and very small p.

1. Very big p. When $p \to 1$, virtually every node is connected to the hub, and hence all nodes should be adjacent to one another. From eqn (9.48) we can easily get

$$\bar{l}_n(p=1) = 1, \tag{9.53}$$

confirming that exactly. This is the worst-case scenario for those who want to contain the spread of harmful information.

2. Very small p. When $p \to 0$, there is virtually no shortcut, and distances between nodes are almost exactly given by m, i.e. their separations on the ring. For the directed model, this means that

$$\lim_{p \to 0} \bar{l}_n(p) = \frac{n}{2}. \tag{9.54}$$

This can be obtained from eqn (9.48) using L'Hôpital's rule, but calculating $\mathcal{P}(l, p)$ explicitly first is much easier in this case.

From these limits, it is not hard to guess that $\bar{l}_n(p)$ decreases with p. It is also intuitive when we think about the physical picture. It is not very straightforward to see this from eqn (9.48) though, since taking derivatives of $\bar{l}_n$ directly is hard. An easier way to see this is putting the model again in the large n limit, where we can approximately drop the $(1-p)^n$ term. Now only the p^{-1} and p^{-2} terms are left, and since $2n+2 \gg 1$, the p^{-1} term dominates and $\bar{l}_n$ indeed decreases monotonically. We can go one step further and see this analytically. We take the continuum limit again, and for $n \to \infty$ we instead study

$$\bar{z}(\rho) \equiv \frac{\bar{l}_n(p)}{n}. \tag{9.55}$$

Note that

$$\lim_{n \to \infty} (1-p)^n = \lim_{n \to \infty} \left(1 - \frac{\rho}{n}\right)^n = e^{-\rho}, \tag{9.56}$$

so we have

$$\begin{aligned}\bar{z}(\rho) &= \frac{2}{np} - \frac{3}{(np)^2} - \lim_{n \to \infty} n^{-1} + \frac{e^{-\rho}}{np}\left(1 - \lim_{n \to \infty} n^{-1} + 3(np)^{-1}\right) \\ &= \frac{2}{\rho} - \frac{3}{\rho^2} + \frac{e^{-\rho}}{\rho}\left(1 + \frac{3}{\rho}\right).\end{aligned} \tag{9.57}$$

Now the range of the variable is $\rho \in \mathbb{R}^*$. This function is much easier to work with.

Now that we have this groundwork, we can add the important practical feature related to real-world human–technology–AI systems and hence control: an additional cost for connections to the hub P. This could be imposed exogenously as a control measure, or it could arise endogenously because of P's limited bandwidth and hence its inevitable slowdown as more connections are made to it. Hence we express the cost $c(\rho)$ as an additional path-length, as mentioned above; however, we stress it could be expressed as a time delay or reduction in flow-rate for harmful material through P, etc. This cost function can take various forms, and we consider some of the simplest cases. Specifically we consider (i) a constant cost c, where c is independent of how many connections that P already has—and hence the cost of using P is considered independent of how much P is likely getting noticed by moderators; (ii) a linear cost $c(\rho) \equiv k\rho$; and (iii) a non-linear cost $c(\rho)$ according to a number of non-linear cost functions. We keep c independent however of l and m so that we satisfy the homogeneous symmetry of the topology. For a general, non-zero cost $c(\rho)$, we can write for this network with directed links:

$$\mathcal{P}(l|m) = \begin{cases} (n-1)^{-1} & l \leqslant c \\ (l-c)p^2(1-p)^{l-c-1} & c < l < m \\ (1-p)^{l-c-1}[1+(l-c-1)p] & l = m \end{cases} \quad (9.58)$$

In the case when $m \leqslant c$, only the first case holds. This first case is new: when the cost of the shortcut is greater than the distance, no shortcuts are taken. On the other hand, the other 2 cases are directly modified from eqn (9.45) by replacing l with $(l-c)$. $(l-c)$ can be understood as an effective shortcut: in order to compensate for the cost, the nodes need to take more shortcuts. The resulting distance distribution is therefore

$$\mathcal{P}(l,p,c) = \begin{cases} (n-1)^{-1} & l \leqslant c \\ (n-1)^{-1}(1-p)^{l-c-1}\left[1+(l-c-1)p+(n-l-1)(l-c)p^2\right] & l > c \end{cases}.$$
(9.59)

Again, we are adding a new phase of 'shortcuts being costly' as well as replacing l with an effective shortcut $(l-c)$. This significantly simplifies our calculation of $\bar{l}_n(p,c)$ since we can reuse much of eqn (9.48). First we write down the expression:

$$\bar{l}_n(p,c) = \sum_{l=1}^{c} l(n-1)^{-1} + \sum_{l=c+1}^{n-1} l\mathcal{P}(l,p,c) \quad (9.60)$$

$$= \frac{c(c+1)}{2(n-1)} + \sum_{l=c+1}^{n-1} l\mathcal{P}(l,p,c). \quad (9.61)$$

The summation is split into 2 parts by the 2 regimes depending on the cost. The first one yields a new constant, previously not seen in the no-cost model. The second summation is more familiar, since we know from eqn (9.59) that

$$\mathcal{P}(l,p,c) = \mathcal{P}(l-c,p) \tag{9.62}$$

for $l > c$. It follows that

$$\sum_{l=c+1}^{n-1} l\mathcal{P}(l,p,c) = \sum_{l-c=1}^{n-c-1} (l-c)\mathcal{P}(l-c,p) + \sum_{l=c+1}^{n-1} c\mathcal{P}(l,p,c)$$

$$= \sum_{l'=1}^{n-c-1} l'\mathcal{P}(l',p) + c\left(1 - \sum_{l=1}^{c} \mathcal{P}(l,p,c)\right)$$

$$= \bar{l}_{n-c}(p) + c\left(1 - \frac{c}{n-1}\right), \tag{9.63}$$

where in the second equality we changed the variables $l' = l - c$ and also evaluate the second summation using the normalization condition of probability $\mathcal{P}(l,p,c)$. This transforms the second summation into a sum of constants. $\bar{l}_{n-c}(p)$ is readily available from eqn (9.48). Now we substitute everything and obtain

$$\bar{l}_n(p,c) = \frac{1}{n-1}\left[\frac{c(c+1)}{2} + c(n-c-1) + \frac{2(n-c)+2}{p} - \frac{3}{p^2} - (n-c)\right.$$
$$\left. + \frac{(1-p)^{n-c}}{p}\left(n-c-2+\frac{3}{p}\right)\right]$$
$$= \frac{1}{n-1}\left[(c-1)\left(n-\frac{c}{2}\right) + \frac{2n+2-2c}{p} - \frac{3}{p^2} + \frac{(1-p)^{n-c}}{p}\left(n-c-2+\frac{3}{p}\right)\right]. \tag{9.64}$$

To see how the cost affects the relationship between $\bar{l}_n$ and p, we look into the partial derivative $\partial \bar{l}_n / \partial p$ to check its monotonicity. But to simplify the calculations, we again take the system to the continuum limit. Now, however, we need an extra variable for the cost. Define

$$\gamma \equiv \frac{c}{n} \tag{9.65}$$

as the cost coefficient of the system. Taking $n \to \infty$ once more, we have

$$\bar{z}(\rho,\gamma) = -\frac{1}{n} - \frac{\gamma^2}{2} + \gamma - \frac{2\gamma}{\rho} + \frac{2}{\rho} - \frac{3}{\rho^2} + \frac{e^{-\rho(1-\gamma)}}{\rho}\left(1 - \gamma + \frac{3}{\rho}\right). \tag{9.66}$$

Recall that when $\gamma = 0$,

$$\frac{\partial \bar{z}}{\partial \rho} < 0 \;\forall \rho \in \mathbb{R}_+, \tag{9.67}$$

and this is exactly the monotonicity that makes the system trivial in terms of control. So the idea is to find γ values that makes $\partial \bar{z}/\partial \rho$ not monotonic. We can choose to approach eqn (9.66) comprehensively with numerical tools, but with some reasonable approximations it can be reduced significantly. We assume that $\rho \gg 1$ and $\gamma \ll 1$: since n is huge, this simply means that the shortcut probability p is not too small and that the cost c is not too great. These are reasonable assumptions. With these, we can take linear approximations and obtain

$$\bar{z}(\rho, \gamma) \approx -\frac{1}{n} + \gamma + \frac{2}{\rho}. \tag{9.68}$$

Taking the partial derivative, we have

$$\frac{\partial \bar{z}}{\partial \rho} \approx \frac{\mathrm{d}\gamma}{\mathrm{d}\rho} - \frac{2}{\rho^2}. \tag{9.69}$$

This means that, as long as we take a $\gamma(\rho)$ function such that $\mathrm{d}\gamma/\mathrm{d}\rho > 2/\rho^2$ for some ρ, $\bar{z}$ is no longer monotonic. In theory $\bar{z}$ can now take any shape. But we stick to the physical meaning of cost and present a few examples where $\gamma(\rho)$ increases as the connections ρ increases.

1. Linear cost function $\gamma \equiv k\rho$ for some constant proportionality constant k. In order for the assumption to hold, we need $k \ll \rho^{-1}$. This cost function adds a constant quantity k to $\partial \bar{z}/\partial \rho$. This results in a minimum in $\bar{z}(\rho, \gamma)$ with respect to ρ, at

$$\rho_c = \sqrt{\frac{2}{k}}. \tag{9.70}$$

When the connections $\rho < \rho_c$, the behaviour of the mean distance remains the same as before, i.e. decreasing connections ρ causes an increase in average distance. For connections larger than the critical value $\rho > \rho_c$ however, reducing connections to the hub shortens the distance, contrary to our intention, and hence accelerates the spread of harmful information. The minimum mean distance is

$$\bar{z}_c(\gamma) \approx 2\sqrt{2k}. \tag{9.71}$$

2. Quadratic cost function $\gamma(\rho) \equiv k\rho^2$. This could be a physically relevant cost function when the cost for using P depends on the number of connected *pairs* created, rather than the number of direct connections. The same approach as before using eqn (9.69) gives

$$\rho_c \approx k^{-1/3}, \tag{9.72}$$

corresponding to

$$\bar{z}_c \approx 3k^{1/3}. \tag{9.73}$$

3. We could furthermore consider a general polynomial $\gamma = k\rho^\alpha$. This gives for the optimal number of connections

$$\rho \approx \left(\frac{2}{\alpha k}\right)^{\frac{1}{1+\alpha}}. \tag{9.74}$$

The corresponding average distance is

$$\bar{z}_c \approx \left(\frac{2}{\alpha} + 2\right)\left(\frac{\alpha k}{2}\right)^{\frac{1}{1+\alpha}}. \tag{9.75}$$

These different cost functions can be explored based on the actual network structure observed empirically in the human–technology–AI system of interest. We leave this for future work.

Exercises

9.1 Review the definition of the heterogeneous rate parameter F, and calculate $F(\tau \to \infty)$ for the distributions given in eqns (9.3) and (9.2).

9.2 Consider a heterogeneous fusion system where we can control the heterogeneous profile according to eqn (9.4).

   (a) Find the defining equations of τ_c for both F_1 and F_2.
   (b) For switching rate $\gamma \gg 1$, plot $\Phi_{1,2}(\tau)$ and find the linear relationship between τ_c and median time τ_1.
   (c) Numerically solve your equations and plot the onset time as a function of τ_1. Check that you obtain something close to Fig. 9.2.

9.3 The control of a multi-species interactive system can also be considered using the phenomenological linear model we discussed in Section 4.6. Though it is not a first-principles theory as for our formal model, the linearity of that model provides us with a tractable means to look at the interactions among species, especially for higher dimensions where the full theory is not straightforward.

In this question we consider a 3-species system and name the species as R, B, G, respectively. According to eqn (4.183) we define the interactions among R, B, G as

$$\frac{dR}{dt} = a_R(R_0 - R) + r_B(B - R)$$
$$\frac{dB}{dt} = a_B(B_0 - B)$$
$$\frac{dG}{dt} = a_G(G_0 - G) + g_R(R - G)$$
$$\quad + g_B(B - G) \tag{9.76}$$

after a critical time $t > t_c$.

   (a) Write down the constant vector $\mathbf{A}$ and the interaction matrix $\mathbf{F}$ as in eqn (4.183). Are the onsets of all 3 species synchronous?
   (b) For this system of ODEs we can analyse its fixed points and see

how they behave in parameter space. The fixed points are values of (R, B, G) for which the system is stationary $(\dot{R}, \dot{G}, \dot{B} = 0)$. Find the fixed point of this system.

(c) Under what conditions is the fixed point stable? How should the stable fixed point be interpreted in the long time limit?

9.4 For the FSS of crowd–anticrowd decision making when $m = 2$, list all strategies. Then find one valid subgroup U and its negation $\bar{U}$. Show that they each contain 4 elements, that their union is an RSS, and that it partitions the FSS.

9.5 When constructing the RSS R for $m_0 + 1$ from the RSS R_0 for m_0 in eqn (9.9), we permuted R_0 into a sequence $\{a_i\}$ and appended a copy of one element in R_0 behind itself to create a new element for R. If we now instead permuted R_0 twice to obtain 2 different sequences $\{a_i\}$ and $\{b_i\}$, can we append b_i behind a_i to create a different subgroup $R' = \{a_i b_i\}$? If yes, prove that R' is indeed a subgroup of the FSS. Is there any condition for this to hold? If not, why?

9.6 For the ring-and-star model where the probability of a shortcut to hub P is p:

(a) Find the probability distribution for the number of shortcuts ν in the network.

(b) Show that $\mathcal{P}(l|m)$ is normalized without explicit calculations.

(c) Calculate explicitly and verify that $\mathcal{P}(l|m)$ is normalized.

9.7 Prove the expression eqn (9.31) of n_Q^Ψ for flat matrix Ψ. You do not need much algebra.

9.8 This exercise is dedicated to simple calculations of the undirected *ring-and-star* model. We assume that the ring is sufficiently large, with $n \gg 1$.

(a) Calculate $\mathcal{P}(1, m)$ for $m \geqslant 2$. How should we treat the 2 nodes both with distance m to our starting node A?

(b) Calculate $\mathcal{P}(2, m)$ for $m \geqslant 3$. Make sure to include all possible routes. Does the exact value of m matter?

(c) Calculate $\mathcal{P}(m, m)$, i.e. for $l = m$ using the same method for the directed model, and express it in terms of l for generality. (**Hint:** you need to consider $(2m + 3)$ nodes, i.e. 2 more nodes with distance $(m+1)$ to A.)

(d) Use $\mathcal{P}(l, l)$ values for both the directed and the undirected models to verify the scaling law at large n.

9.9 Sketch the plot of $\bar{z}(\rho)$, defined by eqn (9.57), for $\rho \geqslant 0$. Pay attention to the limiting behaviours at $\rho \to 0$ and $\rho \to \infty$. Verify that it does indeed decrease monotonically.

10
Final thoughts

Future systems of interacting heterogeneous humans, technology (machinery and software), and AI are an exciting prospect with applications ranging from space missions through to new medical procedures. This book has attempted to lay out the physics of these systems ahead of their arrival, and hence help equip next generation physicists and other scientists with an idea of what to expect, how such systems can be described quantitatively, and what tools could be used to design desired behaviours as well as control and mitigate undesired behaviours. This physics is also interesting in its own right, and opens up the possibility of a unified explanation of the 'bursts' that characterize so many real-world systems (Fig. 1.6, right panel)—including even Gamma Ray Bursts.

Though we see the use of interacting human–technology–AI systems for good as far out-weighing their use for bad, governments and policymakers as well as businesses need to assess the risk of things going wrong—and hence they need to get ahead of surprises and extreme events. This book can serve them in that goal. On that point of bad, nobody knows what and how future AI will be weaponized, or how a multi-dimensional online–offline war may unfold. But knowing the battlefield dynamics clarifies where and how that war will be fought. We refer to our group's laboratory for continual updates on this research: https://donlab.columbian.gwu.edu.

In Chap. 2, we presented patterns in empirical data collected so far from such systems, and hence what to expect in the future. In Chapts 3–6, we presented the new physics that describes this collective behaviour and the empirical patterns. It allows the interacting entities to have intrinsic heterogeneity, but does not consider explicitly any internal dynamics within each entity. In Chap. 7, we added some internal dynamics for each entity in a simple way, by allowing each entity to be in a (small) number of different states to model contagion. Chapter 8 went further by allowing each entity's internal state to exhibit adaptation and decision-making. Chapter 9 looked at how such systems can be controlled.

Based on recent developments in 2024 and 2025, we ourselves see three immediate opportunities for using this book's contents for new research, though readers will probably think of many more:

Opportunity 1: The challenge to understand what goes on 'under the hood' in AI systems (LLMs, etc.) is arguably the most important scientific question that Society now faces. If we can trust what an LLM says, and automatically recognize or avoid hallucinations, we might be able to use its future improved versions to address unsolved problems in medicine, e.g. to understand how cancer develops at the whole-body level, or to understand ageing, or dementia, or immune system diseases. And it may even tell us how to mitigate human conflicts online, offline, and everywhere in between. However, to truly trust a piece of AI, we need to understand how it works—and in particular how it manages to suddenly jump to a generalization (Fig. 2.1(c), grokking). And we need that understanding to be *at least* at the same level of detail that we understand how milk forms a curd (Fig. 1.4). But given the agreement demonstrated in Fig. 2.1(c) between even a simple two-species version of our theory and what is observed in AI grokking, we can now offer a first microscopic explanation for this jump-like emergence of output abilities in Machine-Learning, AI, and new Transformer architectures, i.e. the jumps come from the sudden formation of cohesive units built from several species of circuits within the machine–software system (Chap. 5 theory). Furthermore, a more detailed description of the sequence of cluster assembly—and hence the order of information in phrases for example—can be included using more complex temporal terms in F in Chap. 5. Hence the core theory seems to be ready.

Opportunity 2: Suppose some or many of the entities in Fig. 1.2 each has an internal state driven by a self-contained generative AI system, e.g. each entity could be a single LLM. There are a number of new papers out beginning to explore such systems of interacting agents where each agent is an LLM. It is also something that can be explored by any reader simply by asking ChatGPT to set up agents where each is an LLM, and then telling it what the scenario or 'game' is all about. This opportunity combines most of the main themes in this book, since the goal would be to explore when cohesion does (and does not) arise and also when an anticrowd (and hence potential conflict) forms among the LLM agents. This new field is termed 'Generative Agent Based Modelling'. To form a theory, the same simple idea can be applied from Chap. 8 and part of Chap. 9, where the agents (LLMs) have individual attributes drawn from across some strategy space, e.g. as a result of different training or neural network code. This means that crowd and anticrowd behaviour is expected among a population of interacting LLMs. How that AI subpopulation then interacts with humans, machinery, and other software, is a fascinating question to explore.

Opportunity 3: The emergence of a *quantum* cohesive unit—specifically, quantum entanglement—is the core ingredient for future Quantum Information Processing including Quantum Computing, etc. And there is no reason in principle why such Quantum Information Processing cannot be integrated with the information processing in a human–technology–AI system. Such a quantum cohesive unit (i.e. entanglement) can be generated within a subset of interacting quantum entities such as electrons, atoms, nanostructures, and photons. An example is shown in Fig. 10.1 in which the formation of a 'shock' of entanglement (cohesive unit) emerges in time in quantum systems driven by an external pulsed perturbation (e.g. pulse of light). Moreover, one of our collaborations showed in prior published work (see (Gomez-Ruiz, Rodriguez, Quiroga, and Johnson, 2024) and references therein) that the corresponding equation is somewhat similar to the Burgers'-like equations which play a core role in the theory of Chaps 3–5. But the precise relationship still needs to be established. When this is done, next generation readers and researchers will be in an excellent position to explore this 'future future' world of interacting human–technology–AI–*quantum*.

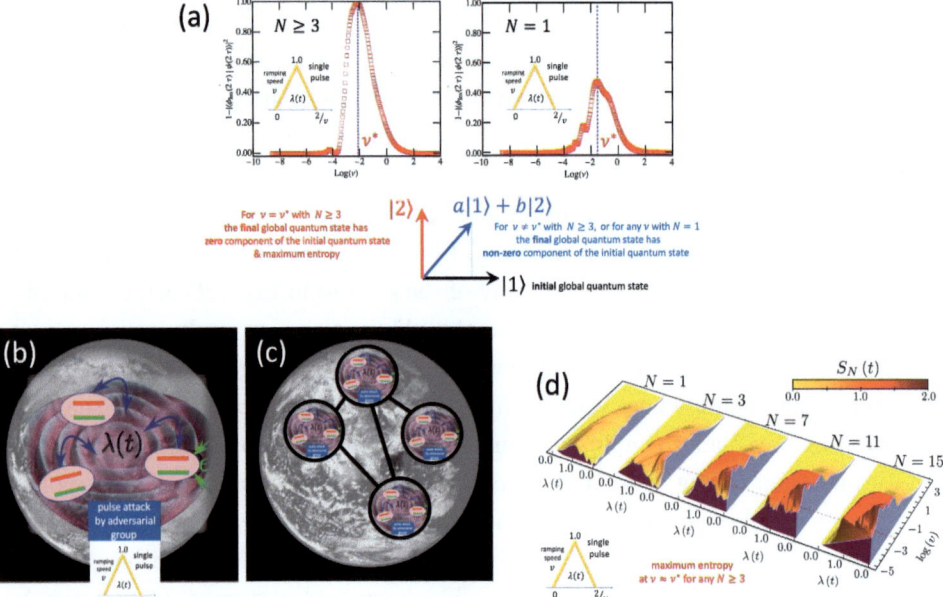

Fig. 10.1: Emergence of a quantum cohesive unit in a system of qubits coupled to a cavity and driven by an external pulse. Such matter–light systems will play a core role in the future quantum technology domain. Source: data from Gomez-Ruiz *et al.* (2024). We are grateful to Prof. Luis Quiroga, Prof. Ferney Rodriguez and Dr Fernando Gomez-Ruiz for their collaboration on this.

Appendix A
Generative equations and their cluster dynamics

In this appendix we list a collection of common ODEs and PDEs and their transformations into mesoscopic interaction ODEs, which are the equations for the clustering dynamics as outlined in Chaps 3 and 4. Depending on the particular human–technology–AI system of interest, one of these clustering dynamical equations could prove useful. The generating functions we are using are

$$\mathcal{D}(x,t) = \sum_k n_k(t) e^{-kx} \tag{A.1}$$

$$\mathcal{E}(x,t) = -\frac{\partial \mathcal{D}}{\partial x} = \sum_k k n_k(t) e^{-kx}. \tag{A.2}$$

Generating functions should include all information of the system (like partition functions in statistical physics). The reason why we list out equations for both $\mathcal{E}$ and $\mathcal{D}$ is that they can give different aggregational behaviours: $\mathcal{E}\mathcal{E}'$ terms correspond to product kernels whereas $\mathcal{D}\mathcal{D}'$ terms correspond to sum kernels.

A.1 One-dimensional

Denote $\mathcal{D}' = \partial \mathcal{D}/\partial x$, $\dot{\mathcal{D}} = \partial \mathcal{D}/\partial t$.

1. Benjamin–Bona–Mahony
   (a) Equation for $\mathcal{D}$:

   $$\dot{\mathcal{D}} + \mathcal{D}\mathcal{D}' - \mathcal{D}''' = 0 \Rightarrow \dot{n}_k - \frac{1}{2} k \sum_{i+j=k} n_i n_j + k^3 n_k = 0$$

   (b) Equation for $\mathcal{E}$:

   $$\dot{\mathcal{E}} + \mathcal{E}\mathcal{E}' - \mathcal{E}''' = 0 \Rightarrow \dot{n}_k - \frac{1}{2} \sum_{i+j=k} ij n_i n_j + k^3 n_k = 0$$

2. Boussinesq
    (a) Equation for $\mathcal{D}$:
    $$\ddot{\mathcal{D}} - \mathcal{D}'' - \mathcal{D}'''' + 3\left(\mathcal{D}^2\right)'' = 0 \Rightarrow \ddot{n}_k + 3k^2 \sum_{i+j=k} n_i n_j - k^2 n_k - k^4 n_k = 0$$

    (b) Equation for $\mathcal{E}$:
    $$\ddot{\mathcal{E}} - \mathcal{E}'' - \mathcal{E}'''' + 3\left(\mathcal{E}^2\right)'' = 0 \Rightarrow \ddot{n}_k + 3k \sum_{i+j=k} ij n_i n_j - k^2 n_k - k^4 n_k = 0$$

3. Burgers
    (a) Equation for $\mathcal{D}$:
    $$\dot{\mathcal{D}} + \mathcal{D}\mathcal{D}' = \nu \mathcal{D}'' \Rightarrow \dot{n}_k - \frac{1}{2}k \sum_{i+j=k} n_i n_j = \nu k^2 n_k$$

    (b) Equation for $\mathcal{E}$:
    $$\dot{\mathcal{E}} + \mathcal{E}\mathcal{E}' = \nu \mathcal{E}'' \Rightarrow \dot{n}_k - \frac{1}{2} \sum_{i+j=k} ij n_i n_j = \nu k^2 n_k$$

4. Generalized Burgers
    (a) Equation for $\mathcal{D}$:
    $$\dot{\mathcal{D}} + \mathcal{D}\mathcal{D}' - \nu \mathcal{D}'' + f(t)\mathcal{D} = 0 \Rightarrow \dot{n}_k - \frac{1}{2}k \sum_{i+j=k} n_i n_j - \nu k^2 n_k + f(t) n_k = 0$$

    (b) Equation for $\mathcal{E}$:
    $$\dot{\mathcal{E}} + \mathcal{E}\mathcal{E}' - \nu \mathcal{E}'' + f(t)\mathcal{E} = 0 \Rightarrow \dot{n}_k - \frac{1}{2} \sum_{i+j=k} ij n_i n_j - \nu k^2 n_k + f(t) n_k = 0$$

5. Fisher's (D is a parameter while $\mathcal{D}$ is the variable)
    (a) Equation for $\mathcal{D}$:
    $$\dot{\mathcal{D}} = D\mathcal{D}'' + \mathcal{D} - \mathcal{D}^2 \Rightarrow \dot{n}_k = (Dk^2 + 1)n_k - \sum_{i+j=k} n_i n_j$$

    (b) Equation for $\mathcal{E}$:
    $$\dot{\mathcal{E}} = D\mathcal{E}'' + \mathcal{E} - \mathcal{E}^2 \Rightarrow \dot{n}_k = (Dk^2 + 1)n_k - \frac{1}{k} \sum_{i+j=k} ij n_i n_j$$

6. Convective Fisher's

(a) Equation for $\mathcal{D}$:

$$\dot{\mathcal{D}} = \frac{1}{2}\mathcal{D}'' + \mathcal{D}(1-\mathcal{D}) - \mu\mathcal{D}\mathcal{D}' \Rightarrow \dot{n}_k = \left(\frac{1}{2}k^2 + 1\right)n_k + \left(\frac{1}{2}\mu k - 1\right)\sum_{i+j=k} n_i n_j$$

(b) Equation for $\mathcal{E}$:

$$\dot{\mathcal{E}} = \frac{1}{2}\mathcal{E}'' + \mathcal{E}(1-\mathcal{E}) - \mu\mathcal{E}\mathcal{E}' \Rightarrow \dot{n}_k = \left(\frac{1}{2}k^2 + 1\right)n_k + \left(\frac{\mu}{2} - \frac{1}{k}\right)\sum_{i+j=k} ijn_i n_j$$

7. Korteweg–de Vries (KdV)
   (a) Equation for $\mathcal{D}$:

$$\dot{\mathcal{D}} + \mathcal{D}''' - 6\mathcal{D}\mathcal{D}' = 0 \Rightarrow \dot{n}_k - k^3 n_k + 3k \sum_{i+j=k} n_i n_j = 0$$

   (b) Equation for $\mathcal{E}$:

$$\dot{\mathcal{E}} + \mathcal{E}''' - 6\mathcal{E}\mathcal{E}' = 0 \Rightarrow \dot{n}_k - k^3 n_k + 3 \sum_{i+j=k} ijn_i n_j = 0$$

8. Cylindrical KdV
   (a) Equation for $\mathcal{D}$:

$$\dot{\mathcal{D}} + \mathcal{D}''' - 6\mathcal{D}\mathcal{D}' + \frac{\mathcal{D}}{2t} = 0 \Rightarrow \dot{n}_k + \left(\frac{1}{2t} - k^3\right)n_k + 3k \sum_{i+j=k} n_i n_j = 0$$

   (b) Equation for $\mathcal{E}$:

$$\dot{\mathcal{E}} + \mathcal{E}''' - 6\mathcal{E}\mathcal{E}' + \frac{\mathcal{E}}{2t} = 0 \Rightarrow \dot{n}_k + \left(\frac{1}{2t} - k^3\right)n_k + 3 \sum_{i+j=k} ijn_i n_j = 0$$

9. Spherical KdV
   (a) Equation for $\mathcal{D}$:

$$\dot{\mathcal{D}} + \mathcal{D}''' - 6\mathcal{D}\mathcal{D}' + \frac{\mathcal{D}}{t} = 0 \Rightarrow \dot{n}_k + \left(\frac{1}{t} - k^3\right)n_k + 3k \sum_{i+j=k} n_i n_j = 0$$

   (b) Equation for $\mathcal{E}$:

$$\dot{\mathcal{E}} + \mathcal{E}''' - 6\mathcal{E}\mathcal{E}' + \frac{\mathcal{E}}{t} = 0 \Rightarrow \dot{n}_k + \left(\frac{1}{t} - k^3\right)n_k + 3 \sum_{i+j=k} ijn_i n_j = 0$$

10. Transitional KdV

(a) Equation for $\mathcal{D}$:
$$\dot{\mathcal{D}} + \mathcal{D}''' - 6f(t)\mathcal{D}\mathcal{D}' = 0 \Rightarrow \dot{n}_k - k^3 n_k + 3f(t)k \sum_{i+j=k} n_i n_j = 0$$

(b) Equation for $\mathcal{E}$:
$$\dot{\mathcal{E}} + \mathcal{E}''' - 6f(t)\mathcal{E}\mathcal{E}' = 0 \Rightarrow \dot{n}_k - k^3 n_k + 3f(t) \sum_{i+j=k} ij n_i n_j = 0$$

11. KdV with variable coefficients

(a) Equation for $\mathcal{D}$:
$$\dot{\mathcal{D}} + bt^m \mathcal{D}''' + at^n \mathcal{D}\mathcal{D}' = 0 \Rightarrow \dot{n}_k - bt^m k^3 n_k - \frac{a}{2} t^n k \sum_{i+j=k} n_i n_j = 0$$

(b) Equation for $\mathcal{E}$:
$$\dot{\mathcal{E}} + bt^m \mathcal{E}''' + at^n \mathcal{E}\mathcal{E}' = 0 \Rightarrow \dot{n}_k - bt^m k^3 n_k - \frac{a}{2} t^n \sum_{i+j=k} ij n_i n_j = 0$$

12. KdV–Burgers

(a) Equation for $\mathcal{D}$:
$$\dot{\mathcal{D}} + 2\mathcal{D}\mathcal{D}' - \nu \mathcal{D}'' + \mu \mathcal{D}''' = 0 \Rightarrow \dot{n}_k - k \sum_{i+j=k} n_i n_j - \nu k^2 n_k - \mu k^3 n_k = 0$$

(b) Equation for $\mathcal{E}$:
$$\dot{\mathcal{E}} + 2\mathcal{E}\mathcal{E}' - \nu \mathcal{E}'' + \mu \mathcal{E}''' = 0 \Rightarrow \dot{n}_k - \sum_{i+j=k} ij n_i n_j - \nu k^2 n_k - \mu k^3 n_k = 0$$

13. Regularized long wave

(a) Equation for $\mathcal{D}$:
$$\dot{\mathcal{D}} + \mathcal{D}' - 6\mathcal{D}\mathcal{D}' - \dot{\mathcal{D}}'' = 0 \Rightarrow (1-k^2)\dot{n}_k - k n_k + 3k \sum_{i+j=k} n_i n_j = 0$$

(b) Equation for $\mathcal{E}$:
$$\dot{\mathcal{E}} + \mathcal{E}' - 6\mathcal{E}\mathcal{E}' - \dot{\mathcal{E}}'' = 0 \Rightarrow (1-k^2)\dot{n}_k - k n_k + 3 \sum_{i+j=k} ij n_i n_j = 0$$

14. Generalized shallow water wave

(a) Equation for $\mathcal{D}$:
$$\mathcal{D}''' + a\mathcal{D}'\dot{\mathcal{D}} + b\dot{\mathcal{D}}\mathcal{D}'' - \dot{\mathcal{D}}' - \mathcal{D}'' = 0$$
$$\Rightarrow \left(1-k^2\right) k \dot{n}_k - k^2 n_k + \frac{1}{2} \sum_{i+j=k} [j(ai+bj)\dot{n}_i n_j + i(aj+bi) n_i \dot{n}_j] = 0$$

When $a = b$,
$$\left(1-k^2\right) \dot{n}_k - k n_k + \frac{a}{2} \sum_{i+j=k} (j\dot{n}_i n_j + i n_i \dot{n}_j) = 0$$

(b) Equation for $\mathcal{E}$:
$$\mathcal{E}''' + a\mathcal{E}'\dot{\mathcal{E}}' + b\dot{\mathcal{E}}\mathcal{E}'' - \dot{\mathcal{E}}' - \mathcal{E}'' = 0$$
$$\Rightarrow (1-k^2)k^2\dot{n}_k - k^3 n_k + \frac{1}{2}\sum_{i+j=k} ij\left[j(ai+bj)\dot{n}_i n_j + i(aj+bi)n_i \dot{n}_j\right] = 0$$

When $a = b$,
$$(1-k^2)k\dot{n}_k - k^2 n_k + \frac{a}{2}\sum_{i+j=k} ij\left(j\dot{n}_i n_j + i n_i \dot{n}_j\right) = 0$$

15. Unnamed example
    (a) Equation for $\mathcal{D}$:
    $$\dddot{\mathcal{D}} + 2\mathcal{D}\dot{\mathcal{D}} - \mathcal{D}'' = 0 \Rightarrow \dddot{n}_k - k^2 n_k + \sum_{i+j=k}(\dot{n}_i n_j + n_i \dot{n}_j) = 0$$

    (b) Equation for $\mathcal{E}$:
    $$\dot{\mathcal{E}} + 2\mathcal{E}\dot{\mathcal{E}} - \mathcal{E}'' = 0 \Rightarrow \dddot{n}_k - k^2 n_k + \frac{1}{k}\sum_{i+j=k} ij\left(\dot{n}_i n_j + n_i \dot{n}_j\right) = 0$$

A.2 Multi-dimensional

Define $\mathcal{D}(r) = \sum_k n_k e^{-k \cdot r}$, a natural extension of the 1D definition. Denote $\mathcal{D}_x = \partial \mathcal{D}/\partial x$, $\mathcal{D}_y = \partial \mathcal{D}/\partial y$, $\dot{\mathcal{D}} = \partial \mathcal{D}/\partial t$, $\mathbf{k} = (k^x, k^y)$ and $k = |\mathbf{k}|$.

1. Kadomtsev–Petviashvili (generalization of KdV). Equation for $\mathcal{D}$ only, since only it is scalar.
$$\frac{\partial}{\partial x}\left(\dot{\mathcal{D}} + \mathcal{D}_{xxx} - 6\mathcal{D}\mathcal{D}_x\right) \pm \mathcal{D}_{yy} = 0$$
$$\Rightarrow -k^x \dot{n}_k + \left[(k^x)^4 \pm (k^y)^2\right] n_k - 3(k^x)^2 \sum_{i+j=k} n_i n_j = 0$$

2. Generalized Kadomtsev–Petviashvili–Burgers
$$\frac{\partial}{\partial x}\left(\dot{\mathcal{D}} + \frac{J}{2t}\mathcal{D} + J_1 \mathcal{D}\mathcal{D}_x + J_2 \mathcal{D}_{xx} + J_3 \mathcal{D}_{xxx}\right) + J_4(t)\mathcal{D}_{yy} = 0$$
$$\Rightarrow -k^x \dot{n}_k - \left(\frac{J}{2t}k^x + J_2(k^x)^3 - J_3(k^x)^4 - J_4(t)(k^y)^2\right)n_k + \frac{J_1}{2}(k^x)^2 \sum_{i+j=k} n_i n_j = 0$$

3. Kuramoto–Sivashinsky
$$\dot{\mathcal{D}} + \nabla^4 \mathcal{D} + \nabla^2 \mathcal{D} + \frac{1}{2}\left|\nabla^2 \mathcal{D}\right|^2 = 0$$
$$\Rightarrow \dot{n}_k + k^4 n_k + k^2 n_k + \frac{1}{2}\sum_{i+j=k} i^2 j^2 n_i n_j = 0$$

4. Lin–Tsien

$$2\dot{\mathcal{D}}_x + \mathcal{D}_x\mathcal{D}_{xx} - \mathcal{D}_{yy} = 0$$

$$\Rightarrow 2k^x \dot{n}_k + \frac{1}{2}k^x \sum_{i+j=k} i^x j^x n_i n_j + (k^y)^2 n_k = 0$$

5. Thomas

$$\mathcal{D}_{xy} + a\mathcal{D}_x + b\mathcal{D}_y + c\mathcal{D}_x\mathcal{D}_y = 0$$

$$\Rightarrow k^x k^y \dot{n}_k - (ak^x + bk^y) n_k + c \sum_{i+j=k} i^x j^y n_i n_j = 0$$

Bibliography

One-Third of U.S. Adults Say Fear of Mass Shootings Prevents Them From Going to Certain Places or Events. https://www.socialworktoday.com/news/dn_081519.shtml.

(2019). Fringe Social Media: Are you digging deep enough? | SMI Aware. https://web.archive.org/web/20201001222412/https://smiaware.com/blog/fringe-social-media-are-you-digging-deep-enough/.

(2021). The Artificial Intelligence Act. https://web.archive.org/web/20230811085634/https://artificialintelligenceact.eu/.

(2022a). AI Experts Predict By 2026, 90% Of Online Content Will Be Generated By Artificial Intelligence. https://idc-a.org/news/industry/AI-Experts-Predict-By-2026-90-Of-Online-Content/127ab0c0-34ba-4c03-8bad-1e4f21923f31.

(2022b). Fringe social media networks sidestep online content rules. https://www.politico.eu/article/fringe-social-media-telegram-extremism-far-right/.

(2023a). Coping with the fear of mass shootings. https://www.siouxlandproud.com/news/local-news/coping-with-the-fear-of-mass-shootings/.

(2023b). Digital Services Act: Commission designates first set of Very Large Online Platforms and Search Engines | Shaping Europe's digital future. https://digital-strategy.ec.europa.eu/en/news/digital-services-act-commission-designates-first-set-very-large-online-platforms-and-search-engines.

(2023c). The haters and conspiracy theorists back on Twitter. *BBC News*. https://www.bbc.com/news/technology-64554381.

(2023d). Strengthened Code of Practice on Disinformation: Signatories to identify ways to step up work one year after launch | Shaping Europe's digital future. https://digital-strategy.ec.europa.eu/en/news/strengthened-code-practice-disinformation-signatories-identify-ways-step-work-one-year-after-launch.

Aldreabi, E., Lee, J. M., and Blackburn, J. (2023). Using Deep Learning to Detect Islamophobia on Reddit. *The International FLAIRS Conference Proceedings*, **36**. https://journals.flvc.org/FLAIRS/article/view/133324.

Ammari, T. and Schoenebeck, S. (2016). "Thanks for your interest in our Facebook group, but it's only for dads": Social Roles of Stay-at-Home Dads. In *Proceedings of the 19th ACM Conference on Computer-Supported Cooperative Work & Social Computing*, CSCW '16, New York, NY, USA, pp. 1363–1375. Association for Computing

Machinery.

Anti-Defamation League (2023). Online hate and harassment in the american experience, 2023. Report, Anti-Defamation League. Accessed June 2025.

Artime, O., Grassia, M., De Domenico, M. et al. (2024). Robustness and resilience of complex networks. *Nat. Rev. Phys.*, **6**, 114–131. https://doi.org/10.1038/s42254-023-00676-y.

Aut, N., Ranaware, S., Ghadge, S., Jadhav, R., and Jagtap, P. (2023). Social Media based Hate Speech Detection using Machine Learning. *International Journal for Research in Applied Science & Engineering Technology*, **11**(V). https://www.ijraset.com/best-journal/social-media-based-hate-speech-detection-using-machine-learning.

Bakshy, E., Messing, S., and Adamic, L. A. (2015). Exposure to ideologically diverse news and opinion on facebook. *Science*, **348**(6239), 1130–1132. Publisher: American Association for the Advancement of Science.

Barabási, A-L. (2016). *Network Science*. Cambridge University Press.

Battiston, F., Amico, E., Barrat, A., Bianconi, G., Ferraz de Arruda, G., Franceschiello, B., Iacopini, I., Kéfi, S., Latora, V., Moreno, Y., Murray, M. M., Peixoto, T. P., Vaccarino, F., and Petri, G. (2021). The physics of higher-order interactions in complex systems. *Nat. Phys.*, **17**, 1093–1098.

Beacken, G., Trauthig, I., and Samuel Woolley, S. (2022). Platforms' Efforts to Block Antisemitic Content Are Falling Short. https://www.cigionline.org/articles/platforms-efforts-to-block-anti-semitic-content-are-falling-short/.

Benninger, M. (2022). 'Fringe' websites radicalized Buffalo shooter, report concludes. https://www.wbng.com/2022/10/18/fringe-websites-radicalized-buffalo-shooter-report-concludes/.

Biever, C. (2023). ChatGPT broke the Turing test — the race is on for new ways to assess AI. *Nature*, **619**(7971), 686–689. Publisher: Nature Publishing Group.

Brown, R. and Livingston, L. (2018). A New Approach to Assessing the Role of Technology in Spurring and Mitigating Conflict: Evidence From Research and Practice. https://jia.sipa.columbia.edu/news/new-approach-assessing-role-technology-spurring-and-mitigating-conflict-evidence-research-and.

Centola, D., Gonzalez-Avella, J., Eguiluz, V., and San Miguel, M. (2007). Homophily, cultural drift, and the co-evolution of cultural groups. *J. Confl. Resol.*, **51**, 905–929.

Chen, E., Lerman, K., and Ferrara, E. (2020). Tracking Social Media Discourse About the COVID-19 Pandemic: Development of a Public Coronavirus Twitter Data Set. *JMIR Public Health and Surveillance*, **6**(2), e19273. Publisher: JMIR Publications Inc., Toronto, Canada.

Cinelli, M., Pelicon, A., Mozetič, I., Quattrociocchi, W., Novak, P. K., and Zollo,

F. (2021). Dynamics of online hate and misinformation. *Scientific Reports*, **11**(1), 22083. Number: 1 Publisher: Nature Publishing Group.

Cosoleto, T. (2023). Surge in young children being targeted by cyber bullies. `https://thewest.com.au/news/social/surge-in-young-children-being-targeted-by-cyber-bullies-c-11223220`.

Crawford, A. and Smith, T. (2023). Illegal trade in AI child sex abuse images exposed. *BBC News*. `https://www.bbc.com/news/uk-65932372`.

Dewey, C. (2022). On fringe social media sites, Buffalo mass shooting becomes rallying call for white supremacists. *Buffalo News*. `https://buffalonews.com/news/local/on-fringe-social-media-sites-buffalo-mass-shooting-becomes-rallying-call-for-white-supremacists/article_74a55388-f61b-11ec-812a-97d8f2646d45.html`.

D'hulst, R and Rodgers, G. J. (1999). The hamming distance in the minority game. *Physica A: Statistical Mechanics and its Applications*, **270**(3), 514–525.

DiResta, R. (2018). The Digital Maginot Line. `https://www.ribbonfarm.com/2018/11/28/the-digital-maginot-line/`.

DisinfoDocket. DisinfoDocket. `https://www.disinfodocket.com/`.

Dodds, P. S., Harris, K. D., Kloumann, I. M., Bliss, C. A., and Danforth, C. M. (2011). Temporal patterns of happiness and information in a global social network: Hedonometrics and Twitter. *PLOS ONE*, **6**(12), 1–1.

Dorogovtsev, S. N. and Mendes, J. F. F. (2000). Exactly solvable small-world network. *Europhysics Letters*, **50**(1), 1.

Douek, E. (2022). Content Moderation as Systems Thinking. *Harvard Law Review*, **136**(2). `https://harvardlawreview.org/print/vol-136/content-moderation-as-systems-thinking/`.

Drake, R. L. (1972). *A General Mathematics Survey of the Coagulation Equation*. Pergamon Press.

Dynamic Online Networks Laboratory. Literature review. Technical report. `https://bpb-us-e1.wpmucdn.com/blogs.gwu.edu/dist/5/3446/files/2022/10/lit_review.pdf`.

Eguíluz, V. M. and Zimmermann, M. G. (2000). Transmission of information and herd behavior: An application to financial markets. *Phys. Rev. Lett.*, **85**, 5659–5662.

Eisenstat, Y. (2023). Hate is surging online — and social media companies are in denial. Congress can help protect users. `https://thehill.com/opinion/congress-blog/4085909-hate-is-surging-online-and-social-media-companies-are-in-denial-congress-can-help-protect-users/`.

Fagan, B. T., MacKay, N. J., and Wood, A. J. (2024). Robustness of steady state and stochastic cyclicity in generalized coalescence-fragmentation models. *The European*

Physical Journal B, **97**, 21.

Flory, P. J. (1953). *Principles of polymer chemistry*. Cornell University Press.

Gelfand, M. J., Harrington, J. R., and Jackson, J. C. (2017). The Strength of Social Norms Across Human Groups. *Perspectives on Psychological Science: A Journal of the Association for Psychological Science*, **12**(5), 800–809.

Gill, P. and Corner, E. (2015). Lone-Actor Terrorist Use of the Internet & Behavioural Correlates. In *Terrorism Online: Politics, Law, Technology and Unconventional Violence* (ed. L. Jarvis, S. Macdonald,, and T.Chen). Routledge.

Gomez-Ruiz, F. J., Rodriguez, F. J., Quiroga, L., and Johnson, N. F. (2024). *Vulnerability of Quantum Information Systems to Collective Manipulation*, Chapter 5. IntechOpen, Rijeka.

González-Bailón, S. et al. (2023). Asymmetric ideological segregation in exposure to political news on Facebook. *Science*, **381**(6656), 392–398. Publisher: American Association for the Advancement of Science.

Gorman, J. C., Demir, M., Cooke, N. J., and Grimm, D. A. (2019). Evaluating sociotechnical dynamics in a simulated remotely-piloted aircraft system: a layered dynamics approach. *Ergonomics*, **62**(5), 629–643.

Guess, A. M. et al. (2023*a*). How do social media feed algorithms affect attitudes and behavior in an election campaign? *Science*, **381**(6656), 398–404. Publisher: American Association for the Advancement of Science.

Guess, A. M. et al. (2023*b*). Reshares on social media amplify political news but do not detectably affect beliefs or opinions. *Science*, **381**(6656), 404–408. Publisher: American Association for the Advancement of Science.

Hart, R. (2023). White Supremacist Propaganda Hit Record Levels In 2022, ADL Says. https://www.forbes.com/sites/roberthart/2023/03/09/white-supremacist-propaganda-hit-record-levels-in-2022-adl-says/.

Hendriks, E. M., Ernst, M. H., and Ziff, R. M. (1983). Coagulation equations with gelation. *J. Stat. Phys*, **31**(3), 519–563.

Hod, S. and Nakar, E. (2002). Self-segregation versus clustering in the evolutionary minority game. *Phys. Rev. Lett.*, **88**, 238702.

Hoffmann, M. and Frase, H. (2023). Adding Structure to AI Harm. https://cset.georgetown.edu/publication/adding-structure-to-ai-harm/.

Horgan, P. and Gill, P. et al. (2013). *From Bomb to Bomb-maker: A Social Network Analysis of the Socio-Psychological and Cultural Dynamics of the IED Process*. Final Report. Office of Naval Research Code 30.

House of Commons Home Affairs Committee (2017). 14th Report - Hate crime: abuse, hate and extremism online. Government Report HC 609, House of Commons, London. https://publications.parliament.uk/pa/cm201617/cmselect/cmha

ff/609/609.pdf.

Hsu, T. (2022). News on Fringe Social Sites Draws Limited but Loyal Fans, Report Finds. *The New York Times*. https://www.nytimes.com/2022/10/06/technology/parler-truth-social-telegram-pew.html.

Hsu, T. and Myers, S. Lee (2023). A.I.'s Use in Elections Sets Off a Scramble for Guardrails. *The New York Times*. https://www.nytimes.com/2023/06/25/technology/ai-elections-disinformation-guardrails.html.

Huang, L, Cooke, N J, Gutzwiller, R S, Berman, S, Chiou, E K, Demir, M, and Zhang, W (2021). Chapter 13 - distributed dynamic team trust in human, artificial intelligence, and robot teaming. In *Trust in Human-Robot Interaction* (ed. C. S. Nam and J. B. Lyons), pp. 301–319. Academic Press.

Huo, F. Y., Manrique, P. D., and Johnson, N. F. (2024). Multispecies Cohesion: Humans, Machinery, AI and Beyond. *Physical Review Letters*, **133**(24).

Jarrett, T. C., Ashton, D. J., Fricker, M., and Johnson, N. F. (2006). Interplay between function and structure in complex networks. *Phys. Rev. E*, **74**, 026116.

Ji, P., Ye, J., Mu, Y., Lin, W., Tian, Y., Hens, C., Perc, M., Tang, Y., Sun, J., and Kurths, J. (2023). Signal propagation in complex networks. *Physics Reports*, **1017**, 1–96. https://doi.org/10.1016/j.physrep.2023.03.005.

Johnson, N., Carran, S., Botner, J., Fontaine, K., Laxague, N., Nuetzel, P., Turnley, J., and Tivnan, B. (2011). Pattern in Escalations in Insurgent and Terrorist Activity. *Science*, **333**(6038), 81–84. Publisher: American Association for the Advancement of Science.

Johnson, N. F. (2017). To slow or not? challenges in subsecond networks. *Science*, **355**(6327), 801–802. Publisher: American Association for the Advancement of Science.

Johnson, N. F. and Hui, P. M. (2003). Crowd-anticrowd theory of collective dynamics in competitive, multi-agent populations and networks. https://arxiv.org/abs/cond-mat/0306516.

Johnson, N. F., Leahy, R., Restrepo, N. J., Velasquez, N., Zheng, M., Manrique, P., Devkota, P., and Wuchty, S. (2019a). Hidden resilience and adaptive dynamics of the global online hate ecology. *Nature*, **573**(7773), 261–265. Number: 7773 Publisher: Nature Publishing Group.

Johnson, N. F., Manrique, P., and Hui, P. M. (2013a). Modeling insurgent dynamics including heterogeneity. *J. Stat. Phys.*, **151**, 395–413.

Johnson, N. F., Manrique, P., Zheng, M., Cao, Z., Botero, J., Huang, S., Aden, N., Song, C., Leady, J., Velasquez, N., and Restrepo, E. M. (2019b). Emergent dynamics of extremes in a population driven by common information sources and new social media algorithms. *Scientific Reports*, **9**(1), 11895. Number: 1 Publisher: Nature

Publishing Group.

Johnson, N. F., Medina, P., Zhao, G., Messinger, D. S., Horgan, J., Gill, P., Bohorquez, J. C., Mattson, W., Gangi, D., Qi, H., Manrique, P., Velasquez, N., Morgenstern, A., Restrepo, E., Johnson, N., Spagat, M., and Zarama, R. (2013*b*). Simple mathematical law benchmarks human confrontations. *Scientific Reports*, **3**(1), 3463. Number: 1 Publisher: Nature Publishing Group.

Johnson, N. F., Sear, R., and Illari, L. (2024). Controlling bad-actor-artificial intelligence activity at scale across online battlefields. *PNAS Nexus*, **3**(1).

Johnson, N. F., Velásquez, N., Restrepo, N. J., Leahy, R., Gabriel, N., El Oud, S., Zheng, M., Manrique, P., Wuchty, S., and Lupu, Y. (2020). The online competition between pro- and anti-vaccination views. *Nature*, **582**(7811), 230–233. Number: 7811 Publisher: Nature Publishing Group.

Johnson, N. F., Zheng, M., Vorobyeva, Y., Gabriel, A., Qi, H., Velasquez, N., Manrique, P., Johnson, D., Restrepo, E., Song, C., and Wuchty, S. (2016). New online ecology of adversarial aggregates: ISIS and beyond. *Science*, **352**(6292), 1459–1463. Publisher: American Association for the Advancement of Science.

Jusup, M., Holme, P., Kanazawa, K., Takayasu, M., Romić, I., Wang, Z., Sunčana, G., Lipić, T., Podobnik, B., Wang, L., Luo, W., Klanjšček, T., Fan, J., Boccaletti, S., and Perc, M. (2022). Social physics. *Physics Reports*, **948**, 1–148.

Kane, E. A., Gershow, M., Afonso, B., Larderet, I., Klein, M., Carter, A. R., de Bivort, B. L., Sprecher, S. G., and Samuel, A. D. T. (2013). Sensorimotor structure of *drosophila* larva phototaxis. *Proceedings of the National Academy of Sciences*, **110**(40), E3868–E3877.

Kilcher, Y. (2023). GPT-4chan Model Card. `https://www.ykilcher.com/gpt-4chan-model-card`.

Krapivsky, P. L., Redner, S., and Ben-Naim, E. (2010). *A Kinetic View of Statistical Physics*. Cambridge University Press.

Kupferschmidt, K. (2023). Studies find little impact of social media on polarization. *Science*, **381**(6656), 367–368. `https://www.science.org/doi/10.1126/science.adj9569`.

Lamensch, M. (2022). To Eliminate Violence Against Women, We Must Take the Fight to Online Spaces. `https://www.cigionline.org/articles/to-eliminate-violence-against-women-we-must-take-the-fight-to-online-spaces/`.

Laws, R., Walsh, A. D., Hesketh, K. D., Downing, K. L., Kuswara, K., and Campbell, K. J. (2019). Differences Between Mothers and Fathers of Young Children in Their Use of the Internet to Support Healthy Family Lifestyle Behaviors: Cross-Sectional Study. *Journal of Medical Internet Research*, **21**(1), e11454. Publisher: JMIR Publications Inc., Toronto, Canada.

Lazer, D. M. J., Baum, M. A., Benkler, Y., Berinsky, A. J., Greenhill, K. M., Menczer, F., Metzger, M. J., Nyhan, B., Pennycook, G., Rothschild, D., Schudson, M., Sloman, S. A., Sunstein, C. R., Thorson, E. A., Watts, D. J., and Zittrain, J. L. (2018). The science of fake news. *Science*, **359**(6380), 1094–1096. Publisher: American Association for the Advancement of Science.

Leahy, R., Restrepo, N. J., Sear, R., and Johnson, N. F. (2022). Connectivity Between Russian Information Sources and Extremist Communities Across Social Media Platforms. *Frontiers in Political Science*, 4.

Lewandowsky, S., Cook, J., Ecker, U., Albarracín, D., Amazeen, M. A., Kendeou, P., Lombardi, D., Newman, E. J., Pennycook, G., Porter, E., Rand, D. G., Rapp, D. N., Reifler, J., Roozenbeek, J., Schmid, P., Seifert, C. M., Sinatra, G. M., Van Der Linden, S., Vraga, E. K., Wood, T. J., Zaragoza, M. S., Fazio, L., Kozyreva, A., and Cook, W. (2020). Debunking Handbook 2020. Technical report, George Mason University. https://www.climatechangecommunication.org/wp-content/uploads/2020/10/DebunkingHandbook2020.pdf.

Liu, X., Li, J.I.A., Watanabe, K., Taniguchi, T., Hone, J., Halperin, B. I., Kim, P., and Dean, C. R. (2022). Crossover between strongly coupled and weakly coupled exciton superfluids. *Science*, **375**(6577), 205–209. https://www.science.org/doi/abs/10.1126/science.abg1110.

Lupu, Y., Sear, R., Velásquez, N., Leahy, R., Restrepo, N. J., Goldberg, B., and Johnson, N. F. (2023). Offline events and online hate. *PLOS ONE*, **18**(1), e0278511. Publisher: Public Library of Science.

Lushnikov, A. A. (2006). Gelation in coagulating systems. *Physica D: Nonlin. Phen.*, **222**(1), 37–53.

Madhusoodanan, J. (2022). Safe space: online groups lift up women in tech. *Nature*, **611**(7937), 839–841. Number: 7937 Publisher: Nature Publishing Group.

Manrique, P. D., Hui, P. M., and Johnson, N. F. (2015). Internal character dictates transition dynamics between isolation and cohesive grouping. *Phys. Rev. E*, **92**, 062803.

Manrique, Pedro D., Huo, Frank. Y., El Oud, Sara, and Johnson, Neil F. (2024). Non-equilibrium physics of multi-species assembly applied to fibrils inhibition in biomolecular condensates and growth of online distrust. *Sci. Rep.*, **14**.

Manrique, P. D., Huo, F. Y., El Oud, S., Zheng, M., Illari, L., and Johnson, N. F. (2023). Shockwavelike Behavior across Social Media. *Physical Review Letters*, **130**(23), 237401. Publisher: American Physical Society.

Manrique, P. D. and Johnson, N. F. (2018). Individual heterogeneity generating explosive system network dynamics. *Phys. Rev. E*, **97**, 032311.

Manrique, P. D., Klein, M., Li, Y. S., Xu, C., Hui, P. M., and Johnson, N. F. (2019).

Getting closer to the goal by being less capable. *Science Advances*, **5**(2), eaau5902.

Manrique, Pedro D., Xu, Chen, Hui, Pak Ming, and Johnson, Neil F. (2016, Aug). Atypical viral dynamics from transport through popular places. *Phys. Rev. E*, **94**, 022304.

Martinez, D. (2014). Klinotaxis as a basic form of navigation. *Frontiers in Behavioral Neuroscience*, **8**.

Miller-Idriss, C. (2020). *Hate in the Homeland: The New Global Far Right*.

Milmo, D. and Hern, A. (2023). Elections in UK and US at risk from AI-driven disinformation, say experts. *The Guardian*. https://www.theguardian.com/technology/2023/may/20/elections-in-uk-and-us-at-risk-from-ai-driven-disinformation-say-experts.

Moon, R. Y., Mathews, A., Oden, R., and Carlin, R. (2019). Mothers' Perceptions of the Internet and Social Media as Sources of Parenting and Health Information: Qualitative Study. *Journal of Medical Internet Research*, **21**(7), e14289. Publisher: JMIR Publications Inc., Toronto, Canada.

Morgan, M. and Kulkarni, A. (2023). Platform-agnostic Model to Detect Sinophobia on Social Media. In *Proceedings of the 2023 ACM Southeast Conference*, ACMSE 2023, New York, NY, USA, pp. 149–153. Association for Computing Machinery. https://dl.acm.org/doi/10.1145/3564746.3587024.

Nanda, N. (2024). Paper replication walkthrough: Reverse-engineering modular addition. https://www.neelnanda.io/mechanistic-interpretability/modular-addition-walkthrough. Accessed: 2024-05-7.

Nanda, N., Chan, L., Lieberum, T., Smith, J., and Steinhardt, J. (2023). Progress measures for grokking via mechanistic interpretability. *International Conference on Learning Representations 2023*. https://arxiv.org/pdf/2301.05217.

Nanda, N. and Lieberum, T. (2024). A mechanistic interpretability analysis of grokking. https://www.alignmentforum.org/posts/N6WM6hs7RQMKDhYjB/a-mechanistic-interpretability-analysis-of-grokking. Accessed: 2024-05-07.

Nelson, A. C., Keener, J. P., and Fogelson, A. L. (2020). Kinetic model of two-monomer polymerization. *Phys. Rev. E*, **101**, 022501.

Nelson, J. (2023). UN Warns of AI-Generated Deepfakes Fueling Hate and Misinformation Online. https://decrypt.co/144281/un-united-nations-ai-deepfakes-hate-misinformation.

Newman, M. (2005). Power laws, Pareto distributions and Zipf's law. *Contemporary Physics*, **46**(5), 323–351. Publisher: Taylor & Francis https://doi.org/10.1080/00107510500052444.

Newman, M. (2018, 03). *Networks*. Oxford University Press.

Niu, X., Doyle, C., Korniss, G., and Szymanski, B. K. (2017). The impact of variable commitment in the naming game on consensus formation. *Sci Rep*, **7**, 41750.

Nyhan, B. et al. (2023). Like-minded sources on Facebook are prevalent but not polarizing. *Nature*, **620**(7972), 137–144. Number: 7972 Publisher: Nature Publishing Group.

Ollagnier, A., Cabrio, E., and Villata, S. (2023). Harnessing Bullying Traces to Enhance Bullying Participant Role Identification in Multi-Party Chats. *The International FLAIRS Conference Proceedings*, **36**. https://journals.flvc.org/FLAIRS/article/view/133191.

Palla, G., Barabási, A-L., and Vicsek, T (2007). Quantifying social group evolution. *Nature*, **446**(7136), 664–667.

Rao, A., Morstatter, F., and Lerman, K. (2022). Partisan asymmetries in exposure to misinformation. *Scientific Reports*, **12**(1), 15671. Number: 1 Publisher: Nature Publishing Group.

Rodrigo, C. M. and Klar, R. (2021). Fringe social networks boosted after mob attack. https://thehill.com/policy/technology/533919-fringe-social-networks-boosted-after-mob-attack/.

Roozenbeek, J., van der Linden, S., Goldberg, B., Rathje, S., and Lewandowsky, S. (2022). Psychological inoculation improves resilience against misinformation on social media. *Science Advances*, **8**(34), eabo6254. Publisher: American Association for the Advancement of Science.

Schweitzer, F. and Andres, G. (2022). Social nucleation: Group formation as a phase transition. *Phys. Rev. E*, **105**, 044301.

Science (2023, July). Volume 381, issue 6656. Published 28 July 2023.

Sear, R., Leahy, R., Restrepo, N. J., and Johnson, N. (2022a). Machine learning reveals adaptive Covid-19 narratives in online anti-vaccination network. In *Proc. of 2021 Conf. of Comp. Social Science Soc. of Americas* (ed. Z. Yang and E. von Briesen), Cham, pp. 164–175. Springer International Publishing.

Sear, R., Restrepo, N. J., and Johnson, N. F. (2022b). Dynamic topic modeling reveals variations in online hate narratives. In *Intell. Comp.* (ed. K. Arai), Cham, pp. 564–578. Springer International Publishing.

Semenov, A., Mantzaris, A. V., Nikolaev, A., Veremyev, A., Veijalainen, J., Pasiliao, E. L., and Boginski, V. (2019). Exploring Social Media Network Landscape of Post-Soviet Space. *IEEE Access*, **7**, 411–426. Conference Name: IEEE Access.

Smith, R., Cubbon, S., and Wardle, C. (2020). Under the surface: Covid-19 vaccine narratives, misinformation and data deficits on social media. Technical report, First Draft.

Soulier, A. and Halpin-Healy, T. (2003). The dynamics of multidimensional secession: Fixed points and ideological condensation. *Phys. Rev. Lett.*, **90**, 258103.

Spagat, M., Johnson, N. F., and Weezel, S. (2018). Fundamental patterns and predictions of event size distributions in modern wars and terrorist campaigns. *PLoS ONE*, **13**(10), e0204639.

Starbird, K. (2019). Disinformation's spread: bots, trolls and all of us. *Nature*, **571**(7766), 449–449. Number: 7766 Publisher: Nature Publishing Group.

Stockmayer, W. H. (1943). Theory of molecular size distribution and gel formation in branched-chain polymers. *J. Chem. Phys.*, **11**(2), 45–55. `https://doi.org/10.1063/1.1723803`.

Stockmayer, W. H. (1944). Theory of molecular size distribution and gel formation in branched polymers II. general cross linking. *J. Chem. Phys.*, **12**(4), 125–131.

United Nations (2023). Common Agenda Policy Brief: Information Integrity on Digital Platforms. Technical report. `https://www.un.org/sites/un2.un.org/files/our-common-agenda-policy-brief-information-integrity-en.pdf`.

Uzogara, E. E. (2023). Democracy Intercepted. *Science*, **381**(6656), 386–387. Publisher: American Association for the Advancement of Science.

van der Linden, S., Leiserowitz, A., Rosenthal, S., and Maibach, E. (2017). Inoculating the Public against Misinformation about Climate Change. *Global Challenges*, **1**(2), 1600008. `https://onlinelibrary.wiley.com/doi/pdf/10.1002/gch2.201600008`.

Velásquez, N., Leahy, R., Restrepo, N. J., Lupu, Y., Sear, R., Gabriel, N., Jha, O. K., Goldberg, B., and Johnson, N. F. (2021). Online hate network spreads malicious COVID-19 content outside the control of individual social media platforms. *Scientific Reports*, **11**(1), 11549. Number: 1 Publisher: Nature Publishing Group.

Verma, A., Sear, R., and Johnson, N. F. (2024). How U.S. presidential elections strengthen global hate networks. *npj Complex*, **1**(18).

Vesna, C-G. and Maslo-Čerkić, Š. (2023). Hate speech online and the approach of the Council of Europe and the European Union. Technical report, University of Rijeka, Tallinn. `https://urn.nsk.hr/urn:nbn:hr:118:377009`.

Wattis, J. A. D. (2006). An introduction to mathematical models of coagulation-fragmentation processes: A discrete deterministic mean-field approach. *Physica D: Nonlinear Phenomena*, **222**(1), 1–20.

Watts, D. J. and Strogatz, S. H. (1998). Collective dynamics of 'small-world' networks. *Nature*, **393**(6684), 440–442.

Wolfe, L. (2023). How the Metaverse is Changing the Gaming Industry. *Medium*. `https://medium.com/@lunawolfe01/how-the-metaverse-is-changing-the-gaming-industry-c6b32dd77cf8`.

Woo, G. (2011). *Calculating Catastrophe*. World Scientific, World Scientific.

Wu, X-Z., Fennell, P. G., Percus, A. G., and Lerman, K. (2018). Degree correlations amplify the growth of cascades in networks. *Physical Review E*, **98**(2), 022321. Publisher: American Physical Society.

Zhao, Z., Calderón, J. P., Xu, C., Zhao, G., Fenn, D., Sornette, D., Crane, R., Hui, P. M., and Johnson, N. F. (2010). Effect of social group dynamics on contagion. *Phys. Rev. E*, **81**, 056107.

Zheng, M., Sear, R.F., Illari, L., Restrepo, N.J., and Johnson, N. F. (2024). Adpative Link Dynamics drive Online Hate Networks and their Mainstream Influence. *npj Complex*, **1**(2).

Ziff, R. M., Hendriks, E. M., and Ernst, M. H. (1982). Critical properties for gelation: A kinetic approach. *Phys. Rev. Lett.*, **49**, 593–595.

Index

Drosophila
  trajectory, 237
Drosophila, 235

abstraction, 31
activation parameter, 150
adaptation, 6, 204
aggregate, 18
aggregation, 29, 31
  system, 32, 33, 69
anti-X, 18, 21, 156, 204, 234
anticorrelated, 241
arms race, 119
assembly path, 143
asynchronous, 145
averaging, 29

bad-actor–AI, 153, 161
  insurgency, 163
bad-actor–vulnerable-mainstream ecosystem, 157
basic reproduction rate, 183, 189
bell-curve, 12
binomial series, 67, 170
bit-string, 206, 244
Black Swan, 7, 12
block diagonal, 102
blockchain-based, 178
boundary, 63
boundary condition, 38, 200
branches, 55, 62
Brownian motion, 37
brute-force, 238
Burgers' equation
  generalized, 101, 144, 163
  inviscid, 48, 51, 53, 57, 79
Burgers'-like, 59, 266

capacitance, 39
causal, 133, 147
central limit theorem, 219
character vector, 41, 208

characteristic curve, 51, 52
characteristic equation
  matrix, 100
  ODE, 61
charge-density, 245
closed system, 59
cluster, 30
  density, 77
  linear, 35
  mass, 31
  profile, 86
  representation, 74
  structure, 31
cluster size distribution, 167, 191
coagulation, 29
coalescence, 29, 31
  rate, 38, 41
  suppression, 40
coarse-grain, 31, 103, 147
  measure of heterogeneity, 41
cohesion, 30
  loss of, 12
cohesive unit, 5, 9, 15, 18, 19, 27, 29, 49, 51, 53, 72, 123, 124, 136, 231, 232
  emergence, 151
  quantum, 265
coin-toss, 239
collective behaviour, 31, 204
collective chemistry, 234
collective process, 154
collision
  frequency, 39
  inelastic, 30, 37
  thermal, 37
combinatorics, 150
common-space model, 181
complex system, 17, 30
complexity, 5
computational efficiency, 151
condensate, 55
configuration class, 210
conflict, 17

offline, 22, 64, 68
online, 17, 64, 68
connection probability, 66, 109, 189
contagion, 180
continuum limit, 33
contour integral, 67
control, 231
control parameter, 183
convergence, 44, 171
convolution, 46
correlation, 30
cost function, 259
counterfactuals, 22
critical point, 13, 139
cross-platform, 105
crowd–anticrowd, 204, 234
cubic equation, 93
cyberspace, 32

data
　collection, 20
　conflict and terrorism, 26
de Moivre–Laplace theorem, 219
decoupled, 83, 102
destructive interactions, 162
diagonalization, 101
diffusion, 37
diffusion equation, 37, 217
diffusion-like, 208
dilute, 63
dilution, 78
dimension, 31
direct product, 241
directed graph, 102, 253
distrust, 1, 159
diverging phenomenon, 51
dynamic network, 155
dynamics, 31
　fluid, 46, 48, 51, 114
　fusion, 98
　fusion–fission, 136, 156, 180, 189
　gel, 142
　internal, 29, 180
　monomer, 77
　viral, 192

ecology, 30
econophysics, 205
eigendirections for gelation, 101
eigenvalue, 100
　maximum, 103

eigenvector, 101
El Farol bar problem, 205
electronic communications, 36, 37
electrostatics, 38
emergent phenomenon, 30, 45, 51, 70, 148
empirical patterns, 15
Encounter Fragmentation, 174
　attrition-based variant, 175
entanglement, 266
epidemic control, 190
epidemiology, 181
equilibrium, 66, 137
　far-from-, 139
Erdős–Rényi, 137
evolution, 6
evolutionary minority game, 208, 224
exponential functions, 57
extremely low latency networks, 5
extremism, 18
EZ model, 64, 164

fat tail, 3, 12, 13
feedback, 5
Fermi surface, 214
finite cutoff, 149, 150
finite population, 149
finite resource, 207
finite systems, 151
fission, 19, 24, 29, 64, 107, 156, 165, 189–191
　complete, 141, 163
　decentralized, 191
　rate, 24, 191
　total, 64
flocking, 30
fluctuation, 14, 152, 174, 175, 239, 243–247, 249, 251
fragmentation, 19, 29, 31, 64
　complete, 64
　rate, 68
　total, 22
frame-independent, 101
functional, 213, 223
fusion, 19, 24, 29
　binary, 31
　co-, 120
　dynamics, 229
fusion–fission, 72, 123, 141, 142, 148, 153, 159, 191
　homogeneous, 141
　rate, 165, 177

Gamma Ray Burst, 10, 136
Gauss' law, 38
Gaussian, 38
   distribution, 12, 219
gel, 9, 29, 49, 53, 55, 124, 177, 232
   curve, 55, 89, 184
   mixed, 84, 162
   phase, 89, 229
gelation, 49, 51, 69, 164, 165, 167, 184, 231, 232
   time, 51
   virtual, 91
gelation onset, 164
   component, 84
   composition, 26, 84–88, 90, 96, 97, 99, 101
   just after, 89
   time, 51, 53, 57, 62, 87, 92, 96
generalizability, 58
generating function, 43–45, 57, 76, 78, 148, 231
generative agent-based modelling, 265
generative PDE, 45–49, 51, 69, 70, 77, 114, 122, 140, 163, 231
Generative Pre-trained Transformer, 154
   ChatGPT, 265
   GPT-2, 156
geometric average, 83
giant connected component, 10, 27, 29, 72, 137, 138, 231
gradient, 51
grand canonical ensemble, 192
graph theory, 137
grokking, 17
group, 241
growth curve, 19, 105, 123, 135
   single species, 128

harmonic mean, 193
hate speech, 156
Heaviside function, 115, 146
heterogeneity, 4, 40, 64
   coarse-graining, 106
   collective, 41
   different types of, 72
   distribution, 233
   of population, 226
heterogeneous, 6, 23, 190
heterophily, 41, 42, 229, 230
homogeneous, 85, 130, 230
   clustering, 39
   in space, 32
   symmetry, 259
homophily, 41, 42, 228, 229
human–machinery–AI, 123, 204
human–technology–AI, 10, 15, 29, 33, 40, 64, 69, 72, 136, 147–150, 227, 231, 235, 259
hydrodynamic, 48
hyperbola, 185
hyperlink, 4, 18, 31, 156, 158, 159, 174, 231

in-built communities, 18
indivisible, 30
induction, 35
inequality, 113
infection route, 188
inflection point, 185
information spread, 180
initial condition, 34, 38, 51, 55–58, 60, 62, 81, 101, 141, 182, 185, 197, 201
   all-monomer, 34, 47, 57, 78, 80, 141
inter-allegiance, 119
internal state, 204
inverse temperature, 45, 150
inviscid fluid, 29
irreducible matrix, 102

kernel, 31
   Brownian, 37
   constant, 34, 46
   exponential, 40
   fragmentation, 108, 110
   generic, 36
   product, 19, 36, 47, 51, 69, 75, 139, 162
   sum, 36, 165, 166
   symmetric, 33, 37
kinetic theory, 137
kinetics, 140
kink, 56
Klein four-group, 241
klinotaxis, 235
Kronecker delta, 59, 112

L'Hôpital's rule, 258
Lambert W-function, 52, 53, 184
Langevin theory, 39, 40
Laplace equation, 38
Laplace transform, 44, 69
large language model, 5, 14, 30, 157, 238, 265
level of approximation, 6, 29
level of magnitude, 32
linear transformation, 130

linearize, 89, 129
localized, 57
logistic equation, 185
logistic function, 150
low energy, 30

macroscopic, 1, 5, 15, 29, 32, 49, 69, 70, 73, 124, 151, 234
  behaviour, 14, 123, 124, 257
  phenomenon, 31, 42
  variable, 42–44, 123, 244
many-body, 6
mapping, 139
mass action, 180
master equations, 32, 33, 42
mean degree, 139
mean-field, 140, 150, 189, 204, 212, 219, 229, 233
  approximations, 204
  average, 219
  limit, 193, 198, 202
  theory, 6, 41, 72, 106, 136, 150
memory storage, 206
mesoscopic, 31, 32, 42, 44, 57, 70, 74, 86, 148
method of characteristics, 51, 59, 80, 140
metric tensor, 75
microcanonical ensemble, 211
minority game, 206
misinformation and disinformation, 1
moment, 34
  first, 43, 50, 76
  second, 48, 50, 84, 99, 126
  tensor, 77
  zeroth, 34, 43, 48, 50
monomer, 35, 40, 50, 58, 59, 63–65, 68, 108, 109, 111, 112, 114, 141, 162, 165, 166, 232
  injection, 58, 59, 63
multi-adversary, 26, 120
multi-body, 29, 72, 123
multi-dimensional, 7, 73, 80, 105, 264
multi-peak, 106, 123, 137
multilayer network, 143

Navier–Stokes, 46
NetLogo, 9, 26
network, 139
  configuration, 150
  node, 18, 19, 24, 137–139, 144, 155, 156, 158, 175, 192, 193, 196, 252
neural network, 265

Newtonian, 29
non-commutativity, 79
non-linear, 9, 33, 44, 48, 51, 167, 182, 259
non-spread condition, 184
normalized, 45

offline–online interplay, 1
one-parameter space, 87
online
  coordinated disinformation campaigns, 31
  extremism, 1
  social networks, 31
  space, 3
online–offline
  complexity, 2
  world, 4, 5
opinions about vaccines, 72
ordinary differential equation, 32, 79, 111, 167, 181, 183–185, 200, 201
orthogonal, 101
oscillation, 183
oscillatory regime of infection, 196

parameter space, 139, 141, 145, 148
partial differentiation, 44
particles, 30
  character of, 41
  collection of, 30
  identical, 40
partition function, 45
Perron–Frobenius theorem, 102, 103, 105
phase diagram, 142, 148
phase transition, 13, 54, 55
  dynamical, 123, 138, 139
  second-order, 139
phenomenology, 151
physical interpretability, 151
polynomial equation, 80, 87
population balance, 181
power series, 67
power-law, 12, 19, 24, 68, 69, 165, 167, 170, 191
prize-to-fine ratio, 224
probability distribution function, 244

quantum information processing, 265
quasi-gel, 84
quasi-static, 228
quasiparticle, 247

random graph, 137–139, 144, 252, 256

degree, 137
single-layer, 142
random walk, 160, 224
re-entrant, 133
peaks, 136
reciprocal space, 44, 53, 76, 148
recursive, 34
Red Queen hypothesis, 160
reduced strategy space, 241
relative motion, 39
renormalization group, 30
rescale, 34, 42
Riccati equation, 59
3-D, 91
matrix, 79
Riemann zeta function, 171
risk, 14, 15, 22, 120

scale, 29
higher, 18
transition, 30
scale free, 13
scale inversely, 139, 148
self-segregation, 225
series, 57
shear interaction, 79
shock, 29, 124, 139
interacting, 114
super-, 119
shock wave, 51, 53, 54, 57, 163
similarity measure, 41
simultaneity, 133
SIR model, 181
SIS model, 181
size distribution, 31, 66
small-world network, 252
smallest positive root, 93, 95, 96, 99
Smoluchowski coagulation equations, 32, 46, 47
Smoluchowski-like, 111
social media, 2, 16, 153, 155
algorithm, 227
communities, 1
platform, 2, 5, 17, 18, 20, 22, 110, 158
users, 1
spatial diffusion, 5
species
abstract notion, 106
as a dynamic notion, 147
bundle, 101
effective, 106
inter-, 74, 119

intra-, 73
linear combinations of, 101
mixed-, 73
multi-, 9, 99, 105
single-, 29
spintronics, 202
spreading threshold, 190
standard deviation, 14
statistical
ensemble, 45, 150
mechanics, 30
steady-state, 24, 66, 68, 108, 170, 183, 186, 212
distribution, 12, 165, 202
Stirling's approximation, 67, 168
stochastic, 207
path, 160
strategy
reinforcement, 204
steering, 231
strategy allocation matrix, 243
strategy space, 208
summation convention, 81, 108, 112
superposition, 135
symmetry, 37

Taylor expansion, 169
time-dependent, 123, 140, 150, 193, 219, 233, 245
timescale, 181
tipping point of contagion, 187, 191, 200
toxic digital content, 154
transcendental, 187
trigonometric, 62

undirected graph, 102

valence electrons, 214
vector space, 41, 72, 76
Venn diagram, 159
Vieta theorem, 93
viral model, 181
viscosity, 40
vulnerable mainstream subsystem, 158

waveform, 53, 58
well-mixed limit, 180

XOR, 241

zero-sum game, 206
zig-zag trajectory, 236, 238, 239